OPTIMIZATION AND STABILITY THEORY FOR ECONOMIC ANALYSIS

Optimization and Stability Theory for Economic Analysis

BRIAN BEAVIS
and
IAN. M. DOBBS

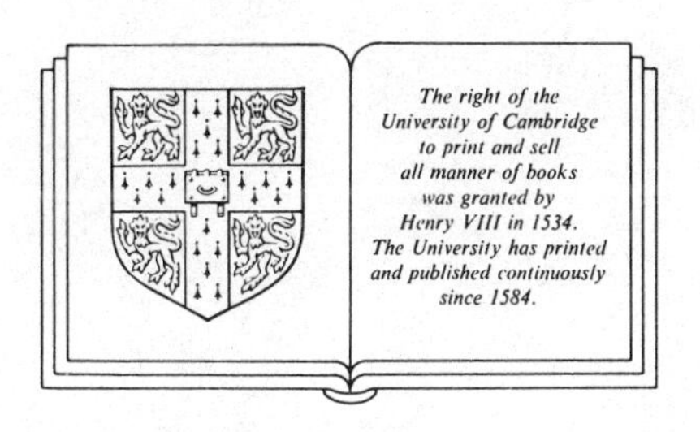
The right of the
University of Cambridge
to print and sell
all manner of books
was granted by
Henry VIII in 1534.
The University has printed
and published continuously
since 1584.

CAMBRIDGE UNIVERSITY PRESS
Cambridge
New York Port Chester Melbourne Sydney

Published by Press Syndicate of the University of Cambridge
The Pitt Building, Trumpington Street, Cambridge CB2 1RP
40 West 20th Street, New York, NY 10011, USA
10 Stamford Road, Oakleigh, Melbourne 3166, Australia

First published 1990

Printed in Great Britain at The University Press, Cambridge

British Library cataloguing in publication data

Beavis, Brian
Optimization and stability theory for economic analysis.
1. Economics. Optimization & stability.
Mathematical models
I. Title II. Dobbs, Ian
330′.0724

Library of Congress cataloguing in publication data

Beavis, Brian.
Optimization and stability theory for economic analysis / Brian Beavis, Ian Dobbs.
p. cm.
Bibliography.
Includes index.
ISBN 0-521-33307-5 (0-521-33605-8 paperback).
1. Mathematical optimization. 2. Economics, Mathematical.
I. Dobbs, Ian. II. Title.
HB143.7.B43 1989
330′.01′51 – dc19 88-31516

ISBN 0 521 33307 5 hard covers
ISBN 0 521 33605 8 paperback

MCP

To Marguerite—B.B.

To my parents, Terry and Robert—I.M.D.

CONTENTS

PREFACE

This book presents a coherent and systematic exposition of the mathematical theory of the problems of optimization and stability. Both of these are topics central to economic analysis since the latter is so much concerned with the optimizing behaviour of economic agents and the stability of the interaction processes to which this gives rise. A basic knowledge of optimization and stability theory is therefore essential for understanding and conducting modern economic analysis. The book is designed for use in advanced undergraduate and graduate courses in economic analysis and should, in addition, prove a useful reference work for practising economists.

Although the text deals with fairly advanced material, the mathematical prerequisites are minimised by the inclusion of an integrated mathematical review designed to make the text self-contained and accessible to the reader with only an elementary knowledge of calculus and linear algebra. We strongly urge the reader to peruse the material contained in the mathematical review before proceeding to the main text. Furthermore, Chapter 1 on convexity is to some extent a reference chapter, and can be regarded as an extension to the mathematical review in that it presents certain fundamental properties of convex sets and functions which are used throughout the text. The reader with a basic knowledge of convexity may begin with Chapter 2, where the theory of static optimization is developed. Chapter 4 on comparative statics and duality can be read immediately after Chapter 2 if so desired. Chapter 3, which deals with equilibrium mathematics is reasonably self-contained, and could be omitted on a first reading, although the material contained therein is widely applied in general equilibrium analysis. Stability theory and dynamic optimization are covered in the remaining chapters, and are best read in order.

Many of the more important theorems in the text are proved, although we often sacrifice elegance in the interests of minimising the mathematical prerequisites. An important feature of the text is the broad range of often extended and detailed economic applications. We believe that one of the best ways of enabling the reader to 'do' modern economic analysis is by working through such extended examples.

The index should enable the reader to locate relevant definitions and theorems. Although the notation used in the text is fairly standard, for convenience, there is a glossary to the symbols introduced in the mathematical review.

The text is an outgrowth of our experience in teaching optimization and stability theory in various second and third year courses in the economics degree programmes at the University of Newcastle upon Tyne. Whether our students would recognize it as such is another matter! Nevertheless, we would like to acknowledge their constructive and destructive comments and insights. We wish to thank Harvey Thompson and several referees for their helpful comments and insights. Naturally they are absolved from any remaining inaccuracies in the text. Finally, thanks are due to Francis Brooke and Patrick McCartan for their guidance, encouragement and editorial assistance throughout the process of writing this manuscript.

BRIAN BEAVIS
IAN M. DOBBS
University of Newcastle upon Tyne

CHAPTER 1

CONVEXITY

1.1 INTRODUCTION

A large part of modern economic analysis, particularly microeconomic analysis which assumes optimizing behaviour on the part of economic agents, relies on theorems dealing with a particular type of set, namely a convex set, and on theorems concerned with special types of functions, namely convex (concave) functions. In Section 1.2 we consider the basic attributes of convex sets and present certain fundamental theorems pertaining to such sets. Convex (concave) functions are discussed in Section 1.3.

1.2 CONVEX SETS

1.2.1 Basic properties

Geometrically, a set is convex if the line segment connecting any two points in the set lies entirely in the set. More formally, we have the following.

DEFINITION 1.1 The **line segment,** $L(x_1,x_2)$, connecting the points $x_1,x_2 \in \mathbb{R}^n$ is the set of points $\{x : x=\lambda x_1+(1-\lambda)x_2, 0 \leqslant \lambda \leqslant 1\}$. A set $X \subset \mathbb{R}^n$ is **convex** if for any two points $x_1,x_2 \in X$,

$\lambda x_1 + (1-\lambda)x_2 \in X$ for all $0 \leqslant \lambda \leqslant 1$. X is **strictly convex** if $\lambda x_1 + (1-\lambda)x_2$ is in the interior of X for all $0 < \lambda < 1$.

Unlike a convex set, the boundary of a strictly convex set cannot contain any linear segments. The null set and any set containing just a single point are by definition convex sets.

DEFINITION 1.2 The point $x \in \mathbb{R}^n$ is a **convex combination** of the points $x_1, \ldots, x_k \in \mathbb{R}^n$ if $x = \sum_{i=1}^k \lambda_i x_i$, where $0 \leqslant \lambda_i \leqslant 1$, $i = 1, \ldots, k$, and $\sum_{i=1}^k \lambda_i = 1$.

THEOREM 1.1 A set X is convex if, and only if, every convex combination of points of X lies in X.

Proof. The sufficiency of the condition is obvious from the definition. We prove the necessity of the condition by induction on the number m of points of X occurring in the convex combination. When $m = 2$, the condition follows from the definition. Assuming that every convex combination of k or fewer points of X yields a point of X, we consider a combination of $k+1$ points. Let $x = \sum_{i=1}^{k+1} \lambda_i x_i$, where $\lambda_i \geqslant 0$ and $\sum_{i=1}^{k+1} \lambda_i = 1$, and $x_i \in X$ for all i. If $\lambda_{k+1} = 1$, then $x = x_{k+1}$, which belongs to X and there is nothing further to prove. Suppose that $\lambda_{k+1} < 1$. In this case $\sum_{i=1}^k \lambda_i = 1 - \lambda_{k+1} > 0$, and we have

$$x = \left(\sum_{i=1}^k \lambda_i\right)\left(\sum_{i=1}^k \lambda_i x_i \Big/ \sum_{i=1}^k \lambda_i\right) + \lambda_{k+1} x_{k+1}.$$

By the induction hypothesis, the point $y = (\sum_{i=1}^k \lambda_i x_i / \sum_{i=1}^k \lambda_i)$ belongs to X. Thus, $x = (1 - \lambda_{k+1})y + \lambda_{k+1} x_{k+1}$ is a convex combination of two points in X and so $x \in X$.

THEOREM 1.2

(i) If X and Y are convex sets then $X \cap Y$ is convex.

(ii) If X and Y are convex sets then $X + Y \equiv \{x + y : x \in X, y \in Y\}$ is convex.

(iii) If X is a convex set and $\mu \geqslant 0$ is a scalar then $\mu X \equiv \{\mu x : x \in X\}$ is convex.

(iv) If $X_i, i=1,\ldots,k$, are convex sets then $X_1 \otimes X_2 \otimes \ldots \otimes X_k$ is convex.

Proof. The above assertions follow easily from the definitions involved. Here we give the proof of (i), leaving the remaining proofs to the reader. Let $x, y \in X \cap Y$. We need to show that $\lambda x + (1-\lambda)y \in X \cap Y$ for all $0 \leqslant \lambda \leqslant 1$. Since $x, y \in X \cap Y$, we have $x, y \in X$ and $x, y \in Y$. But X is convex, so that $\lambda x + (1-\lambda)y \in X$ for all $0 \leqslant \lambda \leqslant 1$. Similarly, Y is convex, so that $\lambda x + (1-\lambda)y \in Y$ for all $0 \leqslant \lambda \leqslant 1$. Hence, $\lambda x + (1-\lambda)y \in X \cap Y$ for all $0 \leqslant \lambda \leqslant 1$.

Note that the union of convex sets is not necessarily convex. The interior of a convex set may be empty as the following example shows.

EXAMPLE 1.1 Consider the convex set $X \subset \mathbb{R}^2$ consisting of the line segment $L(x, y)$, where $x = (a, 0)'$ and $y = (b, 0)'$, with $a < b$; that is, $X = [(x, 0): a \leqslant x \leqslant b]$. Here $\operatorname{int} X = \phi$.

THEOREM 1.3

(i) If X is a convex set then $\operatorname{int} X$ is convex.
(ii) If X is a convex set then $\operatorname{cl} X$ is convex.
(iii) If X is a convex set then $\operatorname{bd}(\operatorname{cl} X) = \operatorname{bd} X$.
(iv) If X is a convex set with $\operatorname{int} X \neq \phi$ then $\operatorname{cl}(\operatorname{int} X) = \operatorname{cl} X$ and $\operatorname{int}(\operatorname{cl} X) = \operatorname{int} X$.

Proof. See Brønsted (1983), p. 22.

DEFINITION 1.3 The **convex hull** of a set $X \subset \mathbb{R}^n$ is the intersection of all convex sets which contain X, and is denoted $\operatorname{conv} X$.

Clearly, X is convex if, and only if, $X = \operatorname{conv} X$. In a natural sense, $\operatorname{conv} X$ is the smallest convex set containing X.

THEOREM 1.4 For any set $X \subset \mathbb{R}^n$, the convex hull of X consists precisely of all convex combinations of points of X.

Proof. Let T denote the set of all convex combinations of elements of X. Since $\operatorname{conv} X$ is convex and $X = \operatorname{conv} X$, Theorem 1.1 implies that $T \subset \operatorname{conv} X$. Conversely, let $x \equiv \sum_{i=1}^{r} \alpha_i x_i$ and

$y=\sum_{i=1}^{s}\beta_i y_i$, where $\alpha_i \geqslant 0$, $\sum_{i=1}^{r}\alpha_i=1$, and $\beta_i \geqslant 0$, $\sum_{i=1}^{s}\beta_i=1$ be two elements of T. Then for any λ, $0 \leqslant \lambda \leqslant 1$, $z \equiv \lambda x+(1-\lambda)y=\lambda\sum_{i=1}^{r}\alpha_i x_i+(1-\lambda)\sum_{i=1}^{s}\beta_i y_i$ is an element of T since each coefficient is between 0 and 1 and $\lambda\sum_{i=1}^{r}\alpha_i+(1-\lambda)\sum_{i=1}^{s}\beta_i=\lambda(1)+(1-\lambda)(1)=1$. Thus T is a convex set containing X and so $\operatorname{conv} X \subset T$. Therefore $\operatorname{conv} X = T$.

Theorem 1.4 implies that a point $x \in \operatorname{conv} X$ is a convex combination of finitely many points of X, but it places no restriction on the number of points of X required to make the combination. Caratheodory's Theorem says that, in an n-dimensional space, the number of points of X in the convex combination never has to be more than $n+1$.

THEOREM 1.5 (Caratheodory's Theorem) If X is a non-empty subset of $\mathbb{R}^n$ then every $x \in \operatorname{conv} X$ can be expressed as a convex combination of $n+1$ or fewer points of X.

Proof. Given a point $x \in \operatorname{conv} X$, Theorem 1.4 implies that $x=\sum_{i=1}^{k}\lambda_i x_i$, where $\lambda_i \geqslant 0$, $\sum_{i=1}^{k}\lambda_i=1$ and $x_i \in X$ for all $i=1,\ldots,k$. Our aim is to show that such an expression exists for x with $k \leqslant n+1$.

If $k>n+1$, then there exist scalars $\alpha_1,\ldots,\alpha_k$, not all zero, such that $\sum_{i=1}^{k}\alpha_i=0$ and $\sum_{i=1}^{k}\alpha_i x_i=\theta$. To see this, note that for $k=n+2$, then the $n-1$ vectors x_2-x_1, $x_3-x_1,\ldots,x_k-x_1$ are linearly dependent, so that there exist scalars $\alpha_2,\ldots,\alpha_k$, not all zero, such that $\alpha_2(x_2-x_1)+\alpha_3(x_3-x_1)+\ldots+\alpha_k(x_k-x_1)=\theta$, or $\sum_{i=1}^{k}\alpha_i x_i=\theta$ where $\alpha_1=-\sum_{i=2}^{k}\alpha_i$. Thus, we have $\sum_{i=1}^{k}\lambda_i x_i=x$ and $\sum_{i=1}^{k}\alpha_i x_i=\theta$. By subtracting an appropriate multiple of the second equation from the first we will obtain a convex combination of fewer than k elements of X which is equal to x.

Since not all of the α_i are zero, we may assume (by reordering subscripts if necessary) that $a_k>0$ and that $\lambda_k\alpha_k^{-1} \leqslant \lambda_i\alpha_i^{-1}$ for all those i for which $\alpha_i>0$. For $1 \leqslant i \leqslant k$, let $\beta_i=\lambda_i-(\lambda_k/\alpha_k)\alpha_i$. Then $\beta_k=0$ and $\sum_{i=1}^{k}\beta_i=\sum_{i=1}^{k}\lambda_i-(\lambda_k/\alpha_k)\sum_{i=1}^{k}\alpha_i=1-0=1$. Furthermore, each $\beta_i \geqslant 0$. Indeed, if $\alpha_i \leqslant 0$ then $\beta_i \geqslant \lambda_i \geqslant 0$. If $\alpha_i > 0$, then $\beta_i = \alpha_i[(\lambda_i/\alpha_i) - (\lambda_k/\alpha_k)] \geqslant 0$. Thus we have

$$\sum_{i=1}^{k-1} \beta_i x_i = \sum_{i=1}^{k} \beta_i x_i = \sum_{i=1}^{k} (\lambda_i - \lambda_k \alpha_i/\alpha_k) x_i$$

$$= \sum_{i=1}^{k} \lambda_i x_i - (\lambda_k/\alpha_k) \sum_{i=1}^{k} \alpha_i x_i = \sum_{i=1}^{k} \lambda_i x_i = x.$$

Hence, we have expressed x as a convex combination of $k-1$ of the points $x_1, \ldots, x_k$. This process may be continued until we have expressed x as a convex combination of at most $n+1$ points of X.

DEFINITION 1.4 The **dimension** of a convex set $X \subset \mathbb{R}^n$, denoted $\dim X$, is the dimension of the affine hull of X.

DEFINITION 1.5 The convex hull of a finite set of points is called a **polytope**. If $X = \{x_1, \ldots, x_{k+1}\}$ and $\dim X = k$, then $\operatorname{conv} X$ is called a **k-dimensional simplex**, and the points $x_1, \ldots, x_{k+1}$ are called *vertices*.

A zero-dimensional simplex is a point; a one-dimensional simplex is a line segment and a two-dimensional simplex is a triangle. From Theorem 1.5 we know that a point x in a simplex $X = \operatorname{conv}\{x_i, \ldots, x_{k+1}\}$ can be expressed as a convex combination of its vertices. We leave it to the reader to show that this representation is unique.

THEOREM 1.6 If X is an open set, then $\operatorname{conv} X$ is open.

Proof. Since X is open, $X \cap \operatorname{bd}(\operatorname{conv} X) = \phi$. But $X \subset \operatorname{conv} X$, so we must have $X \subset \operatorname{int}(\operatorname{conv} X)$. Furthermore, Theorem 1.3(i) implies $\operatorname{int}(\operatorname{conv} X)$ is convex, so $\operatorname{conv} X \subset \operatorname{int}(\operatorname{conv} X)$. On the other hand, $\operatorname{int}(\operatorname{conv} X) \subset \operatorname{conv} X$, so $\operatorname{conv} X = \operatorname{int}(\operatorname{conv} X)$ and hence $\operatorname{conv} X$ is open.

Note that the convex hull of a closed set need not be closed.

THEOREM 1.7 If X is a compact set, then $\operatorname{conv} X$ is compact.

Proof. Let A be the compact subset of $\mathbb{R}^{n+1}$ defined by

$$A = \{(\alpha_1, \ldots, \alpha_{n+1}) : \sum_{i=1}^{n+1} \alpha_i = 1 \text{ and } \alpha_i \geqslant 0, 1 \leqslant i \leqslant n+1\}.$$

The function f defined by

$$f(\alpha_1, \ldots, \alpha_{n+1}, x_{11}, \ldots, x_{1n}, x_{21}, \ldots, x_{2n}, \ldots, x_{n+1,1}, \ldots, x_{n+1,n})$$
$$= \sum_{i=1}^{n+1} \alpha_i (x_{i1}, \ldots, x_{in})$$

is a continuous mapping of $\mathbb{R}^{n+1} \otimes \mathbb{R}^n \otimes \mathbb{R}^n \otimes \ldots \otimes \mathbb{R}^n$ into $\mathbb{R}^n$. By Caratheodory's Theorem, it maps the compact set $A \otimes X \otimes X \otimes \ldots \otimes X$ onto $\operatorname{conv} X$. Therefore, $\operatorname{conv} X$ is also compact.

THEOREM 1.8 Let X_i be a subset of $\mathbb{R}^n$ for $i = 1, \ldots, k$. Then we have

$$\operatorname{conv}\left(\sum_{i=1}^{k} X_i\right) = \sum_{i=1}^{k} \operatorname{conv} X_i.$$

Proof. We leave this for the reader.

1.2.2 Separation of convex sets

DEFINITION 1.6 A **hyperplane** is a set $H(p, \alpha) = \{x \in \mathbb{R}^n : p'x = \alpha\}$, where $p \in \mathbb{R}^n$ is non-zero. The vector p is called the **normal** to the hyperplane $H(p, \alpha)$.

Note that a hyperplane is a level set of a non-identically zero linear function.

DEFINITION 1.7 The hyperplane $H(p, \alpha)$ **separates** two sets A and B if one of the following holds:

(i) $p'x \leqslant \alpha$ for all $x \in A$ and $p'x \geqslant \alpha$ for all $x \in B$,
(ii) $p'x \geqslant \alpha$ for all $x \in A$ and $p'x \leqslant \alpha$ for all $x \in B$.

The hyperplane **strictly separates** two sets A and B if one of the following holds:

(iii) $p'x<\alpha$ for all $x\in A$ and $p'x>\alpha$ for all $x\in B$,

(iv) $p'x>\alpha$ for all $x\in A$ and $p'x<\alpha$ for all $x\in B$.

DEFINITION 1.8 Given a hyperplane $H(p,\alpha)$ the sets

$$\{x:p'x\geqslant\alpha\} \text{ and } \{x:p'x\leqslant\alpha\}$$

are called the **closed half-spaces** determined by the hyperplane $H(p,\alpha)$. If the inequalities are replaced by strict inequalities, the resulting sets are called the **open half-spaces** determined by the hyperplane $H(p,\alpha)$.

Clearly, the hyperplane H separates two non-empty subsets of $\mathbb{R}^n$, A and B, if A is contained in one of the closed half-spaces determined by H and B is contained in the opposite closed half-space. A hyperplane H strictly separates A and B if A is contained in one of the open half-spaces determined by H and B is contained in the opposite open half-space. It should be intuitively clear that, given two non-empty disjoint convex sets in $\mathbb{R}^n$, it is possible to find a hyperplane that separates them, and that, while strict separation requires that the two sets be disjoint, mere separation does not. For example, if two circles in the plane are externally tangent, then their common tangent line separates them, but does not strictly separate them. Although it is necessary that the two sets be disjoint in order to be strictly separated, this condition is not sufficient, even for closed convex sets. For example, in $\mathbb{R}^2$, let $A=\{(x,y):x>0 \text{ and } y\geqslant 1/x\}$ and $B=\{(x,y):x\geqslant 0 \text{ and } y=0\}$. Then A and B are disjoint closed convex sets, but they cannot be strictly separated by a hyperplane (i.e. a line in $\mathbb{R}^2$). Thus the problem of the existence of a hyperplane separating (or strictly separating) two sets is more complex than it might appear to be at first. Before presenting some results concerning the existence of separating hyperplanes we introduce some further necessary terminology.

DEFINITION 1.9 A hyperplane $H(p,\alpha)$ is said to **bound** the set X if either $p'x\geqslant\alpha$ for all $x\in X$ or $p'x\leqslant\alpha$ for all $x\in X$. A hyperplane $H(p,\alpha)$

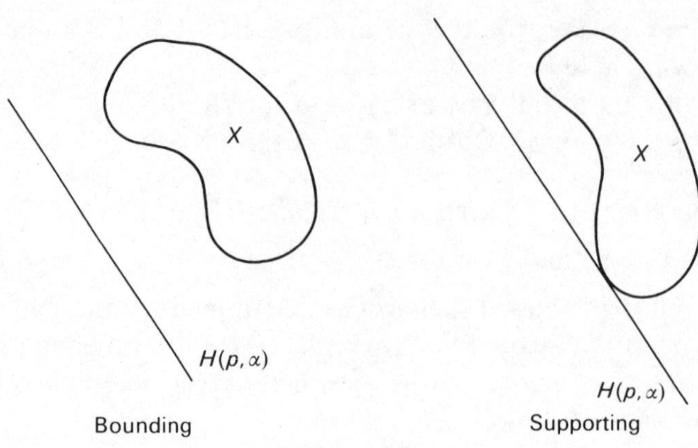

Figure 1.1

$H(p, \alpha)$ **supports** a set X at the point $x \in X$ if $x \in H(p, \alpha)$ and if $H(p, \alpha)$ bounds X.

A bounding hyperplane H to a set X is a hyperplane such that X is completely contained in one of the closed half-spaces determined by the hyperplane H. A supporting hyperplane H for a set X is a bounding hyperplane for X which shares a point in common with the boundary of X.

The above definitions are illustrated in Figure 1.1.

THEOREM 1.9 Given any non-empty closed convex set $X \subset \mathbb{R}^n$ and a point $z \notin X$, then there exists a hyperplane $H(p, \alpha)$ that separates X and $\{z\}$.

Proof. Let $B_\delta(z)$ be a closed ball with centre z such that $B_\delta(z) \cap X \neq \phi$. Let $A = B_\delta(z) \cap X$. Clearly, A is non-empty and compact. Since A is compact and the distance function $d(z, x)$ is continuous, $d(z, x)$ has a minimum in A, from the Weierstrass Theorem. Thus, there exists a $y \in X$ such that $d(z, y) \leqslant d(z, x)$ for all $x \in A$, and thus $d(z, y) \leqslant d(z, x)$ for all $x \in X$. As $z \notin X$ and $y \in X$, $d(z, y) > 0$ (see Figure 1.2). Since X is convex, $y + \lambda(x - y) \in X$ for all $0 \leqslant \lambda \leqslant 1$ and as $d(z, x) \leqslant d(z, y)$ and $d(z, y) > 0$,

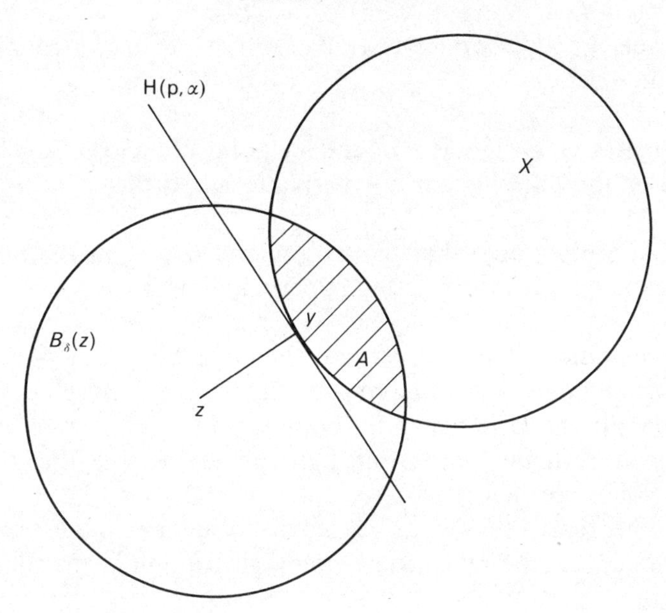

Figure 1.2

we have $\|y+\lambda(x-y)-z\|^2 \geqslant \|y-z\|^2$. Expanding gives $2\lambda(y-z)'(x-y)+\lambda^2\|x-y\| \geqslant 0$ and dividing by $\lambda>0$ and letting $\lambda \to 0^+$ we obtain $(y-z)'(x-y) \geqslant 0$. Define $p \equiv (y-z)$ and $\alpha = p'y$, then we have $p'x \geqslant p'y = \alpha$ for all $x \in X$. Now $p'z = (y-z)'z = -(y-z)'(y-z)+(y-z)'z = -\|y-z\|^2+(y-z)'z = -\|p\|^2+p'z = -\|p\|^2+\alpha < \alpha$. Hence, $p'x \geqslant \alpha > p'z$ for all $x \in X$. Hence, the hyperplane $H(p, \alpha)$ separates X and $\{z\}$.

Theorem 1.9 is sometimes stated in the following form.

THEOREM 1.10 Let X be a non-empty closed, convex set in $\mathbb{R}^n$ not containing the origin. Then there exists a hyperplane that (strictly) separates X and the origin.

Proof. In Theorem 1.9, let $z = \theta$. Then from Theorem 1.9 there exists a $p \in \mathbb{R}^n$ and $\alpha > 0$ such that $p'x \geqslant \alpha$ for all $x \in X$, where p and α are defined as $p = y - z = y$ and $\alpha = p'y$. By choosing a point between y and the origin, instead of y, the inequality $p'x \geqslant \alpha$ can be made strict.

THEOREM 1.11 If x is a boundary point of a closed convex set $X \subset \mathbb{R}^n$ then there exists a hyperplane supporting X at x.

Proof. Simply note that the hyperplane $H(p, \alpha)$ in the proof of Theorem 1.9 is a supporting hyperplane for X.

If the boundary of X is smooth at the given boundary point then there is only one supporting hyperplane passing through the given boundary point. However, if the boundary of X is not smooth, then there can be many supporting hyperplanes passing through the given point (see Figure 1.3).

In Theorem 1.9 X was assumed to be a closed set. This assumption can be relaxed; by doing so we obtain the following result.

THEOREM 1.12 Given any non-empty convex set X in $\mathbb{R}^n$, not necessarily closed, and a point $z \notin X$, there exists a hyperplane $H(p, \alpha)$ that separates X and $\{z\}$.

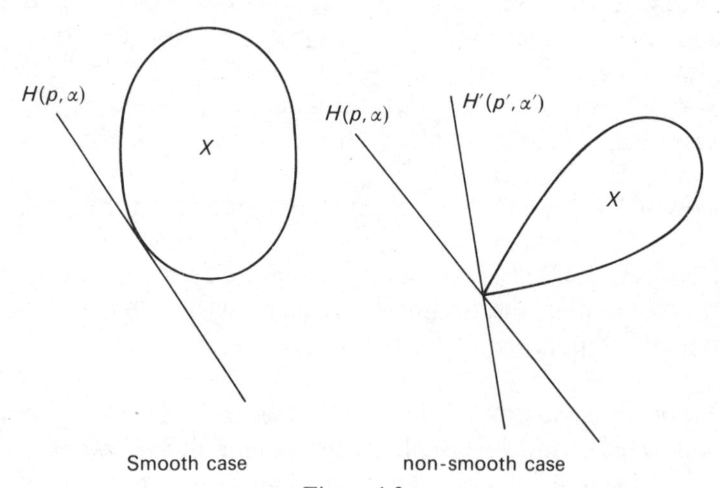

Figure 1.3

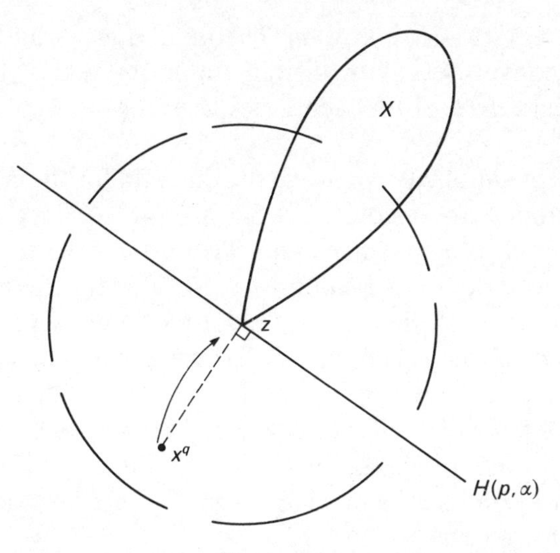

Figure 1.4

Proof. Suppose $z \notin \operatorname{cl} X$. Then from Theorem 1.10 there exists a $p \in \mathbb{R}^n$, $p \neq \theta$ and a scalar α such that $p'x \geqslant \alpha$ for all $x \in \operatorname{cl} X$ and $p'z < \alpha$. Thus $p'x > p'z$ for all $x \in \operatorname{cl} X$ and hence $p'x > p'z$ for all $x \in X$. Suppose now $z \in \operatorname{cl} X$. Since $z \notin X$ by assumption, z is a boundary point of cl X, so that for any neighbourhood containing z there exists a point which is not in cl X. That is, there exists a sequence $\{x^q\}$ such that $x^q \notin \operatorname{cl} X$ and $x^q \to z$. Since $x^q \notin \operatorname{cl} X$ and cl X is non-empty, closed and convex, from Theorem 1.9, there exists a $p^q \in \mathbb{R}^n$, $p^q \neq \theta$ such that $p^{q\prime}x > p^{q\prime}x^q$ for all $x \in \operatorname{cl} X$ (see Figure 1.4). Without loss of generality, we can choose p^q such that $\|p^q\| = 1$. Then the sequence $\{p^q\}$ moves in the unit sphere of $\mathbb{R}^n$. Since the unit sphere is compact, there exists a converging subsequence in the sphere; that is, there exists a subsequence such that $p_s^q \to p$ with $\|p\| = 1$, where $\{p_s^q\}$ corresponds to $\{x_s^q\}$. Take the limit of $p_s^{q\prime}x > p_s^{q\prime}x_s^q$ as $p_s^q \to \infty$. Since an inner product is a continuous function, we have $p'x \geqslant p'z$ for all $x \in \operatorname{cl} X$. Thus, $p'x \geqslant p'z$ for all $x \in X$.

We can use Theorem 1.12 to obtain the following more general result.

THEOREM 1.13 (Minkowski Theorem) Let X and Y be non-empty convex sets with disjoint interiors. Then there exists a hyperplane $H(p,\alpha)$ that separates them.

Proof. For simplicity we prove the theorem for the case where the sets X and Y are disjoint. Let $S=X+(-Y)$. Since X and Y are convex, $-Y$ is also convex and S is thus convex too. Moreover, the origin θ does not belong to S, for if it did there would exist $x^*\in X$, $y^*\in Y$ such that $x^*-y^*=\theta$, or $x^*=y^*$, which contradicts $X\cap Y=\phi$. From Theorem 1.12 there exists a hyperplane that separates S and the origin; that is, there exists a $p\in\mathbb{R}^n$, $p\neq\theta$ such that $p's\geqslant p'\theta=0$ for all $s\in S$. Let $s=x-y$, $x\in X$, $y\in Y$. Then we have $p'x\geqslant p'y$ for all $x\in X$, $y\in Y$. Thus, $\inf_{x\in X}\{p'x\}\geqslant\sup_{y\in Y}\{p'y\}$. Hence we can select α in such a way that $p'x\geqslant\alpha$ for all $x\in X$ and $p'y\leqslant\alpha$ for all $y\in Y$.

Figure 1.5 illustrates the Minkowski Theorem. Note that if X and Y are arbitrary sets, not necessarily convex, then there may not exist a hyperplane that separates them, as Figure 1.6 illustrates.

THEOREM 1.14 Let X and Y be two non-empty disjoint convex

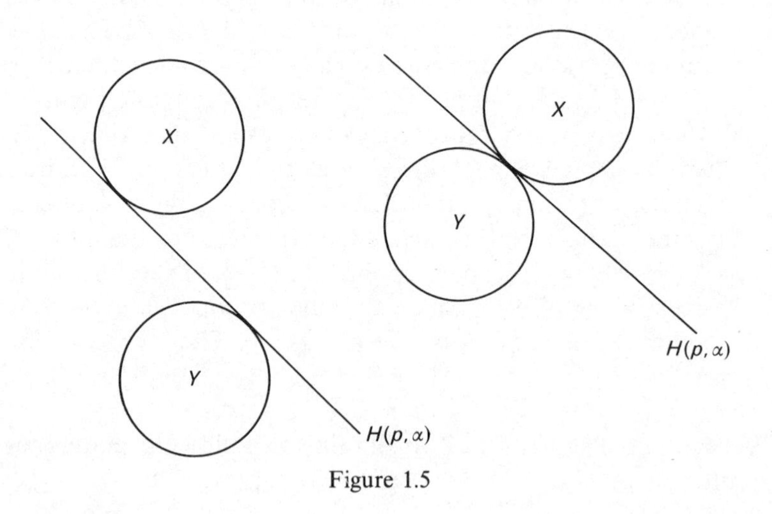

Figure 1.5

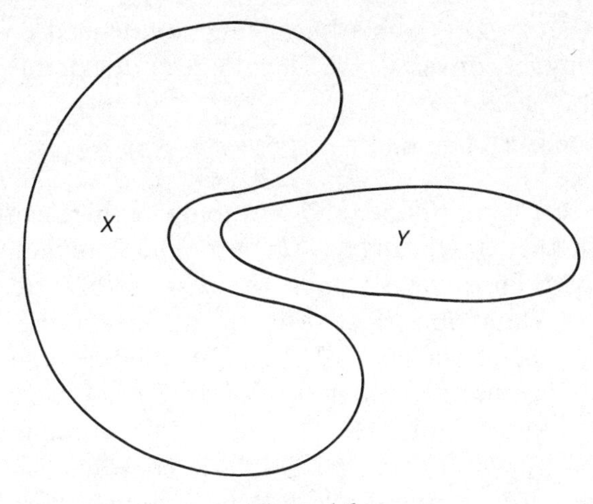

Figure 1.6

sets in $\mathbb{R}^n$ and suppose that Y is compact. Then there exists a hyperplane $H(p, \alpha)$ that strictly separates them.

Proof. Let $S = X + (-Y)$. Since X and Y are convex and Y is compact, S is closed and convex. Furthermore, since $X \cap Y = \phi$, the origin does not belong to S. Thus from Theorem 1.10 there exists a hyperplane which strictly separates S and the origin; that is, there exists a $p \in \mathbb{R}^n$, $p \neq \theta$, such that $p's > p'\theta = 0$ for all $s \in S$. Let $s = x - y$, $x \in X$, $y \in Y$. Then we have $p'x > p'y$ for all $x \in X$, $y \in Y$. Thus, $\inf_{x \in X}\{p'x\} > \sup_{y \in Y}\{p'y\}$. Hence we can select α in such a way that $p'x > \alpha$ for all $x \in X$ and $p'y < \alpha$ for all $y \in Y$.

DEFINITION 1.10 Let X be a convex set. A point $x \in X$ is an **extreme point** of X if x is not an interior point of any line segment in X. The set of all extreme points of X is called the **profile** of X.

The profile of a polygon in $\mathbb{R}^2$ is the set of vertices; a closed half-space and an open convex set have no extreme points.

THEOREM 1.15 Let X be a compact convex subset of $\mathbb{R}^n$. Then X is the convex hull of its profile.

THEOREM 1.16 Let f be a linear function defined on the non-empty compact convex set X. Then there exist extreme points x° and y° of X such that

$$f(x^\circ)=\max_{x\in X} f(x) \qquad \text{and} \qquad f(y^\circ)=\min_{x\in X} f(x).$$

Proof. Since f is continuous and X is compact there exists a point $x'\in X$ such that $f(x')=\max_{x\in X} f(x)=\alpha$. The compactness of X together with Theorem 1.15 imply the existence of extreme points $x_1,\ldots,x_k$ of X and non-negative $\lambda_1,\ldots,\lambda_k$ such that $x'=\sum_{i=1}^k \lambda_i x_i$ with $\sum_{i=1}^k \lambda_i=1$. If none of the extreme points of X satisfies $f(x)=\alpha$, then since α is the maximum of f on X we must have $f(x_i)<\alpha$ for $i=1,\ldots,k$. But this implies that $\alpha=f(x')=\sum_{i=1}^k \lambda_i x_i<\sum_{i=1}^k \lambda_i \alpha=\alpha$. This contradiction implies that some extreme point x° of X must satisfy $f(x^\circ)=\alpha$. The proof for y° is analogous.

An *alternative theorem* states that two systems of linear equations or linear inequalities are so related that exactly one of the two has a solution. Such theorems depend on the separation theorems for convex sets. In Chapter 2 we shall see that much of static optimization theory can be built on such results. Here we shall use the separation theorems to establish two alternative theorems. An extensive list of similar results can be found in Mangasarian (1969); see also Gale (1960).

THEOREM 1.7 (Farkas' Lemma) Let A be a given $m\times n$ real matrix and b a given n vector. The inequality $b'y\geqslant 0$ holds for all vectors y satisfying $Ay\geqslant\theta$ if, and only if, there exists an m vector $x\geqslant\theta$ such that $A'x=b$. Equivalently, $Ay\geqslant\theta, b'y<0$ has a solution if, and only if, $A'x=b$ and $x\geqslant\theta$ has no solution.

Proof. The system $A'x=b$ and $x\geqslant\theta$ has no solution if, and only if, the non-empty closed convex sets $S=\{s:s\in\mathbb{R}^n, s=A'x, x\geqslant\theta\}$ and $T=\{b\}$ are disjoint. If these conditions hold, then, since T is compact, from Theorem 1.14, there exists a hyperplane $H(p,\alpha)$ that strictly separates S and T. That is, there exists a $p\in\mathbb{R}^n$, $p\neq\theta$, and a scalar α such that $p'b<\alpha$ and for every $s\in S$ we have $p's>\alpha$, or, equivalently, for every $p \geqslant \theta$ we have $p'A'x > \alpha$. Letting $x = \theta$, we conclude that $\alpha < 0$, and letting $x = (0, \ldots, x_j, 0, \ldots, 0)$ for $j = 1, \ldots, m$, where x_j is a large positive number, we obtain $p'A' \geqslant \theta$. Hence $Ap \geqslant \theta$ and so p is a solution of $Ay \geqslant \theta$ and $b'y < 0$. Conversely, if $A'x = b$, $x \geqslant \theta$ and $Ay \geqslant \theta$, then $b'y = x'Ay \geqslant \theta$.

THEOREM 1.18 Let A be a given $m \times n$ matrix and b a given n vector. Then exactly one of the following alternatives holds. Either the inequalities $A'x \leqslant b$ have a non-negative solution, or the inequalities $Ay \geqslant \theta$, $b'y < 0$ have a non-negative solution.

Proof. We first show that both alternatives cannot hold. Suppose that $A'x \leqslant b$ has a non-negative solution, so that $A'x + z = b$, where $z \geqslant \theta$, also has a non-negative solution. Also suppose that $Ay \geqslant \theta$ and $b'y < 0$ have a non-negative solution. Then we have $0 > b'y = (A'x + z)'y = x'Ay + z'y$. But $x \geqslant \theta, z \geqslant \theta, y \geqslant \theta$ and $Ay \geqslant \theta$, so that $x'Ay \geqslant 0$ and $z'y \geqslant 0$, and we therefore have $0 > b'y \geqslant 0$, a contradiction.

Suppose then that $A'x \leqslant b$ has no non-negative solution, so that $A'x + z = b$ has no non-negative solution. Using standard block matrix notation we can write $A'x + z = b$ in the form

$$[A' \ I]\begin{bmatrix} x \\ z \end{bmatrix} = b,$$

where $[A' \ I]$ is a block matrix whose left m columns are the columns of A' and whose right n columns are the columns of an $n \times n$ identity matrix I, and x is the column vector whose first m entries are the coordinates of x and whose last n entries are the coordinates of z. Since $[A' \ I]\begin{bmatrix} x \\ z \end{bmatrix} = b$ has no non-negative solution, Farkas Lemma implies the existence of a vector y such that

$$[A' \ I]'y \geqslant \theta \qquad \text{and} \quad b'y < 0.$$

But $[A' \ I]'y = \begin{bmatrix} A \\ I \end{bmatrix} y$, so that $[A' \ I]'y \geqslant \theta$ implies $Ay \geqslant \theta$ and $y \geqslant \theta$, and we have found a non-negative solution to $Ay \geqslant \theta$ and $b'y < 0$.

1.3 CONVEX AND CONCAVE FUNCTIONS

1.3.1 Basic properties

Geometrically, a real-valued function defined on a convex set $X \subset \mathbb{R}^n$ is convex (concave) if the line segment connecting any two points on the surface of the function nowhere lies below (above) the surface. More formally, we have the following.

DEFINITION 1.11 Let f be a real-valued function defined on a convex set $X \subset \mathbb{R}^n$. Then f is **convex** if $f(\lambda x + (1-\lambda)y) \leqslant f(x) + (1-\lambda)f(y)$ for all $x, y \in X$ and all λ, $0 \leqslant \lambda \leqslant 1$: f is **strictly convex** if $f(\lambda x + (1-\lambda)y) < \lambda f(x) + (1-\lambda)f(y)$ for all $x, y \in X$ and all $\lambda, 0 < \lambda < 1$. f is **concave** if $f(\lambda x + (1+\lambda)y) \geqslant \lambda f(x) + (1-\lambda)f(y)$ for all $x, y \in X$ and all $\lambda, 0 \leqslant \lambda \leqslant 1$; f is **strictly concave** if $f(\lambda x + (1-\lambda)y) > \lambda f(x) + (1-\lambda)f(y)$ for all $x, y \in X$ and all $\lambda, 0 < \lambda < 1$.

Clearly, f is (strictly) concave if, and only if, $-f$ is (strictly) convex. Note that a real-valued function f defined on a convex subset of $\mathbb{R}^n$ may be simultaneously convex and concave; for example, $f: X \to \mathbb{R}$, where $X \subset \mathbb{R}^n$, and $f(x) = p'x$ is both convex and concave on X. However, a function f cannot be simultaneously strictly convex and concave.

DEFINITION 1.12 Let f be a real-valued function defined on a convex set $X \subset \mathbb{R}^n$. The **epigraph of** f, denoted epi f, is the subset of $\mathbb{R}^{n+1}$ defined by

$$\text{epi } f = \{(x, \mu) : x \in X, \mu \in \mathbb{R}, \mu \geqslant f(x)\}.$$

The **hypograph of** f, denoted hyp f, is the subset of $\mathbb{R}^{n+1}$ defined by

$$\text{hyp } f = \{(x, \mu) : x \in X, \mu \in \mathbb{R}, \mu \leqslant f(x)\}$$

THEOREM 1.19 Let f be a real-valued function defined on a

convex subset of $\mathbb{R}^n$. Then f is convex if, and only if, epi f is a convex subset of $\mathbb{R}^{n+1}$; f is concave if, and only if, hyp f is a convex subset of $\mathbb{R}^{n+1}$.

Proof. We prove this for the convex case; the proof for the concave case is analogous. Suppose f is convex and let (x_1, μ_1) and (x_2, μ_2) be points in epi f. Then for $0 \leqslant \lambda \leqslant 1$ we have $\lambda\mu_1 + (1-\lambda)\mu_2 \geqslant \lambda f(x_1) + (1-\lambda)f(x_2) \geqslant f(\lambda x_2 + (1-\lambda)x_2)$. Therefore, $\lambda(x_1, \mu_1) + (1-\lambda)(x_2, \mu_2) = (\lambda x_1 + (1-\lambda)x_2, \lambda\mu_1 + (1-\lambda)\mu_2)$ is in epi f and so epi f is convex. Conversely, suppose epi f is a convex subset of $\mathbb{R}^{n+1}$ and let $x_1, x_2 \in X$. Then (x_1, μ_1), (x_2, μ_2) are in epi f, where $\mu_i = f(x_i)$, $i = 1, 2$. Since epi f is convex, we have $\lambda(x_1, \mu_1) + (1-\lambda)(x_2, \mu_2) = (\lambda x_1 + (1-\lambda)x_2, \lambda\mu_1 + (1-\lambda)\mu_2)$ which is in epi f for all $0 \leqslant \lambda \leqslant 1$. But this implies $f(\lambda x_1 + (1-\lambda)x_2) \leqslant \lambda f(x_1) + (1-\lambda)f(x_2)$, so that f is a convex function.

The convexity property of a function is preserved under certain kinds of composition and linear combinations.

THEOREM 1.20 Let f be a real-valued convex function defined on a convex subset X of $\mathbb{R}^n$ and let g be a convex function defined on an interval I in $\mathbb{R}$. Suppose that $f(X) \subset I$ and that g is non-decreasing. The composite function $g \circ f$ is convex on X.

Proof. Since f is convex on X, $f(\lambda x + (1-\lambda)y) \leqslant \lambda f(x) + (1-\lambda)f(y)$ for all $x, y \in X$ and $0 \leqslant \lambda \leqslant 1$. Since g is non-decreasing and convex, we thus obtain $g(f(\lambda x + (1-\lambda)y)) \leqslant g(\lambda f(x) + (1-\lambda)f(y)) \leqslant \lambda g(f(x)) + (1-\lambda)g(f(y))$.

THEOREM 1.21 Let f and g be convex functions defined on the convex set $X \subset \mathbb{R}^n$ and let $\alpha > 0$. Then the functions $f+g$ and αf are convex on X.

Proof. We leave this to the reader.

THEOREM 1.22 Let f be a convex function defined on a convex set $X \subset \mathbb{R}^n$. If the set of points $V = \{x : x \in X, f(x) \leqslant \alpha, \alpha \in \mathbb{R}\}$ is non-empty, V is a convex set.

Proof. Let x, $y \in V$ and consider the point $\lambda x + (1-\lambda)y$ for $0 \leqslant \lambda \leqslant 1$. Since f is convex $f(\lambda x + (1-\lambda)y) \leqslant \lambda f(x) + (1-\lambda)f(y)$. But $f(x) \geqslant \alpha$ and $f(y) \geqslant \alpha$, so that for $0 \leqslant \lambda \leqslant 1$ we have $\lambda f(x) + (1-\lambda)f(y) \leqslant \lambda\alpha + (1-\lambda)\alpha = \alpha$ and hence $f(\lambda x + (1-\lambda)y) \leqslant \alpha$. But this implies $\lambda x + (1-\lambda)y \in V$ for $0 \leqslant \lambda \leqslant 1$ and so V is convex.

Results analogous to those above hold for concave functions, as the reader may verify by replacing f by $-f$. Note that the converse of the last theorem is not necessarily true. A weaker property than the convexity of f, known as quasiconvexity, will suffice to guarantee the convexity of the set V. This is discussed in Section 1.3.2.

In the definition of convexity of functions it is not required that the functions be continuous. Nevertheless, there is a close relationship between convexity and continuity of functions.

THEOREM 1.23 A real-valued convex function defined on an open convex set $X \subset \mathbb{R}^n$ is continuous at every point of X.

Proof. See Rockafellar (1970), p. 82.

Note, however, that a real-valued convex function defined on a closed convex set need not be continuous at every point in its domain as the following example makes clear.

EXAMPLE 1.2 Let $f : X \to \mathbb{R}$, where $X = \{x : x \geqslant 0\}$ and $f(x) = -2$, if $x > 0$, and $f(x) = 0$, if $x = 0$. Clearly, f is a convex function, but f is not continuous at $x = 0$.

Although convex functions defined on open convex sets are continuous at all points in their domain they need not be differentiable at all points in their domain. For example, the function $f : \mathbb{R} \to \mathbb{R}$ where $f(x) = |x|$ is convex and continuous for all $x \in \mathbb{R}$, but it is not differentiable at $x = 0$.

THEOREM 1.24 Let f be a convex function defined on an open convex set $X \subset \mathbb{R}^n$. Then $[f(x+\lambda y)-f(x)]/\lambda$ is a non-decreasing function of λ for $\lambda>0$.

Proof. Let $x \in X$ and let $\mu=\alpha\lambda$ with $0<\alpha\leqslant 1$. Then since f is convex, we have $f(x+\alpha\lambda y)\equiv f((1-\alpha)x+\alpha(x+\lambda y))\leqslant(1-\alpha)f(x)+\alpha f(x+\lambda y)$. It follows that $[f(x+\alpha\lambda y)-f(x)]/\alpha\leqslant f(x+\lambda y)-f(x)$. Therefore, $[f(x+\mu y)-f(x)]/\mu=\{\mu^{-1}[f(x+\alpha\lambda y)-f(x)]\}/\alpha \leqslant[f(x+\lambda y)-f(x)]/\lambda$ for all μ such that $0<\mu\leqslant\lambda$ and all λ and y such that $x+\lambda y \in X$. Since μ is an arbitrary positive number less than or equal to λ, we can see that the required quotient is non-decreasing for $\lambda>0$.

DEFINITION 1.13 Let f be a real-valued function defined on an open set $X \subset \mathbb{R}^n$. The **directional derivative of f at x in the direction of y**, denoted $\delta f(x:y)$, is given by

$$\delta f(x:y)=\lim_{\lambda\to 0+}[f(x+\lambda y)-f(x)]/\lambda$$

for all $x \in X$ and $y \in \mathbb{R}^n$ for which the limit exists ($+\infty$ and $-\infty$ being allowed as limits).

Note that $-\delta f(x:-y)=\lim_{\lambda\to 0-}[f(x+\lambda y)-f(x)]/\lambda$.

THEOREM 1.25 Let f be a convex function defined on an open convex subset X of $\mathbb{R}^n$. Then, given any $y \in \mathbb{R}^n$ and any $x \in X$, f has a directional derivative at x in the direction of y. Moreover, $\delta f(x:y)$ is a positively homogeneous convex function of y, with $\delta f(x:\theta)=0$ and $-\delta f(x:-y)\leqslant\delta f(x:y)$.

Proof. See Rockafellar (1970), p. 214.

DEFINITION 1.14 If $\delta f(x:y)=-\delta f(x:-y)$, then f is said to have a **two-sided directional derivative at x in the direction of y**. If f has a two-sided directional derivative in all directions at x then f is said to be **Gateaux differentiable at x**.

THEOREM 1.26 Let f be a real-valued function defined on $X \subset \mathbb{R}^n$ and let $x^\circ \in X$. If f is differentiable at x° then $\delta f(x^\circ:y)$ exists and is a linear function of y; specifically, $\delta f(x^\circ:y) = f_x(x^\circ)y = \sum_{i=1}^n [\partial f(x^\circ)/\partial x_i]y$.

Proof. Since f is differentiable at x° we have for any $y \neq \theta$

$$0 = \lim_{\lambda \to 0+} [f(x^\circ + \lambda y) - f(x^\circ) - f_x(x^\circ)\lambda y]/\lambda \|y\|$$
$$= [\delta f(x^\circ:y) - f_x(x^\circ)y]/\|y\|.$$

Therefore, $\delta f(x^\circ:y)$ exists and is a linear function of y; $\delta f(x^\circ:y) = f_x(x^\circ)y$, for all y. But $\delta f(x^\circ:y) = \lim_{\lambda \to 0+}[f(x^\circ + \lambda y) - f(x^\circ)]/\lambda$, so that letting $y = e_i$, for $i = 1, \ldots, n$ where e_i is the vector forming the ith row of the $n \times n$ identity matrix, we have

$$f_x(x^\circ)e_i = \lim_{\lambda \to 0+} [f(x^\circ + \lambda e_i) - f(x^\circ)]/\lambda = \partial f(x^\circ)/\partial x_i,$$

$i = 1, \ldots, n.$

Hence, it follows that $f_x(x^\circ) = [\partial f(x^\circ)/\partial x_1, \ldots, \partial f(x^\circ)/\partial x_n]$, so that for any $y = (y_1, \ldots, y_n)'$ we have $\delta f(x^\circ:y) = \sum_{i=1}^n [\partial f(x^\circ)/\partial x_i] y_i$.

The above result clearly shows the relationship between directional derivatives and partial derivatives.

The next concept to be introduced is the notion of a subgradient, which is related to the ordinary gradient in the case of differentiable convex functions, and to directional derivatives in the more general case.

DEFINITION 1.15 Let f be a real-valued convex function defined on $\mathbb{R}^n$. A vector $\xi \in \mathbb{R}^n$ is a **subgradient of f at x°** if $f(x) \geq f(x^\circ) + \xi'(x - x^\circ)$, for all $x \in \mathbb{R}^n$. The set of all subgradients of f at x° is called the **subdifferential of f at x°**, denoted by $\partial f(x^\circ)$. If $\partial f(x^\circ)$ is non-empty, f is said to be **subdifferentiable** at x°. Similarly, let g be a concave function on $\mathbb{R}^n$. Then the vector $\xi \in \mathbb{R}^n$ is a **supergradient of g at x°** if $g(x) \leq g(x^\circ) + \xi'(x - x^\circ)$, for each $x \in \mathbb{R}^n$. The set of all supergradients of g at x°, denoted $\partial g(x^\circ)$, is the

superdifferential of g at x°. If $\partial g(x^\circ)$ is non-empty then g is **superdifferentiable at x°**.

EXAMPLE 1.3 The convex function f defined by $f(x)=|x|$ for all $x\in\mathbb{R}$ is subdifferentiable throughout $\mathbb{R}$, although it is differentiable (in the ordinary sense) only for every $x\in\mathbb{R}$, $x\neq 0$. The set $\partial f(0)$ consists of all points $\xi\in\mathbb{R}$ such that $|x|\geqslant \xi x$, for all $x\in\mathbb{R}$. For $x^\circ\neq 0$, $\partial f(x^\circ)$ consists of the single point $|x^\circ|/x^\circ$. To see this, suppose $\xi\in f(x^\circ)$, $x^\circ\neq 0$. Then ξ must satisfy $|x|\geqslant|x^\circ|+\xi(x-x^\circ)$, for all $x\in\mathbb{R}$. If $x=x^\circ$ this inequality is satisfied by any $\xi\in\mathbb{R}$. Suppose then that $x\neq x^\circ$. If $x-x^\circ>0$, then for the inequality to hold $\xi\geqslant[|x|-|x_1^\circ|]/(x-x^\circ)$ and if $x-x^\circ<0$, then for the inequality to hold $\xi\leqslant[|x|-|x^\circ|]/(x-x^\circ)$. But for ξ to be a subgradient the inequality must hold for all $x\in\mathbb{R}$, in particular for $x=0$, so that for $x^\circ\neq 0$, if ξ is to be a subgradient, we must have $\xi=|x^\circ|/x^\circ$.

THEOREM 1.27 Let f be a real-valued function on an open convex subset X of $\mathbb{R}^n$. Then f is convex if, and only if, f has a subgradient at each point $x^\circ\in X$; that is, for each $x^\circ\in X$ there must exist an $\xi\in\mathbb{R}^n$ such that

$$f(x)\geqslant f(x^\circ)+\xi'(x-x^\circ),\ \text{for all}\ x\in X.$$

Proof. See Bazaraa and Shetty (1976), p. 92.

From the above theorem we immediately have that a real-valued convex function on an open convex subset X of $\mathbb{R}^n$ is subdifferentiable on X.

Subgradients can be characterized by directional derivatives:

THEOREM 1.28 A vector $\xi\in\mathbb{R}^n$ is a subgradient of a real-valued convex function f at a point x° if, and only if, $\delta f(x^\circ\!:\!y)\geqslant\xi' y$ for every direction y.

Proof. Letting $x=x^\circ+\lambda y$, we can turn the subgradient inequality into the condition that $[f(x^\circ+\lambda y)-f(x^\circ)]/\lambda\geqslant\xi' y$, for every y and $\lambda>0$. Since the difference quotient decreases to $\delta f(x^\circ\!:\!y)$ as $\lambda\to 0^+$, this inequality is equivalent to the one in the theorem.

THEOREM 1.29 Let f be a real-valued convex function defined on a convex subset X of $\mathbb{R}^n$. If f is differentiable at x°, then $f_x(x^\circ)$ is the unique subgradient of f at x°. Conversely, if f has a unique subgradient at x°, then f is differentiable at x°.

Proof. Suppose that f is differentiable at x°. Then $\delta f(x^\circ{:}y)$ is a linear function $f_x(x^\circ)y$. By Theorem 1.28, the subgradients at x° are the vectors such that $f_x(x^\circ)y \geqslant \xi' y$, for all y. But this condition is satisfied if, and only if, $\xi' = f_x(x^\circ)$. Thus, $f_x(x^\circ)$ is the unique subgradient of f at x°. The converse is more difficult to establish. A proof can be found in Rockafellar (1970), p. 242.

In the remainder of this section we restrict attention to real-valued differentiable convex functions.

THEOREM 1.30 Let f be a real-valued convex function on an open convex set $X \subset \mathbb{R}^n$. If f is differentiable on X, then f is continuously differentiable on X.

Proof. See Rockafellar (1970), p. 246.

THEOREM 1.31 If f is a real-valued differentiable function on an open convex set $X \subset \mathbb{R}^n$, then f is convex if, and only if,

$$f(x) - f(x^\circ) \geqslant f_x(x^\circ)(x - x^\circ), \quad \text{for all } x^\circ,\ x \in X.$$

Furthermore, f is strictly convex if, and only if, the above holds with strict inequality for all x°, $x \in X$.

Proof. (Note that the above theorem is a special case of Theorem 1.27; we shall, however, provide an alternative proof.)

Suppose f is convex. Then $f(\lambda x + (1-\lambda)x^\circ) \leqslant \lambda f(x) + (1-\lambda)f(x^\circ)$ for all x, $x^\circ \in X$ and all $0 \leqslant \lambda \leqslant 1$, or equivalently,

$$f(x^\circ + \lambda(x - x^\circ)) \leqslant f(x^\circ) + \lambda[f(x) - f(x^\circ)]. \tag{1.1}$$

From Taylor's Theorem the left-side of (1.1) can be expressed as

$$f(x^\circ + \lambda(x - x^\circ)) = f(x^\circ) + \lambda f_x(x^\circ + \lambda\theta(x - x^\circ))(x - x^\circ) \quad \text{with } 0 < \theta < 1. \tag{1.2}$$

Substituting (1.2) into (1.1) and dividing by $\lambda>0$ and taking the limit as $\lambda\to 0^+$ gives $f_x(x^\circ)(x-x^\circ)\leqslant f(x)-f(x^\circ)$, establishing necessity. To prove sufficiency, let $x, x^\circ\in X, x\neq x^\circ$, $x_1\equiv(1-\lambda)x+\lambda x^\circ$, $0<\lambda<1$ and $h\equiv x-x_1$. Then we have $x^\circ=x_1-[(1-\lambda)/\lambda]h$, and since $f(x)-f(x_1)\geqslant f_x(x_1)(x-x_1)$ we thus have

$$f(x)-f(x_1)\geqslant f_x(x_1)h, \tag{1.3}$$

$$f(x^\circ)-f(x_1)\geqslant f_x(x_1)[-(1-\lambda)/\lambda]h. \tag{1.4}$$

Multiplying (1.3) by $(1-\lambda)/\lambda$ and adding the result obtained to (1.4) gives

$$(1-\lambda)f(x)+\lambda f(x^\circ)\geqslant f(x_1)=f((1-\lambda)x+\lambda x^\circ),\quad \text{for } 0<\lambda<1. \tag{1.5}$$

If $\lambda=1$ then (1.5) also holds and reduces to $f(x^\circ)\geqslant f(x^\circ)$. Similarly, if $\lambda=0$, (1.5) reduces to $f(x)\geqslant f(x)$. Hence (1.5) holds for all $0\leqslant\lambda\leqslant 1$, so that f is convex on X. A similar proof can be given for the strict case.

THEOREM 1.32 If f is a real-valued and twice continuously differentiable function on an open convex set $X\subset\mathbb{R}^n$, then f is convex if, and only if, its Hessian matrix, f_{xx}, is positive semidefinite everywhere on X; that is, $z'f_{xx}z\geqslant 0$ for all $z\neq\theta$ and all $x\in X$.

Proof. Since X is convex, any $x\in X$ can be expressed as $x^\circ+z$. From Taylor's Theorem

$$f(x^\circ+z)=f(x^\circ)+f_x(x^\circ)z+\tfrac{1}{2}z'f_{xx}z, \tag{1.6}$$

where f_{xx} is evaluated at $x^\circ+\theta z$, $0<\theta<1$. But from Theorem 1.31, if f is convex then $f(x^\circ+z)-f(x^\circ)\geqslant f_x(x^\circ)z$, so that $z'f_{xx}z\geqslant 0$. Since this is true for all $z\neq\theta, f_{xx}$ is positive semidefinite everywhere on X.

To establish sufficiency, note that from Taylor's Theorem (1.6) holds, so that if f_{xx} is positive semidefinite everywhere on X, we have $f(x^\circ+z)-f(x^\circ)\geqslant f_x(x^\circ)z$ for all $x\in X$, and hence from Theorem 1.31 f is convex on X.

Note that Theorem 1.31 gives necessary and sufficient conditions for f to be convex and necessary and sufficient conditions for f to be strictly convex. In contrast, Theorem 1.32 gives only necessary and sufficient conditions for f to be convex. Sufficient conditions for f to be strictly convex are given in the following theorem.

THEOREM 1.33 If f is a real-valued and twice continuously differentiable function on an open set $X \subset \mathbb{R}^n$ then f is strictly convex if its Hessian matrix, f_{xx}, is positive definite everywhere on X: that is, if $z' f_{xx} z > 0$ for all $z \neq \theta$ and all $x \in X$.

Proof. As for Theorem 1.32.

It is important to realize that Theorem 1.33 does not give necessary conditions for f to be strictly convex. Indeed, f can be strictly convex without its Hessian matrix being positive semidefinite everywhere on X. For example, the function f defined by $f(x) = x^4$ is strictly convex on $X = \mathbb{R}$, but $f''(x) = 0$ for $x = 0$. However, since any strictly convex function is convex, from Theorem 1.32, the Hessian matrix of a strictly convex function must be positive semidefinite everywhere on X.

Results analogous to those given above can be established for concave functions. These are summarized below. To obtain these results all one need do is to replace f by $-f$ in the above proofs.

THEOREM 1.34

(i) Let f be a real-valued differentiable function on an open convex set $X \subset \mathbb{R}^n$. Then f is concave if, and only if,

$$f(x) - f(x^\circ) \leqslant f_x(x^\circ)(x - x^\circ) \text{ for all } x, x \in X.$$

Furthermore, f is strictly concave if, and only if, the above holds with strict inequality for all $x \in X, x \neq x^\circ$.

(ii) Let f be a real-valued twice continuously differentiable function on an open convex set $X \subset \mathbb{R}^n$. Then f is concave if, and only if, its Hessian matrix, f_{xx}, is negative semidefinite everywhere on X; that is, $z' f_{xx} z \leqslant 0$ for all $z \neq \theta$ and all $x \in X$.

Furthermore, if its Hessian matrix is negative definite everywhere on X then f is strictly concave; that is, if $z'f_{xx}z<0$ for all $z\neq\theta$ and all $x\in X$ then f is strictly concave.

1.3.2 Quasiconcave functions

A class of functions with similar properties to those of concave (convex) functions is the class of quasiconcave (quasiconvex) functions. While classes of functions intermediate between those of concavity and quasiconcavity have been considered in the literature on optimization (see for example Avriel, 1976, Chapter 6 and Ponstein, 1967) economists have by and large restricted attention to the classes of concave and quasiconcave functions.

DEFINITION 1.16 A real-valued function f defined on a convex set $X\subset\mathbb{R}^n$ is **quasiconcave** if $f(x)\geqslant f(x^\circ)$ implies $f(\lambda x+(1-\lambda)x^\circ)\geqslant f(x^\circ)$ for all λ, $0\leqslant\lambda\leqslant1$. If $f(x)>f(x^\circ)$ implies $f(\lambda x+(1-\lambda)x^\circ)>f(x^\circ)$ for all λ, $0<\lambda<1$ then f is **strictly quasiconcave**. A real-valued function f is **quasiconvex** if $-f$ is quasiconcave and it is **strictly quasiconvex** if $-f$ is strictly quasiconcave.

THEOREM 1.35 A real-valued function f defined on a convex set $X\subset\mathbb{R}^n$ is quasiconcave if, and only if, for each $\alpha\in\mathbb{R}$, the set $V=\{x:x\in X, f(x)\geqslant\alpha\}$ is convex.

Proof. Let $x,x^\circ\in V$, so that $f(x)\geqslant\alpha$ and $f(x^\circ)\geqslant\alpha$. Suppose f is quasiconcave. Then for $f(x)\geqslant f(x^\circ)$ we have $f(\lambda x+(1-\lambda)x^\circ)\geqslant f(x^\circ)$ for all $0\leqslant\lambda\leqslant1$, so that $f(\lambda x+(1-\lambda)x^\circ)\geqslant\alpha$ and hence V is convex. Conversely, suppose V is convex for all $\alpha\in\mathbb{R}$. Let x, $x^\circ\in X$ and $\alpha^\circ=\min\{f(x), f(x^\circ)\}$. Since V is convex, $\lambda x+(1-\lambda)x^\circ\in V$ for all $0\leqslant\lambda\leqslant1$. Hence, $f(\lambda x+(1-\lambda)x^\circ)\geqslant\alpha^\circ$, so that if V is convex $f(x)\geqslant f(x^\circ)$ implies $f(\lambda x+(1-\lambda)x^\circ)\geqslant f(x^\circ)$ for $0\leqslant\lambda\leqslant1$.

THEOREM 1.36 All concave functions are quasiconcave and all strictly concave functions are strictly quasiconcave.

Proof. If f is concave then $f(\lambda x+(1-\lambda)x^{\circ}) \geqslant \lambda f(x)+(1-\lambda)f(x^{\circ}) = \lambda[f(x)-f(x^{\circ})]+f(x^{\circ})$ for $0 \leqslant \lambda \leqslant 1$. If $f(x) \geqslant f(x^{\circ})$, then for $0 \leqslant \lambda \leqslant 1$ we have $\lambda[f(x)-f(x^{\circ})] \geqslant 0$ and hence $f(\lambda x+(1-\lambda)x^{\circ}) \geqslant f(x^{\circ})$. The strict case can be proved in similar fashion.

It is important to note that the converse of the above theorem is not in general valid. For example, the function f defined on $X=\{x{:}x \geqslant 0\}$ by $f(x)=x^2$ is quasiconcave, but not concave on X; actually, it is strictly convex on X.

THEOREM 1.37 Any monotone increasing function of a quasiconcave function is quasiconcave.

Proof. Let ψ be a monotone increasing transformation. Then ψ does not reverse rankings. Thus, if $f(x) \geqslant f(x^{\circ})$ then $\psi[f(x)] \geqslant \psi[f(x^{\circ})]$ and if $f(\lambda x+(1-\lambda)x^{\circ}) \geqslant f(x^{\circ})$ then $\psi[f(\lambda x+(1-\lambda)x^{\circ})] \geqslant \psi[f(x^{\circ})]$. But if f is quasiconcave $f(x) \geqslant f(x^{\circ})$ implies $f(\lambda x+(1-\lambda)x^{\circ}) \geqslant f(x^{\circ})$ for $0 \leqslant \lambda \leqslant 1$, so that $\psi \circ f$ is quasiconcave.

In economics a minimal property usually required of utility functions is that they be quasiconcave. Since utility functions are unique only up to monotone increasing transformations, Theorem 1.37 ensures that such transformations do not affect the quasiconcavity of the transformed utility function.

As quasiconcavity is a weaker requirement than concavity, some of the properties of concave functions do not carry over to quasiconcave functions. In particular, non-negative linear combinations of quasiconcave functions are not necessarily quasiconcave.

THEOREM 1.38 Let f be a real-valued twice continuously

differentiable function on $\mathbb{R}^n_+$. Define

$$|H^B_r(x)| \equiv \begin{vmatrix} 0 & f_1 & \cdots & f_r \\ f_1 & f_{11} & \cdots & f_{1r} \\ \vdots & \vdots & & \vdots \\ f_r & f_{r1} & \cdots & f_{rr} \end{vmatrix}$$

A sufficient condition for f to be quasiconcave is that the sign of $|H^B_r(x)|$ be the same as the sign of $(-1)^r$ for all $x \in \mathbb{R}^n_+$ and all $r=1,\ldots,n$. A necessary condition for f to be quasiconcave is that $(-1)^r|H^B_r(x)| \geqslant 0$ for all $x \in \mathbb{R}^n_+$ and all $r=1,\ldots,n$.

Proof. See Arrow and Enthoven (1961).

EXAMPLE 1.4 In the theory of production, firms' production functions are frequently assumed to be quasiconcave. Quasiconcavity of production functions does not rule out increasing returns to scale. Letting q denote output and x a vector of inputs, the production function $q=f(x)$, where f is defined on $\mathbb{R}^n_+$, exhibits increasing returns to scale if $f(\lambda x)=\beta f(x)$ and $\beta>\lambda>0$. If f is positively homogeneous of degree s, then f exhibits increasing returns to scale if $s>1$. The Cobb-Douglas production function defined by $f(x)=Ax_1^\alpha x_2^\beta$, $A>0$, $0<\alpha<1$, $0<\beta<1$, exhibits increasing returns to scale if $\alpha+\beta>1$. However, for $\alpha+\beta>1$, the Cobb-Douglas function is not concave. To see this note that $f_1=\alpha q x_1^{-1}$, $f_2=\beta q x_2^{-1}$, $f_{11}=\alpha(\alpha-1)qx_1^{-2}$, $f_{22}=\beta(\beta-1)qx_2^{-2}$ and $f_{12}=f_{21}=\alpha\beta q x_1^{-1}x_2^{-1}$. From Theorem 1.34, concavity of f requires that the Hessian matrix of f be negative semidefinite, so that in particular $f_{11}f_{22}-(f_{12})^2=\alpha\beta q^2x_1^{-2}x_2^{-2}[1-(\alpha+\beta)]\geqslant 0$. But for $\alpha+\beta>1$ this inequality is violated, so that f is not concave for $\alpha+\beta>1$. Nevertheless, the Cobb-Douglas function is quasiconcave even for $\alpha+\beta>1$. We have, for $r=1$, $|H^B_r(x)|=-(\alpha q x_1^{-1})^2<0$ and for $r=2$, $|H^B_r(x)|=\alpha\beta(\alpha+\beta)q^3x_1^{-1}x_2^{-1}>0$, so that the sufficient conditions for quasiconcavity given in Theorem 1.38 are satisfied for all $\alpha+\beta$.

1.3.3 Extremum properties

Convex and concave functions play a central role in the theory of optimization. Their importance in the theory of optimization derives directly from the results given in this section.

DEFINITION 1.17 A real-valued function f has a **global maximum** on a set $S \subset \mathbb{R}^n$ at a point $x^\circ \in S$ if $f(x) \geqslant f(x^\circ)$ for all $x \in S$; if the strict inequality holds for all $x \in S$, $x \neq x^\circ$, then f has a **strict global maximum** at x°. A real-valued function f has a **local maximum** on a set $S \subset \mathbb{R}^n$ at a point $x^\circ \in S$ if there exists some neighbourhood $N(x^\circ)$ such that $f(x) \geqslant f(x^\circ)$ for all $x \in S \cap N(x^\circ)$. If the strict inequality holds for $x \in S \cap N(x^\circ)$, $x \neq x^\circ$, then f has a **strict local maximum** at x°.

By reversing the direction of the inequality signs in the above definition, we obtain definitions of **global minimum**, **local minimum** etc. Clearly, if f has a global (local) maximum at x° then $-f$ has a global (local) minimum at x°. Furthermore, it is evident that a global maximum (minimum) is also a local maximum (minimum), but not vice versa, and that a strict global maximum (minimum) occurs at a unique point.

THEOREM 1.39 If f is a concave or strictly quasiconcave function over the convex set $S \subset \mathbb{R}^n$ then any local maximum of f on S is a global maximum of f on S.

Proof. Since concavity does not imply strict quasiconcavity we need to establish the result for both cases. Suppose f has a local maximum at x° with respect to a neighbourhood N of x°. Suppose that x° is not a global maximum point, so that there exists an $x^* \in S$, $x^* \notin N$ such that $f(x^*) > f(x^\circ)$. Let $B_\delta(x^\circ)$ be an open ball of radius δ with centre x° entirely contained in N. If f is concave $f(\lambda x^* + (1-\lambda)x^\circ) \geqslant \lambda f(x^*) + (1-\lambda) f(x^\circ)$ for $0 \leqslant \lambda \leqslant 1$. Since $f(x^*) > f(x^\circ)$ the right side of the above inequality exceeds $f(x^\circ)$, if $\lambda \neq 0$, so that $f(\lambda x^* + (1-\lambda)x^\circ) > f(x^\circ)$ for $0 < \lambda \leqslant 1$. Similarly, if f is strictly quasiconcave, by definition, we have, for $f(x^*) > f(x^\circ)$, $f(\lambda x^* + (1-\lambda)x^\circ) > f(x^\circ)$ for $0<\lambda<1$. Let $x \equiv \lambda x^* + (1-\lambda)x^\circ$ for $\lambda < \lambda^\circ$ with λ chosen, so that $\lambda^\circ < 1$ and $0<\lambda^\circ<\delta/\|x^* - x^\circ\|$, so that $x \in B_\delta(x^\circ)$ and $x \in S$. Then $f(x) > f(x^\circ)$ for all $0<\lambda<\lambda^\circ$, which contradicts the fact that f has a local maximum at x° with respect to the neighbourhood N.

THEOREM 1.40 A global maximum of a strictly concave or strictly quasiconcave function f over the convex set $S \subset \mathbb{R}^n$ is unique.

Proof. If f is strictly concave it is also strictly quasiconcave so we need only establish the result for strict quasiconcavity. Suppose $f(x^*)$ and $f(x^\circ)$ are global maximum points, x^*, $x^\circ \in S, x^* \neq x^\circ$. From strict quasiconcavity we have for $0<\lambda<1$, $f(\lambda x^* + (1-\lambda)x^\circ) > \min[f(x^*), f(x^\circ)]$, but since $f(\lambda x^* + (1-\lambda)x^\circ) \in S$, as S is convex, this contradicts the fact that $f(x^*)$ and $f(x^\circ)$ are global maxima.

If f is quasiconcave, but not strictly quasiconcave, over the convex set $S \subset \mathbb{R}^n$, then it does *not* follow that any local maximum of f is a global maximum. To see this consider Figure 1.7, where the set S is given by $S = \{x : x \in \mathbb{R}, 0 \leqslant x \leqslant x_1\}$. Here f is an increasing function of a single variable and hence quasiconcave, and it is easily seen that all points in the interval (x_2, x_3) are local maximum points as is the point x_1. However, it is equally apparent that only the point x_1 is a global maximum point.

THEOREM 1.41 If f is a quasiconcave function over the convex set $S \subset \mathbb{R}^n$ then every local maximum point x° is either a global maximum point or $f(x^\circ) = f(x)$ for some $x \in N(x^\circ)$, $x \neq x^\circ$.

Proof. Suppose f has a local maximum at $x^\circ \in S$ with respect to a neighbourhood $N(x^\circ)$ but that x° is not a global maximum point. Then there exists an $x^* \in S$, $x^* \notin N(x^\circ)$ such that $f(x^*) > f(x^\circ)$. Let $B_\delta(x^\circ)$ be an open ball of radius δ with centre x^* entirely contained in $N(x^\circ)$. As f is quasiconcave, $f(\lambda x^* + (1-\lambda)x^\circ) \geqslant f(x^*)$ for

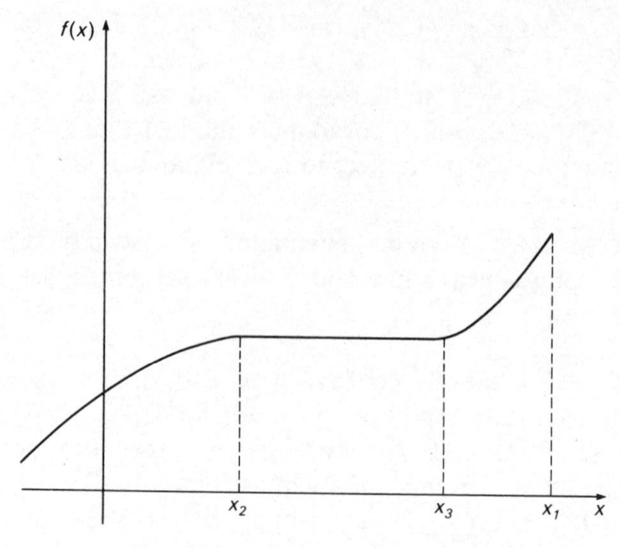

Figure 1.7

$0 \leqslant \lambda \leqslant 1$. Let $x \equiv \lambda x^* + (1-\lambda)x^\circ$ for $\lambda < \lambda^\circ$ with λ° chosen, so that $\lambda^\circ < 1$ and $0 < \lambda^\circ < \delta/\|x^\circ - x^*\|$, so that $x \in B_\delta(x^\circ)$, $x \neq x^\circ$ and $x \in S$. Then $f(x) \geqslant f(x^\circ)$ for all $0 < \lambda < \lambda^\circ$. But, since x° is a local maximum point, $f(x^\circ) \geqslant f(x)$ for all $x \in N(x^\circ)$. Hence if x° is not a global maximum point then $f(x^\circ) = f(x)$ for some $x \in N(x^\circ)$, $x \neq x^\circ$.

THEOREM 1.42 Let f be a concave or quasiconcave function over the convex set $S \subset \mathbb{R}^n$. Let $Y \subset S$ be the set of points at which f takes on its global maximum. Then Y is a convex set.

Proof. If f is concave it is also quasiconcave, so we need only establish the result for quasiconcavity. If a global maximum does not exist, or occurs at a unique point, then Y is trivially convex. Suppose that x^* and x° are global maximum points of f over S, $x^* \neq x^\circ$. Since f is quasiconcave, $f(\lambda x^\circ + (1-\lambda)x^*) \geqslant \min\{f(x^\circ), f(x^*)\}$ for $0 \leqslant \lambda \leqslant 1$. But $f(x^\circ)$ and $f(x^*)$ are global maxima, so $f(\lambda x^\circ + (1-\lambda)x^*) = f(x^*)$, for

$0 \leqslant \lambda \leqslant 1$. Moreover, $\lambda x^\circ + (1-\lambda)x^* \in S$ for $0 \leqslant \lambda \leqslant 1$, as $x^\circ, x^* \in S$ and so Y is convex.

THEOREM 1.43 Let f be a concave function on $\mathbb{R}^n$. Then $\theta \in \partial f(x^\circ)$ if, and only if, f attains its global maximum at x°.

Proof. From the definition of a supergradient, $\theta \in \partial f(x^\circ)$ if, and only if, $f(y) \leqslant f(x^\circ)$ for every $y \in \mathbb{R}^n$; that is, x° is a global maximum point of f.

THEOREM 1.44 Let f be a differentiable concave function on $\mathbb{R}^n$ Then $f_x(x^\circ) = \theta$ if, and only if, f attains its global maximum at x°.

Proof. Recall that $\partial f(x^\circ) = f_x(x^\circ)$ if, and only if, f is differentiable at x°, so that the result follows directly from Theorem 1.43.

Note that the last theorem will generally remain valid if we replace $\mathbb{R}^n$ by an open convex subset $X \subset \mathbb{R}^n$ such that $x^\circ \in X$.

Results analogous to those above hold for convex and quasiconvex functions.

CHAPTER 2

STATIC OPTIMIZATION

2.1 INTRODUCTION

The theory of *static optimization* (or *optimization in finite dimensional spaces*) deals with the problem of determining the values of certain variables, x, called *control variables*, subject to a prescribed set of constraints on their possible values, so as to maximize (minimize) a given function f, called the *objective function*. The control vector x is *feasible* if it satisfies all the constraints of the problem, and the set of all feasible vectors, S, is referred to as the *feasible set*. In general terms the static optimization problem can be expressed as:

$$\underset{x}{\text{maximize}}\; f(x) \text{ subject to } x \in S \subset \mathbb{R}^n. \tag{2.1}$$

As a general rule, we denote the choice or control variables by writing them under the term 'maximize'.

From Definition 1.17 it is clear that if f has a global (local) maximum at $x^\circ \in S$ then $-f$ has a global (local) minimum at $x^\circ \in S$. This observation means that we need only consider either maximization or minimization problems, not both. We have chosen to deal with maximization problems in this chapter. Any minimization problem can be treated as one of maximization by changing the sign of the function to be minimized.

The most fundamental question which can be asked of any

optimization problem is whether a global maximum exists. Obviously, any optimization problem for which the feasible set is empty has neither a global nor a local maximum. However, the static optimization problem (2.1) may not admit a solution (global or local) even when the feasible set is non-empty. For example, the function $f:X \to \mathbb{R}$, where $X = \{x : x \in \mathbb{R}, 0 < x < 1\}$ and $f(x) = x^2$ has no global or local maximum on X. From the Weierstrass Theorem we have that, if the feasible set S is non-empty and compact and the objective function f is continuous on S, then f has a global maximum on S. In the above example, the set X is not compact. Note however that the Weierstrass Theorem gives only sufficient, not necessary, conditions for the existence of a global maximum. For example, if the set X in the above problem is changed to $X = \{x : x \in \mathbb{R}, 0 < x \leqslant 1\}$ then clearly f has a global maximum at $x = 1$, although X is still not compact, as $0 \notin X$.

Assuming that a solution exists for problem (2.1), the problem is to locate it. In solving a practical problem we would have information concerning the specific forms of the objective and constraint functions, the latter defining the feasible set, and we would then try to devise computational procedures to find a solution as efficiently as possible. In economic theory, however, we work only with general functions, usually specifying little more than the signs of their first and second-order derivatives. As a result, the problem is not actually to compute solutions but rather to describe their essential characteristics in the analytically most useful manner. Moreover, when dealing with optimization problems in economics, we are typically concerned to obtain conditions which characterize the essential nature of those points for which a given objective function has a global maximum, rather than in obtaining characterizations of those points for which the objective function has a local maximum. However, virtually all methods for solving static optimization problems identify local rather than global maximum points. Therefore, in applying static optimization theory to economic problems, it is desirable to impose restrictions on the functions concerned so as to ensure that a local maximum point of the objective function is also a global maximum point. This is why in economic theory agents are typically assumed to have objective functions that are concave or strictly quasiconcave, for the results given in Theorems 1.39–44 are then applicable.

2.2 CLASSICAL OPTIMIZATION

2.2.1 Unconstrained optimization

The unconstrained optimization problem is that of choosing values for the control variables x so as to maximize a real-valued function f of those variables; specifically

$$\underset{x}{\text{maximize}}\ f(x) \text{ subject to } x \in S, \tag{2.2}$$

where S is the domain of the objective function and is the entire space $\mathbb{R}^n$ or an open subset of $\mathbb{R}^n$.

From Theorem 1.44 we know that if f is a differentiable concave function on an open convex set $S \subset \mathbb{R}^n$ then f has a global maximum at $x^\circ \in S$ if, and only if, $f_x(x^\circ) = \theta$. The following theorem provides a set of necessary conditions for any twice continuously differentiable function to have a local maximum on an open set $S \subset \mathbb{R}^n$.

THEOREM 2.1 (Necessary conditions) If $f: S \to \mathbb{R}$ is twice continuously differentiable on an open set $S \subset \mathbb{R}^n$ then:

(i) a *first-order* necessary condition for a local maximum to occur at $x^\circ \in S$ is that

$$f_x(x^\circ) = \theta, \tag{2.3}$$

(ii) a *second-order* necessary condition for a local maximum to occur at $x^\circ \in S$ is that

$$z' f_{xx}(x^\circ) z \leqslant 0 \text{ for all } z \in \mathbb{R}^n \tag{2.4}$$

(i.e., $f_{xx}(x^\circ)$ is negative semidefinite).

Proof. Suppose f is twice continuously differentiable on an open set $S \subset \mathbb{R}^n$ and that f has a local maximum at x°. Then $f(x^\circ) \geqslant f(x)$ for all x in some neighbourhood $N(x^\circ) \subset S$. Let $B_\delta(x^\circ)$ be an open ball of radius δ with centre x° entirely contained in $N(x^\circ)$. Every $x \in B_\delta(x^\circ)$ can be expressed as $x = x^\circ + \psi z$, where $\psi > 0$, $|\psi| < \delta$ and $\|z\| = 1$, so that

$$f(x^\circ) \geqslant f(x^\circ + \psi z). \tag{2.5}$$

From Taylor's Theorem

$$f(x^\circ + \psi z) = f(x^\circ) + \psi f_x(x^\circ + \lambda\psi z)z$$

for some λ, $0 < \lambda < 1$, so that from (2.5) we have

$$\psi f_x(x^\circ + \lambda\psi z)z \leqslant 0,$$

and dividing by ψ and taking the limit as $\psi \to 0$ we obtain, from the continuous differentiability of f,

$$f_x(x^\circ)z \leqslant 0. \tag{2.6}$$

Since (2.6) is to hold for all z such that $\|z\| = 1$, it must be that

$$f_x(x^\circ) = \theta. \tag{2.7}$$

Also from Taylor's Theorem we have

$$f(x^\circ + \psi z) = f(x^\circ) + \psi f_x(x^\circ)z + \tfrac{1}{2}\psi^2 z' f_{xx}(x^\circ + \lambda\psi z)z$$

for some λ, $0 < \lambda < 1$, so that from (2.5) and (2.7)

$$\tfrac{1}{2}\psi^2 z' f_{xx}(x^\circ + \lambda\psi z)z \leqslant 0,$$

and dividing by $\frac{1}{2}\psi^2$ and taking the limit as $\psi \to 0$ we obtain

$$z' f_{xx}(x^\circ)z \leqslant 0. \tag{2.8}$$

Since (2.8) is to hold for all z, subject to the restriction on the norm of z, it follows that (2.8) holds for all $z \in \mathbb{R}^n$, since multiplication of z by a scalar does not affect the sign of the quadratic form.

We now consider sufficient conditions for a local maximum. Two sets of sufficient conditions are given below, one of which provides sufficient conditions for a strict local maximum to occur at x°, based on the behaviour of the function at that point, the other providing sufficient conditions for a local (not necessarily strict) maximum to occur at x°, based on the behaviour of f in a neighbourhood of x°.

THEOREM 2.2 (Sufficient conditions) If $f: S \to \mathbb{R}$ is twice continuously differentiable on an open set $S \subset \mathbb{R}^n$ the following conditions are sufficient for f to have a local maximum at x°.

$$f_x(x^\circ) = \theta, \tag{2.9}$$

and for every x in some neighbourhood $N(x^\circ)$ and for every $z \in \mathbb{R}^n$,

$$z' f_{xx}(x) z \leqslant 0. \tag{2.10}$$

Proof. Suppose (2.9) and (2.10) hold but that x° is not a local maximum point. Then there exists an open ball $B_\delta(x^\circ)$ entirely contained in $N(x^\circ)$ with $x \in B_\delta(x^\circ)$, $x \neq x^\circ$ such that $f(x) > f(x^\circ)$. Let $x = x^\circ + \psi z$, where $0 < |\psi| < \delta$ and $\|z\| = 1$. By Taylor's Theorem $f(x) = f(x^\circ) + \psi f_x(x^\circ) z + \frac{1}{2}\psi^2 z' f_{xx}(x^\circ + \lambda \psi z) z$ for some λ, $0 < \lambda < 1$. Our assumptions then lead to $z' f_{xx}(x^\circ + \lambda \psi z) z > 0$, since $x^\circ + \lambda \psi z \in B_\delta(x^\circ)$, but $z' f_{xx}(x^\circ + \lambda \psi z) z > 0$ contradicts (2.10).

THEOREM 2.3 (Sufficient conditions: strict maximum) If $f: S \to \mathbb{R}$ is twice continuously differentiable on an open set $S \subset \mathbb{R}^n$ the following conditions are sufficient for f to have a strict local maximum at x°:

$$f_x(x^\circ) = \theta, \tag{2.11}$$

and

$$z' f_{xx}(x^\circ) z < 0, \text{ for all non-zero vectors } z \in \mathbb{R}^n. \tag{2.12}$$

Proof. Suppose (2.11) and (2.12) hold but that x° is not a strict local maximum point. Then there exists an open ball $B_\delta(x^\circ)$ such that

$$f(x) \geqslant f(x^\circ) \tag{2.13}$$

for some $x \in B_\delta(x^\circ)$, $x \neq x^\circ$. Let $x = x^\circ + \psi z$, where $\|z\| = 1$ and

$0<|\psi|<\delta$. Then from Taylor's Theorem $f(x)=f(x^\circ)+\psi f_x(x^\circ)z+\frac{1}{2}\psi^2 z' f_{xx}(x^\circ+\lambda\psi z)z$, for some $\lambda, 0<\lambda<1$. Hence from (2.13) and (2.11) we have $z' f_{xx}(x^\circ+\lambda\psi z)z \geqslant 0$ and taking the limit as $\psi \to 0$ gives $z' f_{xx}(x^\circ)z \geqslant 0$, which contradicts (2.12).

EXAMPLE 2.1 A price-taking profit maximizing firm produces its output, q, via a technology that can be summarized by a twice continuously differentiable strictly concave production function f defined over $\mathbb{R}^2_{++}$: $q=f(x_1,x_2)$. Let $r=(r_1,r_2)'$ be a vector of input prices with $r \gg \theta$ and $p>0$ be the price of output. The firm's profits are then given by

$$\Pi(x_1,x_2)=pf(x_1,x_2)-r_1x_1-r_2x_2.$$

Since f is strictly concave, Π is strictly concave, so that the conditions

$$pf_1-r_1=0,$$

$$pf_2-r_2=0$$

are necessary and sufficient for profits to be globally maximized. These conditions are that the value of the marginal product of input i, pf_i, be equal to the price of input i, r_i, $i=1,2$.

2.2.2 Equality constrained optimization

We now consider the problem of choosing values for the control variables so as to maximize a given objective function when the choice of values for the control variables is restricted by a set of equations. Specifically, we examine the problem

$$\underset{x}{\text{maximize}}\ f(x) \text{ subject to } g^i(x)=0, \qquad i=1,\ldots,m, \qquad m<n, \tag{2.14}$$

where f and all the g^i are continuously differentiable on an open set $X \subset \mathbb{R}^n$.

The most intuitive method of solution for problem (2.14) involves the elimination of m variables from the problem by use of the constraint equations, thereby converting the problem into an equivalent unconstrained optimization problem involving $n-m$ control variables. The conditions for such an elimination are provided by the Implicit Function Theorem. The actual solution of the constraint equations for m variables in terms of the remaining $n-m$ can often prove a difficult, if not impossible, task. Moreover the method of elimination of variables is seldom applicable to economic problems, as economic theory rarely allows for the specification of particular functional forms. Nevertheless, the theory underlying the method of elimination of variables can be used to obtain analytically useful characterizations of solutions to equality constrained optimization problems.

Using the Implicit Function Theorem, we can derive the following result which gives necessary conditions for a local solution to the equality constrained optimization problem.

THEOREM 2.4 Let f and g^i, $i=1,\ldots,m$, $m<n$, be continuously differentiable on an open set $X \subset \mathbb{R}^n$. Suppose that x° is a local maximum point of f for all $x \in N(x^\circ)$ that also satisfy $g^i(x)=0$, $i=1,\ldots,m$. Also assume that the Jacobian matrix $[\partial g^i(x^\circ)/\partial x_j]$ has rank m. Then there exists a unique vector of real numbers $\lambda^\circ=(\lambda_1^\circ,\ldots,\lambda_m^\circ)'$, commonly termed **multipliers**, such that

$$-f_x(x^\circ)=\sum_{i=1}^{m} \lambda_i^\circ g_x^i(x^\circ)$$

$$\text{i.e. } -[\partial f(x^\circ)/\partial x_j]=\sum_{i=1}^{m} \lambda_i^\circ[\partial g^i(x^\circ)/\partial x_j], \quad \text{for } j=1,\ldots,n. \tag{2.15}$$

In other words, the gradient vector of f at x° is a linear combination of the gradient vectors of the g^i's at this point.

Proof. Clearly, we can always relabel the variables so that the first m columns of the Jacobian matrix $[\partial g^i(x^\circ)/\partial x_j]$ are linearly independent. The set of linear equations

$$\sum_{i=1}^{m} \lambda_i[\partial g^i(x^\circ)/\partial x_j]=-[\partial f(x^\circ)/\partial x_j], \qquad j=1,\ldots,m \tag{2.16}$$

then has a unique solution for the λ_i's denoted λ°. Therefore the first m equations in (2.15) are satisfied. We must now verify that for this choice of the λ_i's the remaining $m-n$ equations in (2.15) are also satisfied. Let $\hat{x} \equiv (x_{m+1}, \ldots, x_n)'$. Then, applying the Implicit Function Theorem to $g^i(x)=0$, $i=1,\ldots,m$, at x°, there exist functions $h^j(\hat{x})$ and an open set $Y \subset \mathbb{R}^{n-m}$ containing x° such that

$$x_j^\circ = h^j(\hat{x}^\circ), \qquad j=1,\ldots,m, \qquad \hat{x}^\circ \in Y, \tag{2.17}$$

and

$$f(x^\circ) = f(h^1(\hat{x}^\circ), \ldots, h^m(\hat{x}^\circ), \hat{x}^\circ). \tag{2.18}$$

As a result of the last expression, it follows from Theorem 2.1 that the first-order partial derivatives of f with respect to $x_{m+1}, \ldots, x_n$ must vanish at x°. Thus,

$$\partial f(x^\circ)/\partial x_j = \sum_{k=1}^{m} [\partial f(x^\circ)/\partial x_k][\partial h^k(x^\circ)/\partial x_j] + [\partial f(x^\circ)/\partial x_j] = 0,$$
$$j=m+1,\ldots,n. \tag{2.19}$$

From the Implicit Function Theorem we have for every $j=m+1,\ldots,n$

$$\sum_{k=1}^{m} [\partial g^i(x^\circ)/\partial x_k][\partial h^k(x^\circ)/\partial x_j] = -[\partial g^i(x^\circ)/\partial x_j],$$
$$i=1,\ldots,m. \tag{2.20}$$

Multiplying each of the equations in (2.20) by $-\lambda_i^\circ$ and adding up, we obtain

$$\sum_{i=1}^{m}\sum_{k=1}^{m} \{-\lambda_i^\circ [\partial g^i(x^\circ)/\partial x_k][\partial h^k(x^\circ)/\partial x_j]\}$$
$$-\sum_{i=1}^{m} \lambda_i^\circ [\partial g^i(x^\circ)/\partial x_j] = 0, \qquad j=m+1,\ldots,n. \tag{2.21}$$

Subtracting (2.21) from (2.19) and rearranging, we get

$$\sum_{k=1}^{m}\left[\{[\partial f(x^\circ)/\partial x_k]+\sum_{i=1}^{m}\lambda_i^\circ[\partial g^i(x^\circ)/\partial x_k]\}[\partial h^k(x^\circ)/\partial x_j]\right]$$
$$+[\partial f(x^\circ)/\partial x_j]+\sum_{i=1}^{m}\lambda_i^\circ[\partial g^i(x^\circ)/\partial x_j]=0,$$
$$j=m+1,\ldots,n. \qquad (2.22)$$

But the first expression on the left side of (2.22) is zero by (2.16), and so

$$\partial f(x^\circ)/\partial x_j+\sum_{i=1}^{m}\lambda_i^\circ[\partial g^i(x^\circ)/\partial x_j]=0, \qquad j=m+1,\ldots,n. \qquad (2.23)$$

This last expression, together with (2.16), yields the result.

The relation between the gradient vector of the objective function and the gradient vectors of the constraint functions at a local maximum point, as expressed in (2.15), leads rather naturally to the formulation of the **Lagrangean function** defined by

$$L(x,\lambda)=f(x)+\sum_{i=1}^{m}\lambda_i g^i(x), \qquad (2.24)$$

where the λ_i are called **Lagrange multipliers**. Theorem 2.4 can be reexpressed in terms of the Lagrangean function as follows.

THEOREM 2.5 (Necessary conditions) Let f and g^i, $i=1,\ldots,m$, $m<n$, be continuously differentiable on an open set $X\subset\mathbb{R}^n$. Suppose that x° is a local maximum point of f for all $x\in N(x^\circ)$ that also satisfy $g^i(x)=0$, $i=1,\ldots,m$. Also assume that the Jacobian matrix $[\partial g^i(x^\circ)/\partial x_j]$ has rank m. Then there exist unique multipliers λ_i°, $i=1,\ldots,m$ such that

$$L_x(x^\circ,\lambda^\circ)=\theta, \qquad (2.25)$$

$$g^i(x^\circ)=0, \qquad i=1,\ldots,m. \qquad (2.26)$$

It is to be emphasized that condition (2.25) is only a necessary condition for a local maximum when the Jacobian matrix

$(\partial g^i(x^\circ)/\partial x_j]$ has rank m. To see this consider the following simple problem

$$\text{maximize } f(x_1,x_2)=x_1x_2 \text{ subject to } g(x_1,x_2)=(1-x_1-x_2)^2=0.$$

Clearly, the optimal solution occurs at $x^\circ=(\frac{1}{2},\frac{1}{2})'$ but we have $g_x(x^\circ)=\theta$. Conditions (2.25) and (2.26) here are

$$x_1^\circ-2\lambda^\circ(1-x_1^\circ-x_2^\circ)=0,$$

$$x_2^\circ-2\lambda^\circ(1-x_1^\circ-x_2^\circ)=0,$$

$$(1-x_1^\circ-x_2^\circ)^2=0.$$

From the first two equations we have $x_1^\circ=x_2^\circ$, and from the last equation $x_1^\circ+x_2^\circ=1$, so $x_1^\circ=x_2^\circ=\frac{1}{2}$, but then the first two equations become $\frac{1}{2}=0$!

Second-order necessary conditions for a local maximum to occur in problem (2.14) are also available:

THEOREM 2.6 (Second-order necessary conditions) Let f and g^i, $i=1,\ldots,m$, $m<n$, be twice continuously differentiable on an open set $X\subset\mathbb{R}^n$. Suppose that x° is a local maximum point of f for all $x\in N(x^\circ)$ that also satisfy $g^i(x)=0$, $i=1,\ldots,m$. Also assume that the Jacobian matrix $[\partial g^i(x^\circ)/\partial x_j]$ has rank m. Then there exist multipliers λ_i°, $i=1,\ldots,m$ such that

$$L_x(x^\circ,\lambda^\circ)=\theta, \tag{2.27}$$

$$g^i(x^\circ)=0, \qquad i=1,\ldots,m, \tag{2.28}$$

$$z'L_{xx}(x^\circ,\lambda^\circ)z\leqslant 0 \tag{2.29}$$

for all $z\in\mathbb{R}^n$ for which $g_x^i(x^\circ)'z=0$, $i=1,\ldots,m$.

Proof. See, for example, Fuss and McFadden (1978), Appendix A.2.

We now consider sufficient conditions for a local maximum in problem (2.14). While sufficient conditions for a local (not necessarily strict) maximum are available in the literature (see

Fiacco, 1968), we consider only sufficient conditions for a strict local maximum, since these are more important in economic applications.

THEOREM 2.7 (Sufficient conditions: strict local maximum) Let f and g^i, $i=1,\ldots,m$, $m<n$, be twice continuously differentiable on an open set $X\subset\mathbb{R}^n$. If there exist vectors $x^\circ\in X$ and $\lambda^\circ=(\lambda_1^\circ,\ldots,\lambda_m^\circ)'$ such that

$$L_x(x^\circ,\lambda^\circ)=\theta, \tag{2.30}$$

and if for every non-zero vector $z\in\mathbb{R}^n$ satisfying

$$g_x^i(x^\circ)z=0, \qquad i=1,\ldots,m \tag{2.31}$$

we have

$$z'L_{xx}(x^\circ,\lambda^\circ)z<0 \tag{2.32}$$

then f has a strict local maximum at x° subject to $g^i(x)=0$, $i=1,\ldots,m$.

Proof. Suppose (2.30)–(2.32) hold and that x° is feasible, but that x° is not a strict constrained local maximum point. Then there exists an open ball $B_\delta(x^\circ)$ and an $\hat{x}\in B_\delta(x^\circ)$, $\hat{x}\neq x^\circ$ such that

$$g^i(\hat{x})=0, \qquad i=1,\ldots,m, \tag{2.33}$$

$$f(\hat{x})>f(x^\circ). \tag{2.34}$$

Let $\hat{x}\equiv x^\circ+\psi z$, where $\|z\|=1$ and $|\psi|<\delta$, $\delta>0$. Then, from (2.33) and Taylor's Theorem, we have for some θ, $0<\theta<1$,

$$g_x^i(x^\circ+\theta\psi z)z=0, \qquad i=1,\ldots,m.$$

Taking the limit as $\psi\to 0$ we obtain

$$g_x^i(x^\circ)z=0. \tag{2.35}$$

Also from Taylor's Theorem, for some $0<\gamma<1$,

$$L(\hat{x},\lambda^\circ)=L(x^\circ,\lambda^\circ)+\psi L_x(x^\circ,\lambda^\circ)z+\tfrac{1}{2}\psi^2 z'L_{xx}(x^\circ+\gamma\psi z,\lambda^\circ)z \tag{2.36}$$

But $L(\hat{x},\lambda^\circ)=f(\hat{x})+\sum_{i=1}^m\lambda_i^\circ g^i(\hat{x})=f(\hat{x})$ from (2.33); also, $L(x^\circ,\lambda^\circ)=f(x^\circ)$, as x° is feasible, and $L_x(x^\circ,\lambda^\circ)=\theta$ from (2.20).

Thus, from (2.36) $f(\hat{x})=f(x^\circ)+\frac{1}{2}\psi^2 z' L_{xx}(x^\circ+\gamma\psi z,\lambda^\circ)z$, so that from (2.34) we have

$$z' L_{xx}(x^\circ+\gamma\psi z,\lambda^\circ)z \geqslant 0,$$

and taking the limit as $\psi \to 0$ we obtain

$$z' L_{xx}(x^\circ,\lambda^\circ)z \geqslant 0. \tag{2.37}$$

Since $\|z\|=1$, and $g_x^i(x^\circ)z=0$, we see that (2.37) contradicts (2.32). Thus, it must be that, if (2.30)–(2.32) hold, then x° must be a strict constrained local maximum.

From the results on constrained quadratic forms given in the Mathematical Review, the above result can be expressed as follows.

THEOREM 2.8 (Sufficient conditions: strict local maximum) Let f and g^i, $i=1,\dots,m$, $m<n$, be twice continuously differentiable on an open set $X \subset \mathbb{R}^n$. Suppose the Jacobian matrix $[\partial g^i(x^\circ)/\partial x_j]$ has rank m and the variables are indexed in such a way that the first m columns of this matrix are linearly independent. If there exist vectors $x^\circ \in X$, $\lambda^\circ=(\lambda_1^\circ,\dots,\lambda_m^\circ)'$ such that

$$L_x(x^\circ,\lambda^\circ)=\theta, \tag{2.38}$$

and if

$$(-1)^r \begin{vmatrix} \dfrac{\partial^2 L(x^\circ,\lambda^\circ)}{\partial x_1^2} & \cdots & \dfrac{\partial L(x^\circ,\lambda^\circ)}{\partial x_1\,\partial x_r} & \dfrac{\partial g^1(x^\circ)}{\partial x_1} & \cdots & \dfrac{\partial g^m(x^\circ)}{\partial x_1} \\ \vdots & & \vdots & \vdots & & \vdots \\ \dfrac{\partial^2 L(x^\circ,\lambda^\circ)}{\partial x_r\,\partial x_1} & \cdots & \dfrac{\partial^2 L(x^\circ,\lambda^\circ)}{\partial x_r^2} & \dfrac{\partial g^1(x^\circ)}{\partial x_r} & \cdots & \dfrac{\partial g^m(x^\circ)}{\partial x_r} \\ \dfrac{\partial g^1(x^\circ)}{\partial x_1} & \cdots & \dfrac{\partial g^1(x^\circ)}{\partial x_r} & 0 & \cdots & 0 \\ \vdots & & \vdots & \vdots & & \vdots \\ \dfrac{\partial g^m(x^\circ)}{\partial x_1} & \cdots & \dfrac{\partial g^m(x^\circ)}{\partial x_r} & 0 & & 0 \end{vmatrix} > 0 \tag{2.39}$$

for $r=m+1,\ldots,n$, then f has a strict local maximum at x° such that $g^i(x^\circ)=0$, $i=1,\ldots,m$.

The above results are concerned with the characterization of constrained local maximum points. Since the necessary conditions for a local maximum are also necessary conditions for a global maximum, Theorems 2.6 and 2.7 provide a set of necessary conditions for a constrained global maximum. Sufficient conditions for a constrained global maximum are given in Section 2.3, where the more general *non-linear programming problem* is treated.

EXAMPLE 2.2 Consider a consumer choosing a vector of commodities x from his consumption set X, where $X=\{x:x\gg\theta, x\in\mathbb{R}^n\}$, so as to maximize his utility function, $U(x)$, subject to a budget constraint $p'x\leqslant m$, where $p\gg\theta$ is a vector of prices for the n commodities and m is the consumer's available income. Typically, the utility function is assumed to be a twice continuously differentiable, strictly quasiconcave function, whose first derivatives, U_i, are all positive for all $x\in X$. This last assumption ensures that the budget constraint will hold with equality. The consumer's problem can thus be expressed as maximize $U(x)$ subject to $g(x)=m-p'x=0$, $x\in X$. Note that since the feasible set is convex and the objective function is strictly quasiconcave any local maximum must be a strict global maximum. Since $g_x(x)\neq\theta$, for all $x\in X$, necessary conditions for a local maximum to occur at $x^\circ\in X$ are

$$L_x(x^\circ,\lambda^\circ)=\theta \tag{i}$$

or equivalently

$$U_j(x^\circ)-\lambda^\circ p_j=0, \qquad j=1,\ldots,n, \tag{ii}$$

and

$$z'U_{xx}(x^\circ,\lambda^\circ)z\leqslant 0 \text{ for all } z\in\mathbb{R}^n \text{ for which } g_x(x^\circ)z=p'z=0. \tag{iii}$$

Condition (iii) is equivalent to the requirement

$$z'U_{xx}(x^\circ)z\leqslant 0 \text{ for all } z\in\mathbb{R}^n \text{ for which } U_x(x^\circ)z=0, \tag{iv}$$

since $L_{xx}(x^\circ,\lambda^\circ)=U_{xx}(x^\circ)$ and $p'z=0$ implies $U_x(x^\circ)z=0$ in view of

(i) and positive λ°. But strict quasiconcavity of $U(x)$ ensures that (iv) holds.

Sufficient conditions for a strict local maximum to occur at x° are that (i) holds and that $z'L_{xx}(x^\circ, \lambda^\circ)z < 0$ for all $z \in \mathbb{R}^n$, $z \neq \theta$, for which $p'z = 0$, the last requirement being equivalent to the condition,

$$z'U_{xx}(x^\circ)z < 0 \text{ for all } z \in \mathbb{R}^n,\ z \neq \theta, \text{ for which } U_x(x^\circ)z = 0. \qquad \text{(v)}$$

Note that strict quasiconcavity of U does not ensure (v), it only guarantees (iv). Thus, if (i) and (v) hold then the consumption vector x°, obtained as the unique solution vector to (i), is the consumption vector which globally maximizes utility. Moreover, if (i) and (v) hold, then the matrix

$$\begin{bmatrix} U_{xx}(x^\circ) & p \\ p' & 0 \end{bmatrix}$$

is non-singular, and, from the Implicit Function Theorem, the consumer's continuously differentiable demand functions $x_i = f^i(p, m)$ can be obtained from (ii) and the budget constraint.

EXAMPLE 2.3 (Optimal commodity taxation) Suppose there are $n+1$ commodities in the economy, the first of which is labour (commodity 0) and the remaining n commodities are consumer goods. Let P_i be the price of commodity i and x_i its quantity. Let labour serve as the numeraire. At the set of prices $P_1, \ldots, P_n$, the consumer will choose his labour supply and consumption of commodities so as to maximize his utility subject to his budget constraint. Suppose the consumer has no exogenous income and that wage income is not subject to taxation. At the set of prices $P_1, \ldots, P_n$, the consumer's labour supply and consumption of commodities are given by the solution to the problem

$$\text{maximize } U(x_0, x_1, \ldots, x_n) \text{ subject to } \sum_{i=0}^{n} P_i x_i = 0. \qquad \text{(i)}$$

Necessary conditions for a solution to (i) are given by

$$\sum_{i=0}^{n} P_i x_i = 0, \qquad \text{(ii)}$$

$$\partial U/\partial x_i - \lambda P_i = 0, \qquad i=0,1,\ldots,n, \tag{iii}$$

where λ is the multiplier associated with (i).

If the sufficient conditions for a strict local maximum hold, then, as in Example 2.2, equations (ii) and (iii) can be solved to yield the consumer's continuously differentiable demand functions and his labour supply function

$$x_i(P_1,\ldots,P_n), \qquad i=0,1,\ldots,n. \tag{iv}$$

Suppose that the consumer goods are subject to indirect taxation and that the government has a fixed amount of tax revenue to collect, R, where the amount of tax to be collected is expressed in terms of labour. Let t_i be the tax on commodity i; then we have

$$\sum_{i=1}^{n} t_i x_i = R, \tag{v}$$

where $t_i \equiv P_i - p_i$, P_i is the price paid by consumers and p_i is the price received by producers. Assume producer prices are given, so that the problem of selecting a tax structure is equivalent to choosing a structure of consumer prices. Furthermore, suppose that the consumer side of the economy can be treated as if there were only a single consumer, whose utility function is

$$U(x_0, x_1, \ldots, x_n), \tag{vi}$$

where $U(.)$ has the usual concavity properties and labour, x_0, is negative.

The Government's optimal commodity taxation problem then becomes that of selecting a tax structure $(t_1,\ldots,t_n)$, or equivalently, a consumer price structure $(P_1,\ldots,P_n)$, which satisfies (v) and maximizes (vi), where the x_i's are consumer's demand functions and his labour supply function, (iv). Formally, the optimal commodity taxation problem is

$$\underset{P_1,\ldots,P_n}{\text{maximize}}\ U(x_0, x_1, \ldots, x_n) \text{ subject to } \sum_{i=1}^{n} t_i x_i = R,$$

where $t_i \equiv P_i - p_i$ and $x_i = x_i(P_1,\ldots,P_n)$, $i=0,1,\ldots,n$.

We have the Lagrangean

$$L(P_1,\ldots,P_n,\mu)=u(x_o,x_1,\ldots,x_n)+\mu\left(\sum_{i=1}^{n} t_i x_i - R\right), \qquad \text{(vii)}$$

from which we obtain the first-order necessary conditions

$$\sum_{i=0}^{n}(\partial U/\partial x_i)(\partial x_i/\partial P_k)+\mu\left[\sum_{i=1}^{n} t_i(\partial x_i/\partial P_k)+x_k\right]=0,$$

$$k=1,\ldots,n. \qquad \text{(viii)}$$

Substituting (iii) into (viii) we get

$$\lambda\sum_{i=0}^{n} P_i(\partial x_i/\partial P_k)+\mu\left[\sum_{i=1}^{n} t_i(\partial x_i/\partial P_k)+x_k\right]=0, \quad k=1,\ldots,n. \qquad \text{(ix)}$$

But from (ii) we have, on differentiating,

$$\sum_{i=0}^{n} P_i(\partial x_i/\partial P_k)+x_k=0, \qquad k=1,\ldots,n, \qquad \text{(x)}$$

so that (ix) can be expressed as

$$\sum_{i=1}^{n} t_i(\partial x_i/\partial P_k)=[(\lambda-\mu)/\mu]x_k, \qquad k=1,\ldots,n. \qquad \text{(xi)}$$

Translating (xi) into an optimal tax structure requires some standard results from consumer theory, namely the Slutsky equation,

$$(\partial x_i/\partial P_k)=(\partial x_i/\partial P_k)|_u - x_k(\partial x_i/\partial I), \qquad \text{(xii)}$$

and the symmetry of the substitution matrix

$$(\partial x_i/\partial P_k)|_u=(\partial x_k/\partial P_i)|_u. \qquad \text{(xiii)}$$

Using (xii) and (xiii), (xi) can be expressed as

$$\sum_{i=1}^{n}(t_i/x_k)(\partial x_k/\partial P_i)|_u=[(\lambda-\mu)/\mu]+\sum_{i=1}^{n} t_i(\partial x_i/\partial I),$$

$$k=1,\ldots,n. \qquad \text{(xiv)}$$

The left side of (xiv) can be interpreted as the relative decrease in demand for commodity k following a tax change, provided that the consumer is compensated so as to stay on the same indifference curve. Since the right side of (xiv) is independent of k, it follows that this proportionate reduction of demand should be the same for all commodities.

2.3 NON-LINEAR PROGRAMMING

2.3.1 Saddle point optimization

The most general form of the static optimization problem is the *non-linear programming problem*

$$\underset{x}{\text{maximize}}\ f(x) \text{ subject to } g^i(x) \geqslant 0, \qquad i=1,\ldots,m, \tag{2.40}$$

where f and the g^i's are real-valued functions defined on an open set $X \subset \mathbb{R}^n$. If f and the g^i's are all concave, problem (2.40) is said to be a *concave programming problem*. If f and the g_i's are all linear functions, problem (2.40) is said to be a *linear programming problem*. The linear programming problem is clearly a concave programming problem. Several computational procedures exist for solving linear programming problems. For a discussion of such techniques the reader is referred to Hadley (1967). In this book we consider only the more general non-linear programming problem.

In the case of concave programming, the set of points S satisfying the constraints is a convex set, and since f is concave any local maximum point is also a global maximum point.

For the non-linear programming problem (2.40), define the associated **Lagrangean function**, as in equation (2.24), as $L(x,\lambda) \equiv f(x) + \sum_{i=1}^m \lambda_i g^i(x)$.

DEFINITION 2.1 The point (x°, λ°) is a **saddlepoint** of the Lagrangean $L(x,\lambda)$ if $L(x,\lambda^\circ) \leqslant L(x^\circ,\lambda^\circ) \leqslant L(x^\circ,\lambda)$, for every $x \in X, \lambda \geqslant \theta$; that is $f(x) + \sum_{i=1}^m \lambda_i^\circ g^i(x) \leqslant f(x^\circ) + \sum_{i=1}^m \lambda_i^\circ g^i(x^\circ) \leqslant f(x^\circ) + \sum_{i=1}^m \lambda_i g^i(x^\circ)$, for every $x \in X$, $\lambda \geqslant \theta$.

The existence of a saddlepoint of $L(x, \lambda)$ and a solution to the non-linear programming problem bear a close theoretical relationship to each other. We proceed to establish this relationship through a number of theorems which are all parts of what are known as the *Kuhn-Tucker theory.*

THEOREM 2.9 If $L(x, \lambda)$ has a saddlepoint (x°, λ°) then $g^i(x^\circ) \geqslant 0$, $i = 1, \ldots, m$ and $\sum_{i=1}^m \lambda_i^\circ g^i(x^\circ) = 0$.

Proof. Suppose (x°, λ°) is a saddlepoint of $L(x, \lambda)$. Then we have $f(x) + \sum_{i=1}^m \lambda_i^\circ g^i(x) \leqslant f(x^\circ) + \sum_{i=1}^m \lambda_i^\circ g^i(x^\circ) \leqslant f(x^\circ) + \sum_{i=1}^m \lambda_i g^i(x^\circ)$. From the right-hand inequality,

$$\sum_{i=1}^m \lambda_i^\circ g_i(x^\circ) \leqslant \sum_{i=1}^m \lambda_i g^i(x^\circ). \tag{2.41}$$

Suppose $g^i(x^\circ) < 0$ for some i. Then whatever may be λ_i° we can make λ_i sufficiently large thereby making $\sum_{i=1}^m \lambda_i g^i(x^\circ)$ small enough to violate (2.41). Hence $g^i(x^\circ) \geqslant 0$, for all i. Now, since $\lambda_i^\circ \geqslant 0$ and $g^i(x^\circ) \geqslant 0$ for all i,

$$\sum_{i=1}^m \lambda_i^\circ g^i(x^\circ) \geqslant 0. \tag{2.42}$$

Also, since (2.41) holds for all $\lambda \geqslant \theta$, it holds for $\lambda = \theta$, and hence

$$\sum_{i=1}^m \lambda_i^\circ g(x^\circ) \leqslant 0. \tag{2.43}$$

Inequalities (2.42) and (2.43) imply $\sum_{i=1}^m \lambda_i^\circ g(x^\circ) = 0$.

Using the above result we can derive sufficient conditions for a global maximum in the non-linear programming problem.

THEOREM 2.10 (Sufficient conditions) If $L(x, \lambda)$ has a saddlepoint (x°, λ°) then f has a global maximum at x° subject to $g^i(x) \geqslant 0$, $i = 1, \ldots, m$.

Proof. From the definition of a saddlepoint and Theorem 2.9 we have $f(x)+\sum_{i=1}^{m}\lambda_i^\circ g^i(x)\leqslant f(x^\circ)$, and since $\lambda_i^\circ\geqslant 0$, $i=1,\ldots,m$, it follows that $f(x^\circ)\geqslant f(x)$ for all x for which $g^i(x)\geqslant 0$, $i=1,\ldots,m$.

Note that Theorem 2.10 does *not* require any concavity assumptions and that it holds whether the Lagrangean is differentiable or not. The converse of the above theorem is not in general true. To see this consider the problem

$$\underset{x}{\text{maximize}}\; f(x)=-x \quad \text{subject to} \quad -x^2\geqslant 0. \tag{2.44}$$

The optimal solution is clearly $x^\circ=0$. If (x°,λ°) is to be a saddlepoint of the Lagrangean for this problem then

$$-x-\lambda^\circ x^2\leqslant -x^\circ-\lambda^\circ x^{\circ 2}\leqslant -x^\circ-\lambda x^{\circ 2}, \text{ for every } x\in\mathbb{R}, \quad \lambda\geqslant 0 \tag{2.45}$$

For $x^\circ=0$, (2.45) becomes

$$x+\lambda^\circ x^2\geqslant 0. \tag{2.46}$$

Clearly, $\lambda^\circ\neq 0$, for if $\lambda^\circ=0$ then (2.46) implies $x\geqslant 0$, but (2.46) must hold for every $x\in\mathbb{R}$ if (x°,λ°) is to be a saddlepoint of $L(x,\lambda)$. Now for any $\lambda^\circ>0$ we can choose $-1/\lambda^\circ<x<0$ and (2.46) will not hold. Thus, there exists no $\lambda^\circ\geqslant 0$ such that (x°,λ°), with $x^\circ=0$, is a saddlepoint of the Lagrangean.

While the converse of Theorem 2.10 generally does not hold, in the case of concave programming it does, *if there exists an* $\hat{x}\in X$ *such that* $g^i(\hat{x})>0$, $i=1,\ldots,m$.

THEOREM 2.11 (Necessary conditions) Let f and g^i, $i=1,\ldots,m$, be concave functions on an open convex set $X\subset\mathbb{R}^n$, and suppose there exists an $\hat{x}\in X$ such that $g^i(\hat{x})>0$, $i=1,\ldots,m$. If f has a global maximum at x° subject to the constraints $g^i(x)\geqslant 0$, $i=1,\ldots,m$, then there exists a $\lambda^\circ\geqslant\theta$ such that (x°,λ°) is a saddlepoint of the Lagrangean $L(x,\lambda)=f(x)+\sum_{i=1}^{m}\lambda_i g^i(x)$.

Proof. Let $z=(z_0,z_1,\ldots,z_m)'\in\mathbb{R}^{m+1}$ and consider a set K_1 defined as

$$K_1\equiv\{z{:}z_0\leqslant f(x)-f(x^\circ)\; z_i\leqslant g^i(x), i=1,\ldots,m\}.$$

Since f and each g^i are concave functions, K_1 is a convex set. Consider another set $K_2 \subset \mathbb{R}^{m+1}$ defined as

$$K_2 = \{z : z \gg \theta\}.$$

K_2 is also a convex set. Furthermore, K_1 and K_2 are disjoint sets because with $z_0 \leqslant f(x) - f(x^\circ) \leqslant 0$ for $z \epsilon K_1$ and $z_0 > 0$ for $z \epsilon K_2$, K_1 and K_2 cannot have any common point. K_1 and K_2 being two disjoint convex sets, from Theorem 1.14 there exists a hyperplane separating them. The point $z = \theta$ is a boundary point of both sets and therefore the separating hyperplane must pass through it. Let the separating hyperplane be $v'z = 0$, $v \neq \theta$ such that

$$v'z \geqslant 0 \text{ for } z \epsilon K_1 \tag{2.47}$$

$$v'z < 0 \text{ for } z \epsilon K_2. \tag{2.48}$$

It is necessary that $v \leqslant \theta$, for if $v \nleqslant \theta$ then there is a component v_k of v such that $v_k > 0$. Let $z^\circ \epsilon K_2$. Then $z^\circ \gg \theta$. Let the kth component of z° be $z_k^\circ = M$, $M > 0$. The term $v_k z_k^\circ$ in $v'z$ will then be positive. By making M sufficiently large, we can make this term dominate over other terms in $v'z$ and thus violate (2.48). So we conclude $v \leqslant \theta$.

Now let $\hat{z} \equiv [f(x) - f(x^\circ), g^1(x), \ldots, g^m(x)]'$, so that $\hat{z} \epsilon K_1$. Then, from (2.47), $v_0 f(x) - v_0 f(x^\circ) + v_1 g^1(x) + \cdots + v_m g^m(x) \geqslant 0$; that is

$$v_0 f(x) + \sum_{i=1}^{m} v_i g^i(x) \geqslant v_0 f(x^\circ). \tag{2.49}$$

But $v_0 \neq 0$; this we show as follows. For $v_0 = 0$, (2.49) reduces to

$$\sum_{i=1}^{m} v_i g^i(x) \geqslant 0. \tag{2.50}$$

Let $\hat{x}$ be a point such that $g^i(\hat{x}) > 0$; such a point exists by assumption. Also $v \leqslant \theta$ and $v \neq \theta$. Hence, for this $\hat{x}$, (2.50) is contradicted. Therefore $v_0 \neq 0$. Dividing (2.49) by v_0 and putting $\lambda_i^\circ = v_i / v_0$, we get

$$f(x) + \sum_{i=1}^{m} \lambda_i^\circ g^i(x) \leqslant f(x^\circ), \tag{2.51}$$

where $\lambda_i^\circ \geqslant 0, i=1,\ldots,m$. Since (2.51) holds for all feasible x, it also holds for x°, and so

$$f(x^\circ)+\sum_{i=1}^{m}\lambda_i^\circ g^i(x^\circ)\leqslant f(x^\circ). \tag{2.52}$$

But $\lambda_i^\circ \geqslant 0$, $i=1,\ldots,m$ and $g^i(x^\circ)\geqslant 0$, $i=1,\ldots,m$, so that $\sum_{i=1}^m \lambda_i^\circ g^i(x^\circ)\geqslant 0$, and hence from (2.52)

$$\sum_{i=1}^{m}\lambda_i^\circ g^i(x^\circ)=0. \tag{2.53}$$

Inequality (2.51) can thus be reexpressed as

$$f(x)+\sum_{i=1}^{m}\lambda_i^\circ g^i(x)\leqslant f(x^\circ)+\sum_{i=1}^{m}\lambda_i^\circ g^i(x^\circ). \tag{2.54}$$

But $\lambda_i \geqslant 0$ and $g^i(x^\circ)\geqslant 0$ for $i=1,\ldots,m$ imply $\sum_{i=1}^m \lambda_i g^i(x^\circ)\geqslant 0$ and so

$$f(x^\circ)+\sum_{i=1}^{m}\lambda_i g^i(x^\circ)\geqslant f(x^\circ)=f(x^\circ)+\sum_{i=1}^{m}\lambda_i^\circ g^i(x^\circ), \quad \text{for} \quad \lambda\geqslant\theta. \tag{2.55}$$

Combining (2.54) and (2.55) gives

$$f(x)+\sum_{i=1}^{m}\lambda_i^\circ g^i(x)\leqslant f(x^\circ)+\sum_{i=1}^{m}\lambda_i^\circ g^i(x^\circ)\leqslant f(x^\circ)+\sum_{i=1}^{m}\lambda_i g^i(x^\circ),$$

for $\lambda\geqslant\theta$, or equivalently, $L(x,\lambda^\circ)\leqslant L(x^\circ,\lambda^\circ)\leqslant L(x^\circ,\lambda)$, $\lambda\geqslant\theta$, which establishes the result.

To sum up the conclusions of Theorems 2.9–2.11, we have shown that a sufficient condition for x° to be a global maximum point of f subject to the constraints $g^i(x)\geqslant 0$, $i=1,\ldots,m$, is that the Lagrangean function $L(x,\lambda)\equiv f(x)+\sum_{i=1}^m \lambda_i g^i(x)$ has a saddlepoint (x°,λ°), $\lambda^\circ\geqslant\theta$. The latter condition becomes a necessary and sufficient condition if f and all g^i are concave functions and there exists an $\hat{x}\in X$ such that $g^i(\hat{x})>0$, $i=1,\ldots,m$. The condition that there exists an $\hat{x}\in X$ such that $g^i(\hat{x})>0$, $i=1,\ldots,m$ is known as *Slater's constraint qualification.* It can be shown that Slater's constraint qualification is not required for those g^i's which are linear. Thus, in the case of a concave programming problem with linear constraints, the saddlepoint condition is a necessary and sufficient condition for a constrained global maximum. The role of constraint qualifications is considered in more detail in Section 2.3.2.

The above results do not require that f and all g^i be differentiable functions. If they are, then some further necessary and sufficient conditions for a solution to the non-linear programming problem can be derived.

Suppose (x°, λ°) is a saddlepoint of the Lagrangean $L(x,\lambda)=f(x)+\sum_{i=1}^{m} \lambda_i g^i(x)$, where f and all g^i are continuously differentiable functions on an open set $X \subset \mathbb{R}^n$. Then, from Theorem 2.9, $g^i(x^\circ) \geqslant 0$, $i=1,\ldots,m$, and $\sum_{i=1}^{m} \lambda_i^\circ g^i(x^\circ)=0$. But, from the definition of a saddlepoint, $\lambda_i^\circ \geqslant 0$, $i=1,\ldots,m$, so that we must have $\lambda_i^\circ g^i(x^\circ)=0$, $i=1,\ldots,m$. Furthermore, if (x°, λ°) is a saddlepoint, $L(x^\circ, \lambda^\circ) \geqslant L(x, \lambda^\circ)$ for all $x \in X$, so that x° is a global maximum point of $L(x, \lambda^\circ)$ and, since X is an open set, $L_x(x^\circ, \lambda^\circ)=\theta$. Thus, if $L(x, \lambda)$ has a saddlepoint at (x°, λ°) and f and all g^i are continuously differentiable on an open set $X \subset \mathbb{R}^n$, then

$$
\begin{aligned}
&L_x(x^\circ, \lambda^\circ)=0, \\
&g^i(x^\circ) \geqslant 0, \qquad i=1,\ldots,m, \\
&\lambda_i^\circ g^i(x^\circ)=0, \qquad i=1,\ldots,m, \\
&\lambda_i^\circ \geqslant 0, \qquad i=1,\ldots,m.
\end{aligned}
\tag{2.56}
$$

Conditions (2.56) are the *Kuhn-Tucker* (K–T) *conditions*. The conditions $\lambda_i^\circ g^i(x^\circ)=0$, $i=1,\ldots,m$, are termed the *complementary slackness conditions*.

In general the K–T conditions are not sufficient conditions for the existence of a saddlepoint of the Lagrangean. However, if they hold at (x°, λ°) *and* if f and all g^i are concave functions on $X \subset \mathbb{R}^n$ then the Lagrangean function has a saddlepoint at (x°, λ°). To see this, suppose f and all g^i are concave and the K–T conditions hold at (x°, λ°). From the concavity of f and the g^i, $L(x, \lambda^\circ)=f(x)+\sum_{i=1}^{m} \lambda_i^\circ g^i(x)=f(x)+\lambda^{\circ\prime} g(x)$, where

$g(x) \equiv [g^1(x), \ldots, g^m(x)]'$, is concave in x, since $\lambda^\circ \geqslant \theta$. Hence, $L(x, \lambda^\circ) - L(x^\circ, \lambda^\circ) \leqslant L_x(x^\circ, \lambda^\circ)(x - x^\circ)$ for all $x \in X$. But the right-side of this inequality is zero from the K–T conditions, and so

$$L(x, \lambda^\circ) \leqslant L(x^\circ, \lambda^\circ) \text{ for all } x \in X. \tag{2.57}$$

Now consider $L(x^\circ, \lambda) = f(x^\circ) + \lambda' g(x^\circ)$. Clearly, $L(x^\circ, \lambda)$ is linear in λ and hence convex in λ, so that $L(x^\circ, \lambda) - L(x^\circ, \lambda^\circ) \geqslant L_\lambda(x^\circ, \lambda^\circ)(\lambda - \lambda^\circ)$. But $L_\lambda(x^\circ, \lambda^\circ) = g(x^\circ)$, and as $g^i(x^\circ) \geqslant 0$, $i = 1, \ldots, m$, and $\lambda_i^\circ g^i(x^\circ) = 0$, $i = 1, \ldots, m$, it follows that for $\lambda \geqslant \theta$,

$$L(x^\circ, \lambda) \geqslant L(x^\circ, \lambda^\circ). \tag{2.58}$$

Combining (2.57) and (2.58) gives $L(x, \lambda^\circ) \leqslant L(x^\circ, \lambda^\circ) \leqslant L(x^\circ, \lambda)$, for $x \in X$, $\lambda \geqslant \theta$.

We can combine the above results relating saddlepoints to the K–T conditions with those given in Theorems 2.10 and 2.11 to obtain the following.

THEOREM 2.12 (Kuhn-Tucker sufficiency condition) If f and g^i, $i = 1, \ldots, m$, are continuously differentiable concave functions on an open convex set $X \subset \mathbb{R}^n$ then the K–T conditions

$$L_x(x^\circ, \lambda^\circ) = \theta,$$

$$g^i(x^\circ) \geqslant 0, \; i = 1, \ldots, m,$$

$$\lambda_i^\circ g^i(x^\circ) = 0, \qquad i = 1, \ldots, m,$$

$$\lambda_i^\circ \geqslant 0, \qquad i = 1, \ldots, m,$$

are sufficient conditions for f to have a global maximum at x° subject to the constraints $g^i(x) \geqslant 0$, $i = 1, \ldots, m$.

THEOREM 2.13 (Kuhn-Tucker necessary conditions) If f and g^i, $i = 1, \ldots, m$, are continuously differentiable concave functions on an open convex set $X \subset \mathbb{R}^n$, and if there exists an $\hat{x} \in X$ such that $g^i(\hat{x}) > 0$, $i = 1, \ldots, m$, then the K–T conditions are necessary conditions for f to have a global maximum at x° subject to the constraints $g^i(x) \geqslant 0$, $i = 1, \ldots, m$.

In the case of concave programming, the K–T conditions are thus sufficient for a global maximum; moreover, when Slater's constraint qualification holds and the problem is one of concave programming, the K–T conditions are also necessary conditions for a global maximum.

EXAMPLE 2.4 Consider the problem

$$\underset{x}{\text{maximize}}\ f(x) = -(x_1+1)^2 - (x_2-2)^2$$

$$\begin{aligned}\text{subject to } g^1(x) &= 2 - x_1 \geqslant 0,\\ g^2(x) &= 1 - x_2 \geqslant 0,\\ g^3(x) &= x_1 \geqslant 0,\\ g^4(x) &= x_2 \geqslant 0.\end{aligned}$$

Here f is strictly concave and all g^i are linear and thus concave. The K–T conditions are therefore necessary and sufficient conditions for a global maximum. Given the strict concavity of f, the global maximum is unique and the K–T conditions will thus be satisfied at a unique point x°.

The Lagrangean function for the problem is

$$\begin{aligned}L(x,\lambda) = {}& -(x_1+1)^2 - (x_2-2)^2\\ &+ \lambda_1(2-x_1) + \lambda_2(1-x_2) + \lambda_3 x_1 + \lambda_4 x_2\end{aligned}$$

and the K–T conditions are

$$-2(x_1+1) - \lambda_1 + \lambda_3 = 0, \tag{i}$$

$$-2(x_2-2) - \lambda_2 + \lambda_4 = 0, \tag{ii}$$

$$\lambda_1(2-x_1) = 0, \qquad \lambda_2(1-x_2) = 0, \tag{iii}$$

$$\lambda_3 x_1 = 0, \qquad \lambda_4 x_2 = 0, \tag{iv}$$

$$2 - x_1 \geqslant 0, \qquad 1 - x_2 \geqslant 0, \tag{v}$$

$$x_1 \geqslant 0, \qquad x_2 \geqslant 0, \qquad \lambda_i \geqslant 0, \qquad i = 1, \ldots, 4. \tag{vi}$$

Consider the possibility that $x_1^\circ > 0$, $x_2^\circ > 0$. Then $\lambda_3^\circ = \lambda_4^\circ = 0$ from (iv), and then from (i) $\lambda_1^\circ = -2(x_1^\circ + 1)$; but $x_1^\circ > 0$, which implies $\lambda_1^\circ < 0$, which violates (vi). Thus, x_1° and x_2° cannot both be positive. Consider then $x_1^\circ = 0$, $x_2^\circ > 0$. Then $\lambda_4^\circ = 0$ from (iv) and from (ii) $\lambda_2^\circ = -2(x_2^\circ - 2)$. But $\lambda_2^\circ(1 - x_2^\circ) = 0$, so that $-2(x_2^\circ - 2)(1 - x_2^\circ) = 0$ which implies $x_2^\circ = 2$ or $x_2^\circ = 1$. However, from (v) $x_2^\circ \leqslant 1$. Thus, $x_2^\circ = 1$ and $\lambda_2^\circ = 2$. For $x_1^\circ = 0$, $\lambda_1^\circ = 0$ from (iii) and hence from (i) $\lambda_3^\circ = 2$. The K–T conditions are thus satisfied for $x_1^\circ = 0$, $x_2^\circ = 1$, $\lambda_1^\circ = 0$, $\lambda_2^\circ = \lambda_3^\circ = 2$, $\lambda_4^\circ = 0$, and the point $x^\circ = (0, 1)'$ is the unique global maximum point.

So far we have required that $X \subset \mathbb{R}^n$ and that X be an open set. In many applications of non-linear programming the control variables are restricted to be non-negative. This was so in the above example. As we saw in that example, we can take account of the non-negativity restrictions by introducing the explicit constraints $g^j(x) = x_j \geqslant 0$, $j = 1, \ldots, n$. However, the non-negativity restrictions, $x \geqslant \theta$, can be handled more directly by restricting X to be the non-negative orthant of $\mathbb{R}^n$. Theorems 2.9–2.13 continue to hold when X is so restricted, except that the condition $L_x(x^\circ, \lambda^\circ) = \theta$ in the K–T conditions needs modification to allow for the possibility of x° being a boundary point of X. In the case where X is the non-negative orthant of $\mathbb{R}^n$ the appropriate K–T conditions are

$$\begin{aligned}
&\partial L(x^\circ, \lambda^\circ)/\partial x_j \leqslant 0, \qquad j = 1, \ldots, n, \\
&x_j^\circ[\partial L(x^\circ, \lambda^\circ)/\partial x_j] = 0, \qquad j = 1, \ldots, n, \\
&x_j^\circ \geqslant 0, \qquad j = 1, \ldots, n, \\
&g^i(x^\circ) \geqslant 0, \qquad i = 1, \ldots, m, \\
&\lambda_i^\circ g^i(x^\circ) = 0, \qquad i = 1, \ldots, m, \\
&\lambda_i^\circ \geqslant 0, \qquad i = 1, \ldots, m.
\end{aligned} \tag{2.59}$$

To see this, if (x°, λ°) is a saddlepoint, then $L(x^\circ, \lambda^\circ) \geqslant L(x, \lambda^\circ)$ for all $x \in X$. When X is restricted to the non-negative orthant of $\mathbb{R}^n$, and x° is an interior point of X, $(x^\circ \gg \theta)$, then x° is a local maximum of $L(x, \lambda^\circ)$ with respect to some neighbourhood of x°, so that $L_x(x^\circ, \lambda^\circ) = \theta$ continues to hold. Suppose, however, that x° is a boundary point of X. Then, from Taylor's Theorem we have $L(x,\lambda^\circ)=L(x^\circ,\lambda^\circ)+\psi L_x(x^\circ+\psi\gamma z,\lambda^\circ)z$, for some $0<\gamma<1$, where $\psi>0$ and $z\geqslant\theta$. $L(x^\circ,\lambda^\circ)\geqslant L(x,\lambda^\circ)$ for all $x\in X$, then implies $\psi L_x(x^\circ+\psi\gamma z,\lambda^\circ)'z\leqslant 0$, and dividing by ψ and taking the limit as $\psi\to 0$ gives $L_x(x^\circ,\lambda^\circ)z\leqslant 0$. Thus, $L_x(x^\circ,\lambda^\circ)\leqslant\theta$ for $x^\circ\geqslant\theta$ with $z\geqslant\theta$ as $L_x(x^\circ,\lambda^\circ)=\theta$ for $x\gg\theta$. These conditions are equivalent to the first three parts of the K–T conditions.

2.3.2 Constraint qualifications

If f or one of the g^i's are non-concave we can still write down the K–T conditions and look for a solution to them. However, the conditions are no longer necessary and a solution so obtained may not yield a maximum, as the following examples illustrate.

EXAMPLE 2.5 Consider the problem

$$\text{maximize } f(x)=-x_2$$

$$\text{subject to } (1-x_1)^3-x_2\geqslant 0,$$

$$x_1\geqslant 0, \qquad x_2\geqslant 0.$$

The feasible set for this problem is shown in Figure 2.1, from which it is easily seen that $x^\circ=(x_1^\circ,0)'$, where $0\leqslant x_1^\circ\leqslant 1$ are all global maximum points. Here the constraint function $g(x)=(1-x_1)^3-x_2$ is non-concave. The K–T conditions for this problem (using 2.59) are

$$-3\lambda^\circ(1-x_1^\circ)^2\leqslant 0, \qquad -(1+\lambda^\circ)\leqslant 0,$$

$$-3x_1^\circ\lambda^\circ(1-x_1^\circ)^2=0, \qquad -x_2^\circ(1+\lambda^\circ)=0,$$

$$(1-x_1^\circ)^3-x_2^\circ\geqslant 0,$$

$$\lambda^\circ[(1-x_1^\circ)^3-x_2^\circ]=0,$$

$$\lambda^\circ\geqslant 0, \qquad x_1^\circ\geqslant 0, \qquad x_2^\circ\geqslant 0,$$

where the associated Lagrangean is $L(x,\lambda)=-x_2+\lambda[(1-x_1)^3-x_2]$. It is easily verified that $0\leqslant x_1^\circ\leqslant 1$, $x_2^\circ=0$ satisfy the above K–T conditions.

For this problem the K–T conditions are satisfied at all global

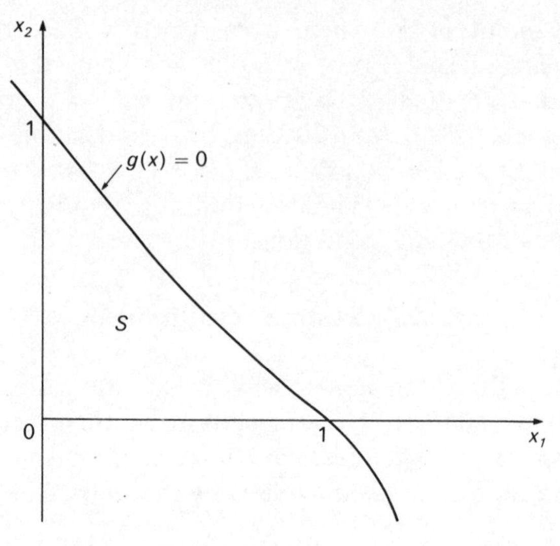

Figure 2.1

maximum points even though the constraint function is non-concave.

EXAMPLE 2.6 Consider the problem of Example 2.5 modified so that the objective function is altered to $f(x) = x_1$, the constraints remaining as before. From Figure 2.1 it is easily seen that $x^\circ = (1, 0)'$ is the optimal solution. The K–T conditions for this problem are

$$1 - 3\lambda^\circ(1 - x_1^\circ)^2 \leqslant 0, \qquad -\lambda^\circ \leqslant 0,$$

$$x_1^\circ[1 - 3\lambda^\circ(1 - x_1^\circ)^2] = 0, \qquad -x_2^\circ\lambda^\circ = 0,$$

$$\lambda^\circ[(1 - x_1^\circ)^3 - x_2^\circ] = 0, \qquad (1 - x_1^\circ)^3 - x_2^\circ \geqslant 0,$$

$$\lambda^\circ \geqslant 0, \qquad x_1^\circ \geqslant 0, \qquad x_2^\circ \geqslant 0.$$

At the optimal point $x_1^\circ = 1$, $x_2^\circ = 0$ the first inequality is clearly violated, so that in this case the K–T conditions are not satisfied at the optimal point.

Much effort has been devoted to the determination of conditions under which the K–T conditions are necessary conditions for a solution to the non-linear programming problem. We show below that if a certain set is empty at the optimal point x° then the K–T conditions necessarily hold at x°. We then present some sufficient conditions, commonly referred to as *constraint qualifications*, which ensure that this set is empty at the optimal point. We do not engage in an extended discussion of the various constraint qualifications and their interrelationships which have been discovered in the literature. For extensive discussion of the various constraint qualifications the reader is referred to surveys by Bazaraa and Shetty (1976), Bazaraa, Goode and Shetty (1972), Gould and Tolle (1971) and Takayama (1985). We present only those constraint qualifications that have proved to be particularly useful within the context of economic optimization problems.

Consider the problem maximize $f(x)$ subject to $g^i(x) \geqslant 0$, $i=1,\ldots,m$, where f and all g^i are continuously differentiable functions on an open set $X \subset \mathbb{R}^n$. Let $S \equiv \{x : x \in X, g^i(x) \geqslant 0, i=1,\ldots,m\}$ denote the feasible set. Then f has a local maximum at x° subject to the constraints if $f(x^\circ) \geqslant f(x)$ for all $x \in N(x^\circ) \cap S$. Let $B_\delta(x^\circ)$ be an open ball entirely contained within $N(x^\circ)$. Every $x \in B_\delta(x^\circ)$ can be expressed as $x = x^\circ + z$, with $\|z\| < \delta$, where $z \neq \theta$, if, and only if, $x \neq x^\circ$. A vector $z \neq \theta$ is a *feasible direction vector from* x° if there exists an $\sigma > 0$ such that $x^\circ + \gamma z \in B_\sigma(x^\circ) \cap S$ for all γ, $0 \leqslant \gamma \leqslant \sigma / \|z\|$. We are interested in feasible direction vectors from x° since, if f has a constrained local maximum at x°, we must have $f(x^\circ) \geqslant f(x^\circ + \gamma z)$ for sufficiently small γ. For z to be a feasible direction vector from x° we must have $g^i(x^\circ + \gamma z) \geqslant 0$, $i=1,\ldots,m$. If at x° the kth constraint is non-binding (i.e., $g^k(x^\circ) > 0$), then for sufficiently small γ, $g^k(x^\circ + \gamma z) \geqslant 0$, so that we need only consider those constraints which are binding at x°. Define the set

$$B(x^\circ) \equiv \{i : g^i(x^\circ) = 0\}. \tag{2.60}$$

We now show that if z is a feasible direction vector from x° then $g^i_x(x^\circ)z \geqslant 0$ for all $i \in B(x^\circ)$. Suppose $g^k_x(x^\circ)z < 0$ for some $k \in B(x^\circ)$ and z is a feasible direction vector from x°. From Taylor's Theorem, $g^k(x^\circ + \gamma z) = g^k(x^\circ) + \gamma g^k_x(x^\circ)z + o(\|z\|)$. For $k \in B(x^\circ)$, $g^k(x^\circ) = 0$

and, since $g_x^k(x^\circ)z<0$, it follows that for sufficiently small γ, $g^k(x^\circ+\gamma z)<0$, which contradicts the fact that z is a feasible direction vector from x°. Thus, a necessary condition for z to be a feasible direction vector from x° is that $g_x^i(x^\circ)z \geqslant 0$ for all $i \epsilon B(x^\circ)$. (Note however that $g_x^i(x^\circ)z \geqslant 0$ for all $i \epsilon B(x^\circ)$ is not a sufficient condition for z to be a feasible direction vector from x°).

Define the sets

$$Z^1(x^\circ) \equiv \{z : g_x^i(x^\circ) \geqslant 0 \text{ for all } i \epsilon B(x^\circ), \text{ and } f_x(x^\circ)z \leqslant 0\}, \tag{2.61}$$

$$Z^2(x^\circ) \equiv \{z : g_x^i(x^\circ) \geqslant 0 \text{ for all } i \epsilon B(x^\circ), \text{ and } f_x(x^\circ)z > 0\}, \tag{2.62}$$

$$Z^3(x^\circ) \equiv \{z : g_x^i(x^\circ) < 0 \text{ for at least one } i \epsilon B(x^\circ)\}. \tag{2.63}$$

Clearly, the sets $Z^1(x^\circ)$, $Z^2(x^\circ)$, $Z^3(x^\circ)$ are mutually disjoint and exhaustive, and for z to be feasible direction vector from x°, $z \epsilon Z^1(x^\circ) \cup Z^2(x^\circ)$. Now, if f has a constrained local maximum at x°, then, from Taylor's Theorem, for sufficiently small γ, we must have $f_x(x^\circ)z \leqslant 0$ for all feasible direction vectors from x°. Thus, if $Z^2(x^\circ)$ contains some feasible direction vectors, then x° could not be a constrained local maximum point. However, x° may be a constrained local maximum point when $Z^2(x^\circ) \neq \phi$, provided that the only vectors contained in $Z^2(x^\circ)$ are infeasible direction vectors from x°. In Example 2.6 this is precisely the situation. There, $Z^2(x^\circ) \neq \phi$ at the optimal point $x^\circ = (1,0)'$. To see this, note that $Z^2(x^\circ) = A \cap C$, where $A \equiv \{z : g_x^i(x^\circ)z \geqslant 0 \text{ for } i \epsilon B(x^\circ)\}$ and $C \equiv \{z : f_x(x^\circ)z > 0\}$. Here the binding constraints are $g^1(x) = (1-x_1)^3 - x_2$ and $g^2(x) = x_2$, so that A is the set of all $z = (z_1, z_2)'$ which satisfy the inequalities

$$z_1[\partial g^1(x^\circ)/\partial x_1] + z_2[\partial g^1(x^\circ)/\partial x_2] \geqslant 0,$$

$$z_1[\partial g^2(x^\circ)/\partial x_1] + z_2[\partial g^2(x^\circ)/\partial x_2] \geqslant 0.$$

That is, $A = \{z = (z_1, z_2)' : z_2 = 0\}$. C is the set of all z which satisfy

$$z_1[\partial f(x^\circ)/\partial x_1] + z_2[\partial f(x^\circ)/\partial x_2] > 0,$$

so that $C = \{z = (z_1, z_2)' : z_1 > 0\}$. Consequently,

$$Z^2(x^\circ) = \{z = (z_1, z_2)' : z_1 > 0, z_2 = 0\} \neq \phi.$$

However, $Z^2(x^\circ)$ contains only infeasible direction vectors, since, for $z \epsilon Z^2(x^\circ)$, $g^1(x^\circ + \gamma z) = (-\gamma z_1)^3 < 0$, as $\gamma > 0$ and $z_1 > 0$.

Now, if $Z^2(x^\circ) = \phi$, the only feasible direction vectors from x° must be contained in $Z^1(x^\circ)$. We now show that if $Z^2(x^\circ) = \phi$ then $z \epsilon Z^1(x^\circ)$ implies that the K–T conditions necessarily hold at x°. Given the definition of $Z^1(x^\circ)$, $z \epsilon Z^1(x^\circ)$ implies, from Farkas' Lemma (Theorem 1.7), that there exist scalars $\lambda_i^\circ \geqslant 0$, $i \epsilon B(x^\circ)$, with strict inequality holding for at least one i such that $-f_x(x^\circ) = \sum_{i \epsilon B(x^\circ)} \lambda_i^\circ g_x^i(x^\circ)$. If we define $\lambda_i^\circ = 0$ for all $i \notin B(x^\circ)$ then $z \epsilon Z^1(x^\circ)$ implies $f_x(x^\circ) + \sum_{i=1}^m \lambda_i^\circ g_x^i(x^\circ) = \theta$. But this is just the condition that $L_x(x^\circ, \lambda^\circ) = \theta$, where $L(x, \lambda) = f(x) + \sum_{i=1}^m \lambda_i g^i(x)$ is the Lagrangean function. Also, since $\lambda_i^\circ \geqslant 0$, $i \epsilon B(x^\circ)$, and $\lambda_i^\circ = 0$, $i \notin B(x^\circ)$, we have $\lambda_i^\circ g^i(x^\circ) = 0$, $i = 1, \ldots, m$. Thus, if x° is a constrained local maximum point and if $Z^2(x^\circ) = \phi$, then necessarily the K–T conditions must hold. In Example 2.5 the K–T conditions hold at the optimal point even though f is not concave, since $Z^2(x^\circ) = \phi$. There, $Z^2(x^\circ) = A \cap D$, where $A = \{z = (z_1, z_2)' : z_2 = 0\}$ and D is the set of all z which satisfy

$$z_1[\partial f(x^\circ)/\partial x_1] + z_2[\partial f(x^\circ)/\partial x_2] > 0,$$

so that $D = \{z = (z_1, z_2)' : z_2 > 0\}$, and hence $Z^2(x^\circ) = \phi$. We have already seen that in the case of Example 2.6, $Z^2(x^\circ) \neq \phi$, and that the K–T conditions do not hold at the optimal point x°.

If it can be shown that $Z^2(x^\circ) = \phi$ at a constrained local maximum point x° then the K–T conditions are necessary conditions. However, it is often difficult to verify *directly* that $Z^2(x^\circ)$ is indeed empty at a constrained local maximum point x°. Consequently, other more directly verifiable conditions (constraint qualifications) have been developed which ensure that $Z^2(x^\circ) = \phi$; some of these are presented below.

DEFINITION 2.2 **Arrow–Hurwicz–Uzawa constraint qualification** (AHUCQ) Let f and g^i, $i = 1, \ldots, m$, be continuously differentiable on an open convex set $X \subset \mathbb{R}^n$. Let x° be a point in the feasible set, S. Define the sets

$$C = \{i : i \epsilon B(x^\circ) \text{ and } g^i(x) \text{ is convex on } X\},$$

$$N = \{i : i \in B(x^\circ) \text{ and } g^i(x) \text{ is non-convex on } X\}.$$

The AHUCQ requires that there exists a $z \in \mathbb{R}^n$ such that

$$g^i_x(x^\circ)z \geqslant 0 \text{ for all } i \in C,$$

$$g^i_x(x^\circ)z > 0 \text{ for all } i \in N.$$

Note that $C \cup N = B(x^\circ)$.

THEOREM 2.14 Let f and g^i, $i = 1, \ldots, m$, be continuously differentiable on an open convex set $X \subset \mathbb{R}^n$. If f has a local maximum at x° subject to the constraints $g^i(x) \geqslant 0$, $i = 1, \ldots, m$, and if the AHUCQ holds at x°, then $Z^2(x^\circ) = \phi$ and the K–T conditions necessarily hold at x°.

Proof. See Arrow, Hurwicz and Uzawa (1961) or Takayama (1985).

The AHUCQ as expressed above is not in a form which can easily be applied. We therefore present some conditions which imply that the AHUCQ is met.

THEOREM 2.15 The AHUCQ holds at x° if *one* of the following is true:

(i) the functions g^i, $i = 1, \ldots, m$, are convex,

(ii) the functions g^i, $i = 1, \ldots, m$ are concave and there exists an $\hat{x} \in X$ such that $g^i(\hat{x}) > 0$, $i = 1, \ldots, m$ (*Slater's constraint qualification*),

(iii) the vectors $g^i_x(x^\circ)$ for all $i \in B(x^\circ)$ are linearly independent (*Rank condition*),

(iv) the feasible set $S = \{x : g^i(x) \geqslant 0, i = 1, \ldots, m, x \in X\}$ is convex and possesses an interior point, and $g^i_x(x^\circ) \neq \theta$ for all $i \in B(x^\circ)$.

Proof. See Arrow, Hurwicz and Uzawa (1961) or Takayama (1985).

In the case of Example 2.4 the reader may verify that all four of the above conditions are in fact satisfied, whereas in Example 2.6 none are satisfied. In Example 2.5, $x^\circ=(x_1^\circ,0)'$, where $0\leqslant x_1^\circ\leqslant 1$, are all globally optimal points. From Figure 2.1 it is easily seen that none of the conditions (i), (ii) and (iv) hold. However, for $x^\circ=(0,0)'$, the binding constraints are $g^1(x)=x_1$, $g^2(x)=x_2$, so that $g_x^1(\theta)=(1,0)'$, $g^2(\theta)=(0,1)'$. Clearly, $g_x^1(\theta)$ and $g_x^2(\theta)$ are linearly independent, so that the Rank condition (iii) is met. For $0<x_1^\circ<1$, $x_2^\circ=0$ there is only one binding constraint, $g^2(x)=x_2$, and the Rank condition is clearly satisfied, as $g_x^2(x^\circ)\neq\theta$. At the remaining optimal point, $x^\circ=(1,0)'$, the Rank condition does not hold since the binding constraints in this case are $g^2(x)=x_2$ and $g^3(x)=(1-x_1)^3-x_2$, so that $g_x^2(x^\circ)=(0,1)'$ and $g_x^3(x^\circ)=(0,-1)'$. Thus, for this example, the AHUCQ does not hold at the optimal point $x^\circ=(1,0)'$. However, we have already seen that at $x^\circ=(1,0)'$, $Z^2(x^\circ)=\phi$.

It is important to note that the constraint qualifications in Theorem 2.15 are only *sufficient* conditions for $Z^2(x^\circ)=\phi$, so that $Z^2(x^\circ)$ may indeed be empty even though none of the above constraint qualifications are met.

The relationships between a solution to the non-linear programming problem, a saddlepoint of the Lagrangean function and the K–T conditions are summarized in Figure 2.2.

Necessary conditions for a local maximum in the non-linear programming problem which do not require that $Z^2(x^\circ)=\phi$ are available. These are the *John conditions*, which are much weaker

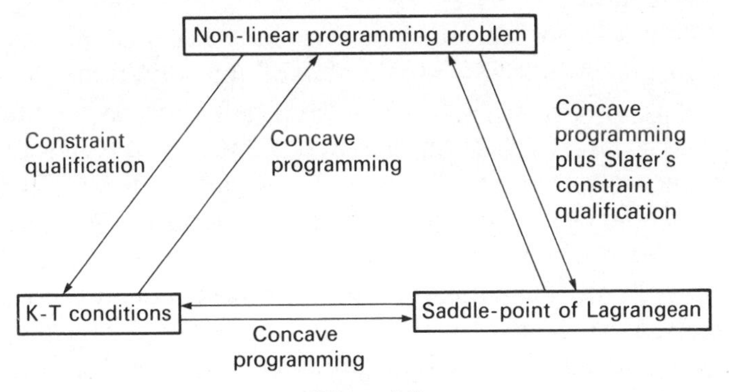

Figure 2.2

than the K–T conditions. If we define the *weak Lagrangean function* by

$$\hat{L}(x,\lambda)=\lambda_0 f(x)+\sum_{i=1}^{m}\lambda_i g^i(x)$$

then it can be shown (see John, 1948 or Avriel, 1976) that if f has a local maximum at x° subject to the constraints $g^i(x)\geqslant 0, i=1,\ldots,m$, then there exist multiplier $\lambda_i^\circ\geqslant 0$, $i=0,1,\ldots,m$, not all zero, such that

$$\begin{aligned}&\hat{L}_x(x^\circ,\lambda^\circ)=\theta,\\ &\lambda_i^\circ g^i(x^\circ)=0, \qquad i=1,\ldots,m,\\ &g^i(x^\circ)\geqslant 0, \qquad i=1,\ldots,m.\end{aligned} \tag{2.64}$$

Consider the problem in Example 2.6. The weak Lagrangean for this problem is $\hat{L}=\lambda_0 x_1+\lambda_1[(1-x_1)^3-x_2]+\lambda_2 x_1+\lambda_3 x_3$ and the John conditions are

$$\lambda_0^\circ-3\lambda_1^\circ(1-x_1^\circ)^2+\lambda_2^\circ=0, \qquad -\lambda_1^\circ+\lambda_2^\circ=0,$$

$$\lambda_1^\circ[(1-x_1^\circ)^3-x_2^\circ]=0, \qquad \lambda_2^\circ x_1^\circ=0, \qquad \lambda_3^\circ x_2^\circ=0,$$

$$x_1^\circ\geqslant 0, x_2^\circ\geqslant 0, \qquad \lambda_1^\circ\geqslant 0, \qquad i=0,1,2,3,$$

with strict inequality for at least one i,

$$(1-x_1^\circ)^3-x_2^\circ\geqslant 0.$$

Letting $\lambda_0^\circ=0$, $\lambda_1^\circ=1$, $\lambda_2^\circ=0$, $\lambda_3^\circ=1$, we observe that the John necessary conditions are satisfied at the optimal point $x^\circ=(1,0)'$. This example also illustrates the main weakness of the John conditions; substituting $\lambda_0^\circ=0$, the above John conditions are in fact satisfied at $x^\circ=(1,0)'$ for *any* differentiable objective function whether it has a local maximum at that point or not. As a screening device, the John conditions are not especially effective and this explains the fact that they are rarely used. It is important that λ_0° be non-zero for the solution not to be independent of the objective function f. In any meaningful economic problem one would expect the solution to be dependent on the objective function. When $Z^2(x^\circ)=\phi$, it can be shown that $\lambda_0^\circ\neq 0$ and hence λ_0° is guaranteed to be positive, so that it can be taken to be unity (by dividing all $m+1$

multipliers by λ_0°). The various constraint qualifications therefore ensure that $\lambda_0^\circ \neq 0$ in the John conditions.

We now present some simple applications of non-linear programming in economics.

EXAMPLE 2.7 (Profit maximizing firm) In the neoclassical theory of the firm, the firm's problem is that of maximizing profits, $\Pi = pq - rx$, subject to a non-negativity restriction, $x \geqslant \theta$, and the firm's technological capabilities as described by a production function, $q \leqslant f(x)$, where q denotes output, p the price of output, x a vector of inputs and r a vector of corresponding input prices. Typically, the production function is assumed to be twice-continuously differentiable and concave with $f(\theta) = 0$. Clearly, there will be some $(\hat{q}, \hat{x})$ which satisfies $q < f(x), x \gg \theta$ and $q > 0$, so that Slater's constraint qualification will hold. If the firm is a price-taker in all markets, the firm's profit function is linear in q and x, and hence concave in (q, x), so the K–T conditions are necessary and sufficient for profits to be globally maximized.

The Lagrangean for the firm's problem is

$$L(q, x, \lambda, \mu) = pq - r'x + \lambda[f(x) - q] + \sum_{i=1}^{n} \mu_i x_i,$$

and the K–T conditions are

$$p - \lambda^\circ = 0, \tag{i}$$

$$-r_i + \lambda^\circ f_i(x^\circ) + \mu_i^\circ = 0, \qquad i = 1, \ldots, n, \tag{ii}$$

$$f(x^\circ) - q^\circ \geqslant 0, \tag{iii}$$

$$\lambda^\circ[f(x^\circ) - q^\circ] = 0, \tag{iv}$$

$$\lambda^\circ \geqslant 0, \qquad x_i^\circ \geqslant 0, \qquad \mu_i^\circ \geqslant 0, \qquad i = 1, \ldots, n, \tag{v}$$

$$\mu_i^\circ x_i^\circ = 0, \qquad i = 1, \ldots, n. \tag{vi}$$

From (i), $\lambda^\circ > 0$, since p, the price of output, is positive, so that, from (iv), $f(x^\circ) = q^\circ$; that is, the firm will produce the maximum level of output which its technology allows with the input level x° – the firm will use its inputs efficiently. Suppose $x_j^\circ > 0$, $x_k^\circ > 0$;

then, from (vi), $\mu_j^\circ = \mu_k^\circ = 0$, and, since $\lambda^\circ > 0$, from (ii), we have $r_j/r_k = f_j(x^\circ)/f_k(x^\circ)$. That is, when the inputs x_j and x_k are used in production, they will be employed in a manner such that the ratio of their marginal products equals the input price ratio. Suppose now that $x_j^\circ = 0$ and $x_k^\circ > 0$; then $\mu_i^\circ \geqslant 0$ and $\mu_k^\circ = 0$, and, as $\lambda^\circ > 0$, (ii) implies $r_k/r_i \leqslant f_k(x^\circ)/f_i(x^\circ)$ and $r_i \geqslant p f_i(x^\circ)$, the latter inequality meaning that the firm will not employ input x_i in production when the price of that input exceeds the value of its marginal product $(p f_i(x^\circ))$.

EXAMPLE 2.8 (Peak-load pricing) Consider the following simple peak-load pricing model in which the objective is to maximize social welfare, W, as measured by the sum of producer's and consumers' surplus, subject to the restriction that output in any period not exceed capacity. Formally the problem is

$$\text{maximize } W(x, y) = \sum_{j=1}^{n} \int_0^{x_j} p_j(\tau_j)\, \mathrm{d}\tau_j - \sum_{j=1}^{n} b_j x_j - cy \tag{i}$$

$$\text{subject to} \quad x_j \leqslant y, \qquad j = 1, \ldots, n, \tag{ii}$$

where W is defined on the positive orthant of $\mathbb{R}^{n+1}$. Here x_j denotes output in period j; $p_j = p_j(x_j)$ is the price of output in period j, where demands are assumed to be independent across periods with $\mathrm{d}p_j/\mathrm{d}x_j < 0$, for all j; b_j is the operating cost in period j; the b_j are assumed to be constant and equal to b in all periods (for simplicity); y is capacity; and c is the constant cost of a unit of capacity. (Note that since W is defined over the positive orthant of $\mathbb{R}^{n+1}$, we are assuming that capacity is strictly positive, along with output, in every period.)

We first show that the objective function W is concave. Since

$$\mathrm{d}\left[\int_0^{x_j} p_j(\tau_j)\mathrm{d}\tau_j\right] \Big/ \mathrm{d}x_j = p_j(x_j), \quad \partial W/\partial x_j = p_j(x_j) - b, \quad \partial W/\partial y = -c,$$

we have $\partial^2 W/\partial x_j^2 = \mathrm{d}p_j/\mathrm{d}x_j$, for all j, $\partial^2 W/\partial y^2 = 0$, $\partial^2 W/\partial x_j \partial x_i = 0$, for $j \neq i$, and $\partial^2 W/\partial y\, \partial x_j = 0$ for all j. Hence, the Hessian matrix of

$W(x, y)$ is

$$H^w = \begin{bmatrix} dp_1/dx_1 & 0 & \dots & 0 & 0 \\ 0 & dp_2/dx_2 & \dots & 0 & 0 \\ \vdots & \vdots & & \vdots & \vdots \\ - & - & - & dp_n/dx_n & 0 \\ 0 & 0 & \dots & 0 & 0 \end{bmatrix}$$

Here $|H^w| = 0$ and all principal minors of H^w, denoted $|\tilde{H}_k^w|$, satisfy $(-1)^k|\tilde{H}_k^w| \geqslant 0$, so that H^w is negative semidefinite everywhere; hence W is concave.

The constraint functions given by (ii) are all linear, so the AHUCQ holds, and the feasible set is convex. Since W is concave, the K–T conditions are necessary and sufficient for W to be globally maximized, given the capacity restrictions. The K–T conditions are obtained from the Lagrangean

$$L(x, y, \lambda) = \sum_{j=1}^{n} \int_0^{x_j} p_j(\tau_j) d\tau_j - \sum_{j=1}^{n} bx_j - cy + \sum_{j=1}^{n} \lambda_j (y - x_j).$$

and are as follows:

$$p_j - b - \lambda_j^\circ = 0, \quad j = 1, \dots, n, \tag{iii}$$

$$-c + \sum_{j=1}^{n} \lambda_j^\circ = 0, \tag{iv}$$

$$y^\circ - x_j^\circ \geqslant 0, \quad j = 1, \dots, n, \tag{v}$$

$$\lambda_j^\circ (y^\circ - x_j^\circ) = 0, \quad j = 1, \dots, n, \tag{vi}$$

$$\lambda_j^\circ \geqslant 0, \quad j = 1, \dots, n. \tag{vii}$$

By definition, a peak-period is one in which capacity is fully utilized (i.e., $y = x_j$), so that from (vi) and (vii) $\lambda_j^\circ \geqslant 0$ in a peak-period. In an off-peak period, spare capacity exists (i.e., $y - x_j > 0$), so that $\lambda_j^\circ = 0$ in an off-peak period. Using this, together with (iii), we see that social welfare maximization requires that the following pricing rule be adopted.

$p_j = b + \lambda_j^\circ$ for j a peak-period,

$p_j = b$ for j an off-peak period

(i.e., off-peak users pay only marginal operating costs and make no contribution to capacity costs).

If the above pricing rule is adopted then the enterprise will just break even. To see this, let $A = \{j : y^\circ - x_j^\circ = 0\}$ and $B = \{j : y^\circ - x_j^\circ > 0\}$. The enterprise's total revenue, if it follows the above pricing rule, is

$$\sum_{j \in A} p_j x_j^\circ + \sum_{j \in B} p_j x_j^\circ = \sum_{j \in A} (b + \lambda_j^\circ) x_j^\circ + \sum_{j \in B} b x_j^\circ$$

$$= \sum_{j \in A \cup B} b x_j^\circ + \sum_{j \in A} \lambda_j^\circ x_j^\circ = \sum_{j=1}^{n} b x_j^\circ + c y^\circ, \qquad \text{(viii)}$$

since $j \in A \cup B$ is $j = 1, \ldots, n$, and for $j \in A$, $y^\circ = x_j^\circ$ and $\sum_{j \in A} \lambda_j^\circ = \sum_{j=1}^{n} \lambda_j^\circ = c$, from (iv). Thus, the right-side of (viii) is just the enterprise's total costs. Hence, the enterprise just breaks even.

2.3.3 Quasiconcave programming

In many non-linear programming problems in economics it is assumed that the relevant functions are continuously differentiable and concave. Moreover, in such problems, it is often possible to meaningfully relax the assumption of concavity to that of quasiconcavity. A non-linear programming problem of the form maximize$_x f(x)$ subject to $g^i(x) \geqslant 0$, $i = 1, \ldots, m$, and $x \geqslant \theta$, where the objective and constraint functions are all quasiconcave is referred to as a *quasiconcave programming problem.*

The K–T conditions are necessary conditions for any objective function f to have a constrained local maximum when f and all g^i are continuously differentiable and the AHUCQ holds. In the case of quasiconcave programming, condition (ii) of Theorem 2.15 (Slater's constraint qualification) and condition (iv) of the same theorem are especially useful constraint qualifications. With these constraint qualifications, we have (cf. Theorem 2.15), the following result:

THEOREM 2.16 (Necessary conditions: quasiconcave programming) Let g^i, $i=1,\ldots,m$, be continuously differentiable quasiconcave functions and suppose there exists an $\hat{x} \gg \theta$ such that $g^i(\hat{x})>0, i=1,\ldots,m$. Then the AHUCQ holds at x° provided that either of the following conditions is satisfied:

(i) g^i is concave for $i=1,\ldots,m$,
(ii) $g^i_x(x^\circ)\neq\theta$ for all $i\in B(x^\circ)$,

and the K–T conditions are necessary conditions for *any* continuously differentiable objective function to have a local maximum at x° subject to the constraints $g^i(x)\geqslant 0$, $i=1,\ldots,m$, $x\geqslant\theta$.

That the quasiconcavity of f and g^i, $i=1,\ldots,m$, do *not* imply that the K–T conditions are sufficient conditions for a global maximum is easily seen by considering the following simple problem, due to Arrow and Enthoven (1961); maximize $f(x)=(x-1)^3$ subject to $g(x)=2-x\geqslant 0$, $x\geqslant 0$. Here, the objective function is quasiconcave and the constraint function is linear and thus quasiconcave. The K–T conditions for this problem are

$$3(x^\circ-1)^2-\lambda^\circ\leqslant 0,$$

$$x^\circ[3(x^\circ-1)^2-\lambda^\circ]=0,$$

$$2-x^\circ\geqslant 0,$$

$$\lambda^\circ(2-x^\circ)=0,$$

$$\lambda^\circ\geqslant 0, \qquad x^\circ\geqslant 0.$$

If $x^\circ=1$, $\lambda^\circ=0$ then the K–T conditions are satisfied, yet clearly the constrained local maximum occurs at $x^\circ=2$. (See Figure 2.3 where the feasible set is the line segment $[0,2]$.) Note also that the K–T conditions are also satisfied at $x^\circ=2$, $\lambda^\circ=3$, as they must be, since the AHUCQ holds given the linearity of the constraint function.

In order to obtain sufficient conditions for a solution to the quasiconcave programming problem we need to introduce a new concept.

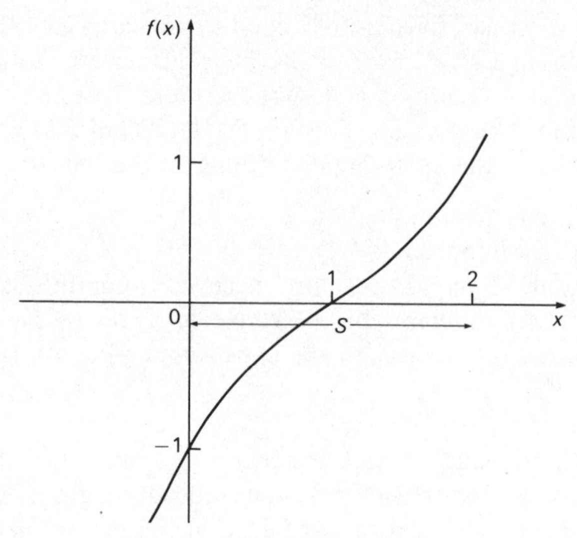

Figure 2.3

DEFINITION 2.3 Given the feasible set $S=\{x:x\geqslant\theta,\ g^i(x)\geqslant 0,\ i=1,\ldots,m\}$, we call the ith component of x, x_i, a **relevant variable** if there exists an $\hat{x}\in S$ such that $\hat{x}_i>0$. That is, a relevant variable is one which can take on a positive value without violating the constraints.

The following theorem provides a set of *four* sufficient conditions for a global solution to the quasiconcave programming problem.

THEOREM 2.17 (Sufficient conditions: quasiconcave programming) Let f and g^i, $i=1,\ldots,m$, be continuously differentiable quasiconcave functions, defined for $x\geqslant\theta$. Let (x°,λ°) satisfy the K–T conditions and let any *one* of the following conditions be satisfied:

(i) $\partial f(x^\circ)/\partial x_i<0$ for at least one relevant variable x_i,

(ii) $\partial f(x^\circ)/\partial x_i>0$ for at least one relevant variable x_i,

(iii) $f_x(x^\circ)\neq\theta$ and f is twice-continuously differentiable in a neighbourhood of x° (i.e., all second-order derivatives of f exist at x°, although they may all be zero),

(iv) f is concave.
Then f has a global maximum at x° subject to the constraints $g^i(x) \geqslant 0$, $i = 1, \ldots, m$, $x \geqslant \theta$.

Proof. See Arrow and Enthoven (1961).

Note that only one of the above four conditions need be satisfied for f to have a constrained global maximum at x° if the K–T conditions hold at (x°, λ°). In many economic applications, assumptions are made which ensure that condition (ii) will be met.

EXAMPLE 2.9 (Utility maximization) The standard problem in consumer theory is that of a consumer choosing a bundle of commodities x from a consumption set X, where X is the non-negative orthant of $\mathbb{R}^n$, so as to maximize a utility function $U(x)$ subject to a budget constraint $p'x \leqslant m$, where m is the consumer's income and $p \gg \theta$ is a vector of commodity prices. Typically, the utility function is taken to be twice-continuously differentiable and strictly quasiconcave, and a non-satiation requirement is imposed, namely that $U_i > 0$ for $x_i > 0$ for some (possibly all) i. The consumer's problem is thus one of quasiconcave programming. Condition (ii) of Theorem 2.17 and the hypothesis of Theorem 2.16 are easily seen to be satisfied, so the following K–T conditions are necessary and sufficient for a strict global maximum:

$$U_i(x^\circ) - \lambda^\circ p_i \leqslant 0, \qquad i = 1, \ldots, n, \tag{i}$$

$$x_i^\circ [U_i(x^\circ) - \lambda^\circ p_i] = 0, \qquad i = 1, \ldots, n, \tag{ii}$$

$$\lambda^\circ (m - p'x^\circ) = 0, \tag{iii}$$

$$m \geqslant p'x^\circ, \qquad x^\circ \geqslant \theta, \qquad \lambda^\circ \geqslant 0. \tag{iv}$$

Since $U_i(x^\circ) > 0$ for at least one i and $p_i > 0$ for all i, then from (i) $\lambda^\circ > 0$, so that from (iii) $m = p'x^\circ$, i.e., non-satiation implies that the budget constraint will hold with equality. Since $\lambda^\circ > 0$, then from (i) we have $U_i(x^\circ)/U_j(x^\circ) \leqslant p_i/p_j$, with equality holding when x_i° and x_j° are both positive. Thus, the consumer's utility maximizing choice of consumption bundle is such that the marginal rate of substitution between any pair of purchased commodities is equal to the ratio of the prices of those commodities, while the marginal rate of substitution between a commodity that is purchased and one that is not does not exceed the ratio of the prices of the two commodities.

In cases where the objective and constraint functions do not possess appropriate concavity or quasiconcavity properties, sufficient conditions for a local solution to the general non-linear programming problem are available. These are rarely employed in economic optimization problems, where a global rather than a local solution is required. They can be found in Fiacco (1968) or McCormick (1967).

CHAPTER 3

EQUILIBRIUM MATHEMATICS

3.1 INTRODUCTION

In this chapter we consider some important mathematical concepts which are frequently used to establish the existence of various types of equilibrium notion in the economics literature. A wide variety of equilibrium notions occur in the economics literature but the confines of space dictate that we consider only some of the more important notions. In Section 3.2, the essential mathematical concepts are presented. These are employed in Section 3.3 to establish the existence of a competitive equilibrium in a private ownership economy with a finite number of agents and commodities, and in Section 3.4, to establish the existence of a Nash equilibrium in n-person non-cooperative games.

3.2 EQUILIBRIUM MATHEMATICS

In most economic models the actions taken by an agent are determined by the values of those variables which constitute his economic 'environment'. If those values uniquely determine the action to be taken by the agent then we can work with functions (point–point mappings). However, when the action to be taken by the agent is not uniquely determined, there is a set of possible actions and we need to work with correspondences (point-set mappings).

3.2.1 Continuity of correspondences

Intuitively, the concept of continuity for a mapping expresses the idea that points 'close' to each other in the domain of the mapping are mapped into points which are 'close' to each other in the range of the mapping. In this section we show how the intuitive notion of continuity can be made precise in the case of mappings which are correspondences. *We assume throughout this section that the sets X and Y are subsets of some Euclidean space.*

DEFINITION 3.1 Let $F{:}X \to Y$ be a correspondence and let $F(x) \subset Y$ be the image set of the point $x \in X$. F is **upper hemicontinuous at the point $x_0 \in X$**, if, for every open set V containing $F(x_0)$, there exists a neighbourhood $N(x_0)$ such that $F(x) \subset V$ for every $x \in N(x_0)$. F is **upper hemicontinuous** if it is upper hemicontinuous at every point $x \in X$.

The above definition is illustrated in Figures 3.1–3.3. Consider the correspondences $F{:}X \to Y$ whose graphs are shown in Figures 3.1–3.3. In Figures 3.1–3.2, F is upper hemicontinuous at x_0, since it is evident that, for every open interval V such that $F(x_0) \subset V$, we can find some interval $(x_0-\delta, x_0+\delta)$, $\delta>0$ such that for all x, $x_0-\delta<x<x_0+\delta$, we have $F(x) \subset V$. One such open interval V, with $F(x_0) \subset V$, is shown together with a $N(x_0)$ such that for all $x \in N(x_0)$ we have $F(x) \subset V$. In Figure 3.3, F is not upper hemicontinuous at x_0. Here we can find an interval $(x_0-\delta, x_0+\delta)$ and an open set V such that $F(x_0) \subset V$, but $F(x) \not\subset V$ for all x, $x_0-\delta<x<x_0+\delta$.

EXAMPLE 3.1 Let $F{:}X \to R$, where $X=[0,2]$ and

$$F(x)=\begin{cases} [-1+0.25x, x^2-1], & \text{for } 0 \leqslant x < 1 \\ [-1,1], & \text{for } x=1 \\ [-1+0.25x, x^2-1], & \text{for } 1<x \leqslant 2. \end{cases}$$

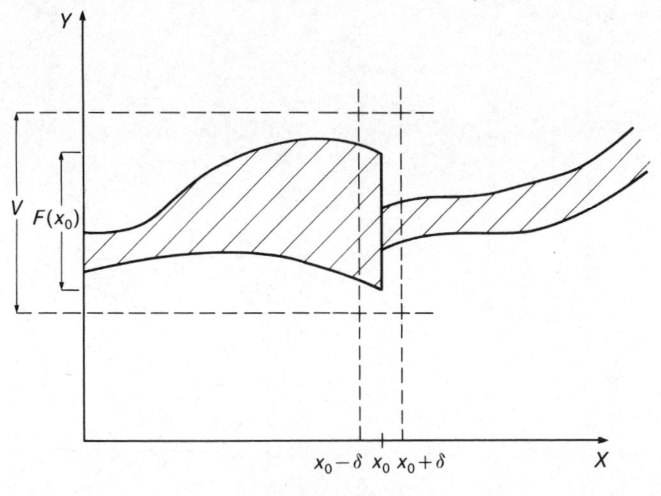

Figure 3.1

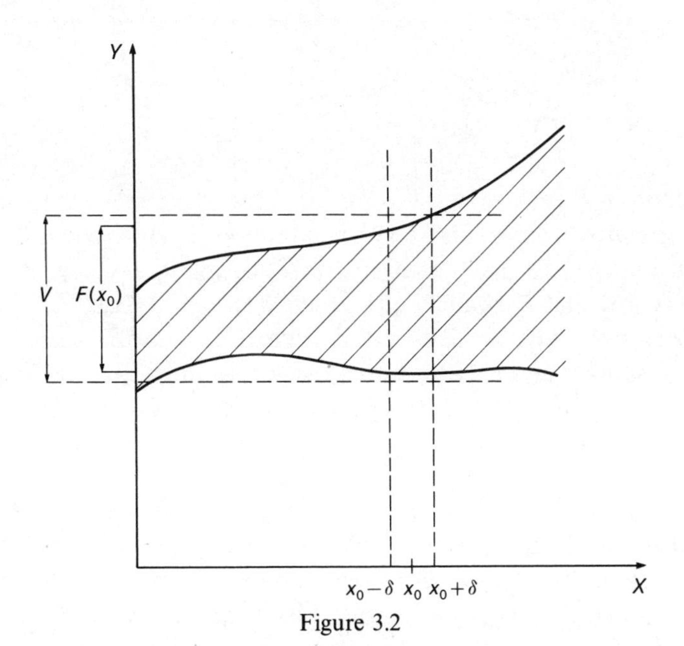

Figure 3.2

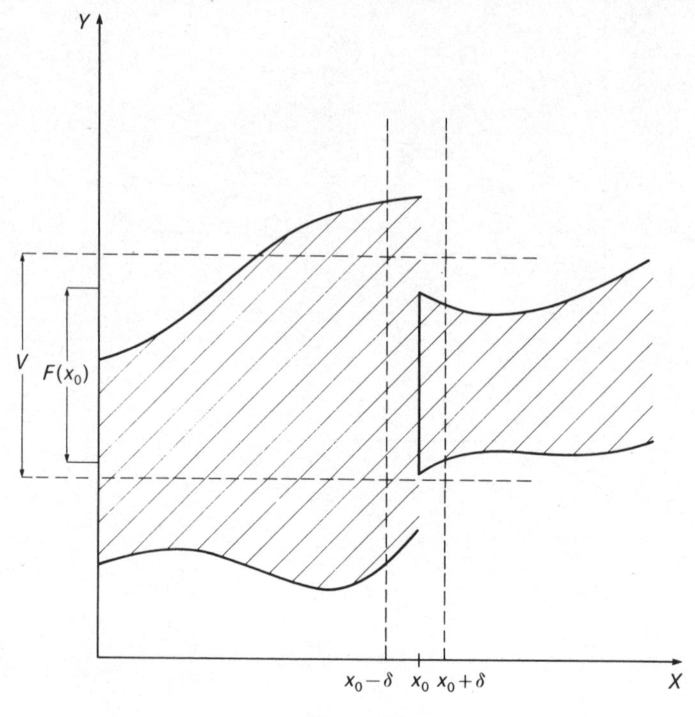

Figure 3.3

The graph of F is shown in Figure 3.4. Clearly, at $x=1$, F is upper hemicontinuous, since, for every open interval V such that $F(1)\subset V$, we can find a $\delta>0$ such that $F(x)\subset V$ for all x, $1-\delta<x<1+\delta$. Suppose now that F is redefined so that $F(1)=[-0.75, -0.25]$. The graph of F now becomes as shown in Figure 3.5. F is no longer upper hemicontinuous at $x=1$, since there exists an open interval V such that $F(1)\subset V$, but there is no $\delta>0$ such that $F(x)\subset V$ for all x, $1-\delta<x<1+\delta$.

DEFINITION 3.2 The correspondence $F:X\rightarrow Y$ is **lower hemicontinuous at the point $x_0\in X$** if, for every open set V in Y, with $F(x_0)\cap V$ non-empty, there exists a neighbourhood $N(x_0)$ such that $F(x)\cap V$ is non-empty for every $x\in N(x_0)$. The

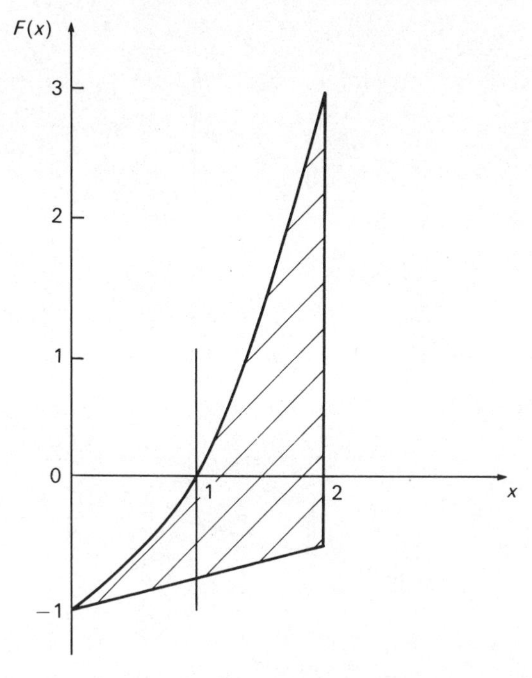

Figure 3.4

correspondence F is **lower hemicontinuous** if it is lower hemicontinuous at every point $x \in X$.

The above definition is illustrated in Figures 3.6–3.8. Consider the correspondences $F:X \rightarrow Y$ whose graphs are shown in Figures 3.6–3.7. In both cases F is lower hemicontinuous at x_0, since it is evident that, for every open interval V such that $F(x_0) \cap V$ is non-empty, we can find a neighbourhood of x_0, $N(x_0)$ such that for all $x \in N(x_0)$, $F(x) \cap V$ is non-empty. One such open interval V, with $F(x_0) \cap V$ non-empty is shown, and a $N(x_0)$ illustrated for which $F(x) \cap V$ is non-empty for all $x \in N(x_0)$. In Figure 3.8, F is not lower hemicontinuous at x_0. Here there exists an open interval V with $F(x_0) \cap V$ non-empty, but there is no neighbourhood of x_0 such that $F(x) \cap V$ is non-empty for all x in that neighbourhood. Indeed, given the open interval V illustrated in Figure 3.8, for any small

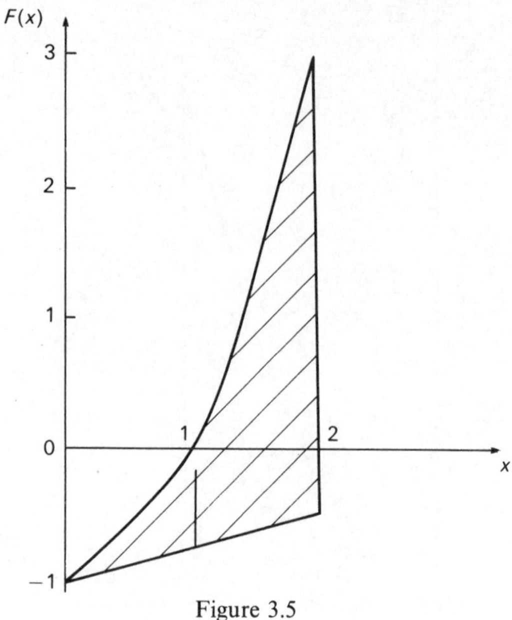

Figure 3.5

neighbourhood of x_0, $F(x)\cap V$ is a null set for every x, $x_0<x<x_0+\delta$.

Consider the correspondences defined in Example 3.1 and whose graphs are illustrated in Figures 3.4–3.5. The correspondence whose graph is shown in Figure 3.4 is not lower hemicontinuous at $x=1$, but the correspondence whose graph is given in Figure 3.5 is lower hemicontinuous at $x=1$.

DEFINITION 3.3 A correspondence $F:X\to Y$ is **continuous at** $\boldsymbol{x_0\in X}$ if it is both lower and upper hemicontinuous at x_0. If F is continuous for all $x\in X$ then F is a **continuous correspondence.**

The correspondences illustrated in Figures 3.2 and 3.7 are continuous correspondences.

DEFINITION 3.4 The correspondence $F:X\to Y$ is **closed at** $\boldsymbol{x\in X}$ if whenever $x^n\to x$, $y^n\to F(x^n)$ and $y^n\to y$ then $y\in F(x)$. A correspondence F is **closed** if it is closed at every point of its domain; that is, if its graph $\{(x,y)\in X\otimes Y:y\in F(x)\}$ is closed in

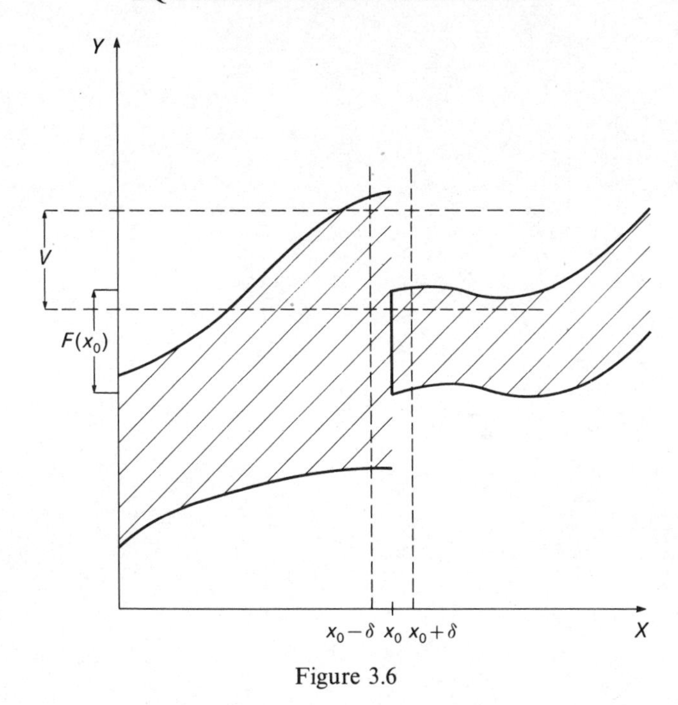

Figure 3.6

$X \otimes Y$. The correspondence F is **open** if the graph of F is open in $X \otimes Y$.

In general, a correspondence may be closed without being upper hemicontinuous, and vice-versa. For example, the correspondence $F:\mathbb{R} \to \mathbb{R}$ defined by $F(x) = (0, 1)$ is upper hemicontinuous but not closed.

DEFINITION 3.5 A correspondence $F:X \to Y$ is **compact-valued** if $F(x)$ is a compact subset of Y for every $x \in X$.

THEOREM 3.1 Let $X \subset \mathbb{R}^m$, $Y \subset \mathbb{R}^k$ and $F:X \to Y$ be a correspondence. If the range of the correspondence is compact and $F(x)$ is a closed set for each $x \in X$, then F is closed if, and only if, it is upper hemicontinuous.

Proof. See Border (1985), p. 56.

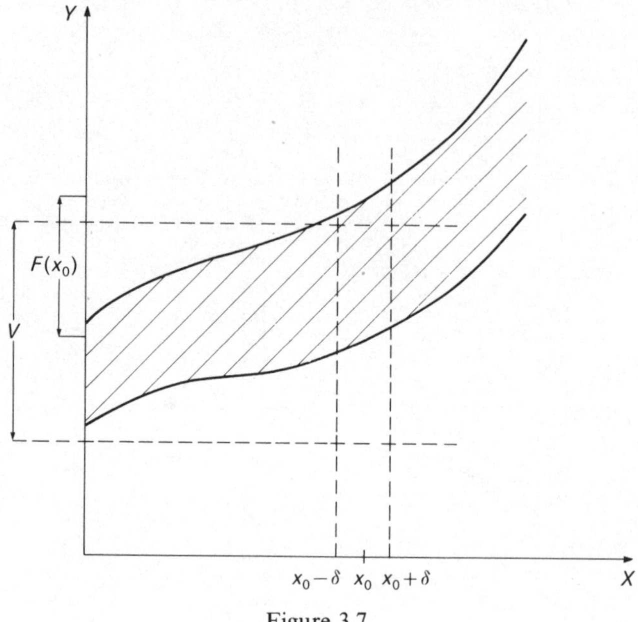

Figure 3.7

In most economic applications the correspondences are compact-valued, and in such cases the definition of an upper hemicontinuous correspondence has an equivalent formulation in terms of sequences.

THEOREM 3.2 The compact-valued correspondence $F:X \to Y$ is upper hemicontinuous at $x \in X$ if, and only if, for every sequence $\{x_n\}$ converging to x and every sequence $\{y_n\}$ with $y_n \in F(x_n)$, there exists a converging subsequence of $\{y_n\}$ whose limit belongs to $F(x)$.

Proof. See Border (1985), p. 57, or Hildenbrand and Kirman (1976), Appendix III.

The following is an equivalent formulation of the definition of lower hemicontinuity in terms of sequences. Note that, unlike the case of upper hemicontinuity, no compactness assumption is needed.

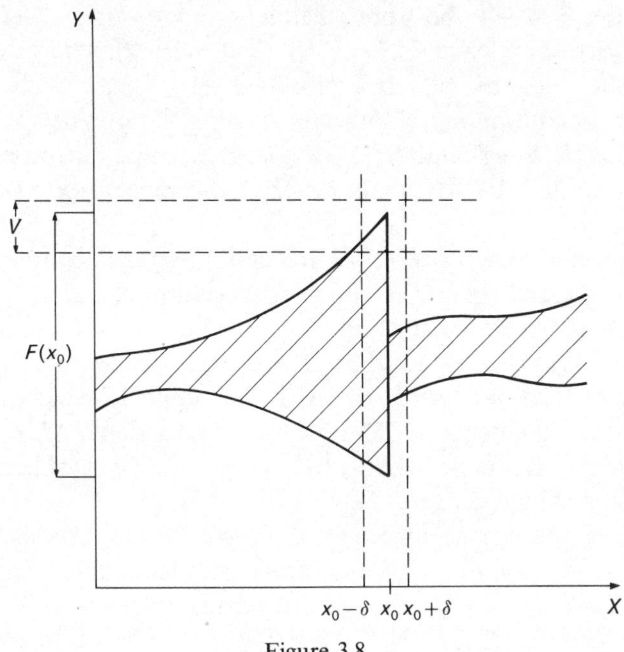

Figure 3.8

THEOREM 3.3 The correspondence $F:X \rightarrow Y$ is lower hemicontinuous at $x \in X$ if, and only if, for every sequence $\{x_n\}$ converging to x and every $y \in F(x)$, there exists a sequence $\{y_n\}$ converging to y with $y_n \in F(x_n)$.

Proof. See Hildenbrand and Kirman (1976), Appendix III.

Theorem 3.4 below gives some important properties of upper hemicontinuous correspondences. Proofs of the various results can be found in Border (1985), Chapter 11, or Hildenbrand and Kirman (1976), Appendix III.

THEOREM 3.4

(i) Let $F:X \rightarrow Y$ be upper hemicontinuous and compact valued and let $K \subset X$ be compact. Then the image $F(K) = \cup_{x \in K} F(x)$ is a compact set.

(ii) Let $F:X \to Y$ be upper hemicontinuous at x. Define the correspondence $G:X \to Y$ by $G(x) = \text{cl}\, F(x)$, where $\text{cl}\, F(x)$ is the closure of the subset $F(x)$. Then G is upper hemicontinuous. (Note the converse is not true.)

(iii) Let $F:X \to Y$ and $G:X \to Y$ and define the correspondence $H:X \to Y$ by $H(x) = F(x) \cap G(x)$. Suppose $F(x) \cap G(x) \neq \phi$. Then

(a) If F and G are upper hemicontinuous at x and closed-valued then H is upper hemicontinuous at x;

(b) If G is closed at x and F is upper hemicontinuous at x and $F(x)$ is compact then H is upper hemicontinuous at x.

(iv) Let $F:X \to Y$ and $G:Y \to Z$ be upper hemicontinuous correspondences. Define the correspondence $H:X \to Z$ by $H(x) = G(F(x))$, where $G(F(x)) = \cup_{y \in F(x)} G(y)$. Then H is upper hemicontinuous.

(v) Let the correspondences $F_i:X \to Y_i$, $i = 1, \ldots, k$, be upper hemicontinuous at x and compact-valued and let $Z = Y_1 \otimes \cdots \otimes Y_k$. Define the correspondence $G:X \to Z$ by $G(x) = F_1(x) \otimes \cdots \otimes F_k(x)$. Then G is upper hemicontinuous at x and compact valued.

(vi) Let the correspondences $F_i:X \to Y_i$, $i = 1, \ldots, k$, be upper hemicontinuous at x and compact-valued. Define the correspondence G by $G(x) = \sum_{i=1}^{k} F_i(x)$. Then G is upper hemicontinuous at x and compact-valued.

(vii) Let $F:X \to Y$ be upper hemicontinuous at x and compact-valued and let Y be convex. Define the correspondence G by $G(x) = \text{conv}\, F(x)$, where $\text{conv}\, F(x)$ is the convex hull of $F(x)$. Then G is upper hemicontinuous at x.

Similar results for lower hemicontinuous correspondences are given in Theorem 3.5. Note, however, that there are some important differences between some of the results for lower hemicontinuous correspondences and the corresponding results for upper hemicontinuous correspondences. Again for proofs of the statements in Theorem 3.5 the reader is referred to Border (1985). Chapter 11, and Hildenbrand and Kirman (1976), Appendix III.

THEOREM 3.5

(i) Let $F:X \to Y$ and define the correspondence $G:X \to Y$ by $G(x) = \text{cl}\, F(x)$, where $\text{cl}\, F(x)$ is the closure of the subset $F(x)$. Then F is lower hemicontinuous at x if, and only if, G is lower hemicontinuous at x.

(ii) Let $F:X \to Y$ and $G:X \to Y$ and define the correspondence $H:X \to Y$ by $H(x) = F(x) \cap G(x)$. Suppose $F(x) \cap G(x) \neq \phi$. Then if F is lower hemicontinuous at x and if G has an open graph then H is lower hemicontinuous at x.

(iii) Let $F:X \to Y$ and $G:Y \to Z$ be lower hemicontinuous correspondences. Define the correspondence $H:X \to Z$ by $H(x) = G(F(x))$, where $G(F(x)) = \cup_{y \in F(x)} G(y)$. Then H is lower hemicontinuous.

(iv) Let the correspondences $F_i:X \to Y_i$, $i = 1, \ldots, k$, be lower hemicontinuous at x and let $Z = Y_1 \otimes \cdots \otimes Y_k$. Define the correspondence $G:X \to Z$ by $G(x) = F_1(x) \otimes \cdots \otimes F_k(x)$. Then G is lower hemicontinuous at x.

(v) Let the correspondences $F_i:X \to Y_i$, $i = 1, \ldots, k$, be lower hemicontinuous at x. Define the correspondence G by $G(x) = \sum_{i=1}^{k} F_i(x)$. Then G is lower hemicontinuous at x.

(vi) Let $F:X \to Y$ be lower hemicontinuous at x and Y be convex. Define the correspondence G by $G(x) = \text{conv}\, F(x)$, where $\text{conv}\, F(x)$ denotes the convex hull of $F(x)$. Then G is lower hemicontinuous at x.

One of the most useful theorems employed in mathematical economics is the Maximum Theorem which deals with the case where a continuous real-valued function is being maximized over a compact set which varies continuously with some parameter vector. The set of solutions is an upper hemicontinuous correspondence with compact values, and the value of the maximized function varies continuously with the parameters. Typical applications of the Maximum Theorem in economics are the establishment that supply and demand correspondences are upper hemicontinuous, under appropriate circumstances (see Section 3.3).

THEOREM 3.6 (Maximum Theorem) Let $X \subset \mathbb{R}^m$, $Y \subset \mathbb{R}^k$ and let

$\gamma: X \to Y$ be a compact-valued correspondence. Let $f: Y \to R$ be a continuous function. Define the correspondence $\mu: X \to Y$ by $\mu(x) = \{y \in \gamma(x) : y$ maximizes f on $\gamma(x)\}$, and the function $g: X \to \mathbb{R}$ by $g(x) = f(y)$ for $y \in \mu(x)$. If γ is continuous at x, then μ is closed and upper hemicontinuous at x and g is continuous at x. Furthermore μ is compact-valued.

Proof. Note that, since γ is compact-valued, μ is non-empty and compact-valued. It suffices to show that μ is closed at x, for then $\mu = \gamma \cap \mu$ and Theorem 3.4 implies that μ is upper hemicontinuous at x. Let $x^n \to x$, $y^n \in \mu(x^n)$, $y^n \to y$. We wish to show $y \in \mu(x)$ and $g(x^n) \to g(x)$. Since γ is upper hemicontinuous and compact-valued, Theorem 3.1 implies $y \in \gamma(x)$. Suppose $y \notin \mu(x)$. Then there is a $z \in \gamma(x)$ with $f(z) > f(y)$. Since γ is also lower hemicontinuous at x, from Theorem 3.3 there is a sequence $\{z^n\}$ such that $z^n \to z$, $z^n \in \gamma(x^n)$. Since $z^n \to z$, $y^n \to y$ and $f(z) > f(y)$, the continuity of f implies that eventually $f(z^n) > f(y^n)$, contradicting $y^n \in \mu(x^n)$. Now $g(x^n) = f(y^n) \to f(y) = g(x)$, so g is continuous at x.

3.2.2 Fixed-point theorems

A basic role in the proof of existence of various types of equilibria has been played by fixed-point theorems. In each case there is some kind of continuous mapping of a set into itself, and the aim is to demonstrate that at least one point of the set remains invariant under the mapping.

DEFINITION 3.6 A **fixed-point of a function** $f: X \to Y$, where $X \cap Y \neq \phi$, is a point $x^\circ \in X$ such that $x^\circ = f(x^\circ)$. A **fixed-point of a correspondence** $F: X \to Y$, where $X \cap Y \neq \phi$, is a point $x^\circ \in X$ such that $x^\circ \in F(x^\circ)$.

THEOREM 3.7 (Brouwer Fixed-Point Theorem) Let X be a compact convex subset of $\mathbb{R}^n$ and let $f: X \to X$ be a continuous function. Then f has a fixed-point.

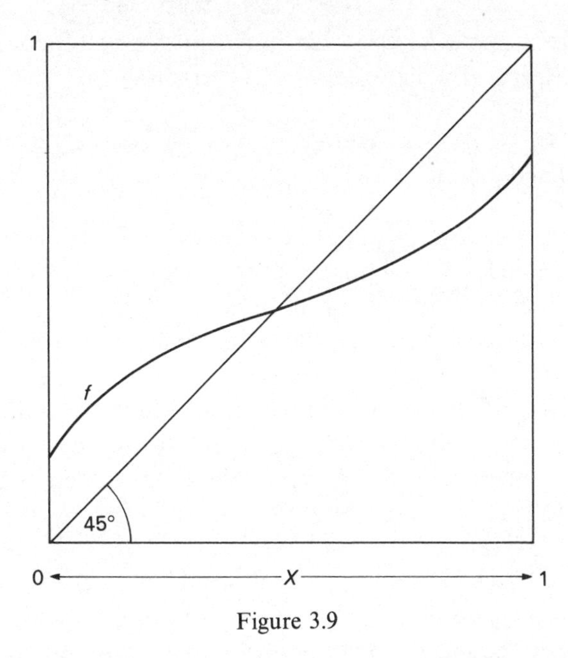

Figure 3.9

The proof of Brouwer's Theorem is rather lengthy and is therefore omitted. Good elementary proofs may be found in Border (1985), Klein (1973) and Tompkins (1964). The theorem is however easy to establish when f is a continuous function of a real variable. Let $X=[0,1]$ and consider Figure 3.9. If f crosses the 45° line, then clearly f has a fixed-point. Suppose then that f lies everywhere strictly above the 45° line. This is impossible since $f(1) \leqslant 1$ by the assumption that $f: X \to X$. Similarly, suppose f lies everywhere strictly below the 45° line. Again this is impossible since $f(0) \geqslant 0$ by the assumption that $f: X \to X$. Hence f must have a fixed-point.

Note that while Brouwer's Theorem guarantees the existence of a fixed-point it does not say anything about the uniqueness of the fixed-point. It is very easy to draw a continuous function in Figure 3.9 which cuts the 45° line more than once.

Recall that a metric space (X,d) is *complete* if every Cauchy sequence in it converges to some point in the space. Also recall that a function $f:(X,d)\to(X,d)$ is a *contraction* if there exists $r<1$ such that, for all $x, y\in X$, $d[f(x), f(y)]<rd(x,y)$.

THEOREM 3.8 A contraction on a complete metric space has a unique fixed-point.

Although no assumptions about convexity or finite-dimensionality on X are needed for the above theorem, we note that contractions are rather special kinds of continuous functions.

Brouwer's Theorem refers only to functions and, in economics, as we have already noted, correspondences are frequently encountered, so the Brouwer Theorem is of limited applicability. A generalization of Brouwer's Theorem to the case where the mappings concerned are correspondences has been provided by Kakutani (1941).

THEOREM 3.9 (Kakutani's Fixed-Point Theorem) Let $X\subset\mathbb{R}^n$ be compact and convex and let $\mu:X\to X$ be a closed or upper hemicontinuous correspondence with non-empty convex, compact values. Then μ has a fixed-point.

Proofs of Kakutani's Theorem are available in Border (1985), Hildenbrand and Kirman (1976) and Klein (1973). Generalizations of Kakutani's Theorem may be found in Border (1985), Chapter 15.

EXAMPLE 3.2 A price vector p° is an equilibrium price vector if $z(p^\circ)\leqslant\theta$, $p^\circ\geqslant\theta$, and if $z_k(p^\circ)<0$ then $p_k^\circ=0$, where z is a vector of aggregate excess demands. Suppose the aggregate excess demand functions are continuous, homogeneous of degree zero $[z_k(\lambda p)=z_k(p)$ for all $\lambda>0]$, and Walras' Law holds $[\sum_k z_k(p)p_k=0$ for all $p\geqslant\theta]$. Then an equilibrium price vector exists, as we shall now show.

Since the excess demand functions are homogeneous we can restrict attention to price vectors in the unit simplex $P\equiv\{p\in\mathbb{R}^n:\sum_k p_k=1,\ p_k\geqslant 0 \text{ for all } k\}$. P is closed, bounded and convex. Now consider the continuous mapping $f:P\to P$ defined by $f_k(p)=\{p_k+\max[0,z_k(p)]\}/\{1+\sum_k\max[0,z_k(p)]\}$, for all k. By

Brouwer's Theorem f has a fixed-point; there exists a $p^\circ \in P$ such that $p^\circ = f(p^\circ)$, i.e., $p_k^\circ = \{p_k^\circ + \max[0, z_k(p^\circ)]\}/\{1 + \sum_k \max[0, z_k(p^\circ)]\}$, for all k. Cross-multiplying each of these equations we obtain $p_k^\circ\{\sum_k \max[0, z_k(p^\circ)]\} = \max[0, z_k(p^\circ)]$ for all k, and multiplying by $z_k(p^\circ)$ and summing over all k gives $\sum_k z_k(p^\circ)p_k^\circ\{\sum_k \max[0, z_k(p^\circ)]\} = \sum_k z_k(p^\circ)\{\max[0, z_k(p^\circ)]\}$. But from Walras' Law $\sum_k z_k(p^\circ)p_k^\circ = 0$, so that $\sum_k z_k(p^\circ)\max[0, z_k(p^\circ)] = 0$. Each term in the sum is greater than or equal to zero since each term is either 0 or $[z_k(p^\circ)]^2$. But if any term were strictly greater than zero the inequality could not hold. Hence, we have $z_k(p^\circ) \leqslant 0$ for all k. Finally, we need to show that if $z_k(p^\circ) < 0$ then $p_k^\circ = 0$. Suppose $z_k(p^\circ) < 0$. From Walras' Law $\sum_j z_j(p^\circ)p_j^\circ = \sum_{j, j \neq k} z_j(p^\circ)p_j^\circ + z_k(p^\circ)p_k^\circ = 0$. But $z_j(p^\circ) < 0$ and $p_j^\circ \geqslant 0$ for all j, so that $\sum_{j, j \neq k} z_j(p^\circ)p_j^\circ \leqslant 0$. Hence, we must have $z_k(p^\circ)p_k^\circ \geqslant 0$, otherwise Walras' Law would be violated. But if $z_k(p^\circ) < 0$ then $p_k^\circ \leqslant 0$. But $p_k^\circ \geqslant 0$. Hence, $p_k^\circ = 0$.

3.3 EXISTENCE OF COMPETITIVE EQUILIBRIUM

3.3.1 Introduction

In this section, we show how the mathematical concepts introduced in Section 3.2 can be used to establish the existence of a competitive equilibrium in a private ownership economy with a finite number of agents and commodities. The modern analysis of the existence of competitive equilibrium is largely the result of work by Arrow and Debreu (1954), Debreu (1959), Gale (1955), McKenzie (1959) and Nikaido (1956). Our exposition is based on Debreu (1959) and Klein (1973). Not all the assumptions we shall make are necessary for proving the existence of a competitive equilibrium, but we are concerned here with merely illustrating the nature of an otherwise complex exercise. (Indeed, our treatment of the existence of competitive equilibrium is quite crude given the current state of the art. Recent research on the existence of competitive equilibrium is reviewed in Debreu, 1982, McKenzie, 1981 and Sonnenschein, 1977.)

3.3.2 Finite private ownership economy

Consider an economy with a finite number of commodities (indexed by $i=1,\ldots,n$), a finite number of firms (indexed by $f=1,\ldots,F$), and a finite number of consumers (indexed by $h=1,\ldots,H$).

A commodity is any good or service completely identified by three characteristics: (a) physical properties, (b) location of delivery, (c) time of delivery. A commodity bundle is an n-tuple of real numbers in the commodity space $\mathbb{R}^n$, with the ith component representing the quantity in which commodity i is included.

Consumer h is represented by a consumption possibility set $X_h \subset \mathbb{R}^n$, totally preordered by a preference-indifference relation $\lesssim_h$, together with a vector of initial endowments $\bar{x}_h \in \mathbb{R}^n$. (A relation τ is reflexive if $a\tau a$ for all a, and τ is transitive if $a\tau b$ and $b\tau c$ implies $a\tau c$, for any a, b, c; τ is a preordering if τ is reflexive and transitive, and τ is a total preordering if it is a preordering and, for any a, b, either $a\tau b$ or $b\tau a$.) A point $x_h \in X_h$ is a consumption plan, where positive components denote inputs (consumption proper) and negative components denote outputs (supplies, e.g., labour services). The aggregate consumption possibility set of the economy is $X \equiv \sum_{h=1}^{H} X_h$; an aggregate consumption plan is $x=\sum_{h=1}^{H} x_h$, where clearly $x \in X$; the total initial resources are given by $\bar{x}=\sum_{h=1}^{H} \bar{x}_h$, where $\bar{x}_h$ is the resource endowments of consumer h.

Firm f is represented by a production possibility set $Y_f \subset \mathbb{R}^n$. A production plan is a point $y_f \in Y_f$, where negative components represent inputs (raw materials, labour) and positive components outputs (production proper). More than one component of a production plan may be positive, so that joint production is allowed. The production possibility set for the economy is $Y=\sum_{f=1}^{F} Y_f$; an aggregate production plan is $y=\sum_{f=1}^{F} y_f$, where clearly $y \in Y$.

With each commodity there is associated a non-negative real number, its price. We represent a normalized price system by an n-tuple $p \in P=\{p: p \geqslant \theta, \sum_{i=1}^{n} p_i = 1\}$. Given the above sign conventions regarding inputs and outputs, the profit of firm f from producing y_f at prices p is $\pi_f = p' y_f$.

A private ownership economy is one in which consumers own resources and firms, so that consumer h has a claim on the profit of firm f given by $\alpha_{hf} \geqslant 0$ and owns $\bar{x}_h$. $\sum_{h=1}^{H} \alpha_{hf} = 1$ for all f. Under these conditions the wealth of consumer h is $w_h = p'\bar{x}_h + \sum_{f=1}^{H} \alpha_{hf} p' y_f$. A point $w \in \mathbb{R}^H$ is a wealth distribution. If p and w are given, we refer to $(p, w) \in \mathbb{R}^{H+n}$ as a price-wealth pair.

DEFINITION 3.7 A **private ownership economy** is a system

$$E = \{(X_h, \precsim_h), (Y_f), \bar{x}, (\alpha_{hf})\}$$

with commodity space $\mathbb{R}^n$, resources $\bar{x} = \sum_{h=1}^{H} \bar{x}_h$, and $h = 1, \ldots, H$; $f = 1, \ldots, F$.

3.3.3 Properties of demands and supplies

We impose the following conditions on production sets. For all f:

A1. Y_f is compact and convex,

A2. $Y_f \cap \mathbb{R}^n_+ = \{\theta\}$ (impossibility of free production),

A3. $Y_f \cap (-Y_f) = \{\theta\}$ (irreversibility),

where $\mathbb{R}^n_+$ is the non-negative orthant of $\mathbb{R}^n$.

Note that A2 implies $\theta \in Y_f$ and together with convexity, A1, this rules out increasing returns to scale. Given A1–A3 it can be shown that: (A1′) Y is compact and convex, (A2′) $Y \cap \mathbb{R}^n_+ = \{\theta\}$, and (A3′) $Y \cap (-Y) = \{\theta\}$.

It is assumed that firm f chooses a production plan which maximizes its profit relative to $p \in P$; that is, firm f chooses $y_f \in Y_f$ which is a maximum for the set pY_f. In general, this determines a point-set mapping $s_f : P \to Y_f$ defined by $s_f(p) = \{y_f : y_f \in Y_f, p' y_f = \max pY_f\} : s_f$ is the supply correspondence of firm f. For the economy as a whole, the aggregate supply correspondence is defined by $S(p) = \sum_{f=1}^{F} s_f(p)$.

A profit function for firm f is a mapping $\pi_f : P \to \mathbb{R}$ defined by $\pi_f(p) = \max pY_f$.

Given the above assumptions on firm production sets we have:

R1. s_f is upper hemicontinuous and π_f is continuous for each f,

R2. S is upper hemicontinuous,

R3. $s_f(p)$ and $S(p)$ are convex sets.

TO ESTABLISH (R1), define the function $g:Y \to \mathbb{R}$ by $g(y_f)=p'y_f$, which is clearly continuous for given p. Define the correspondence $\gamma:P \to Y_f$ by $\gamma(p)=Y_f$. As a constant mapping γ is trivially a continuous correspondence. Moreover, γ is compact-valued since Y_f is compact. Firm f's supply correspondence is $s_f(p)=\{y_f:y_f \in Y_f, p'y_f=\max pY_f\}$, which can be expressed as $s_f(p)=\{y_f \in \gamma(p): y_f$ maximizes g on $\gamma(p)\}$. Firm f's profit function is $\pi_f(p)=\max pY_f$, which can be expressed as $\pi_f(p)=g(y_f)$ for $y_f \in s_f(p)$. Then, from the Maximum Theorem, we have s_f is closed and upper hemicontinuous and π_f is continuous. Furthermore s_f is compact-valued. To see (R2), note that since $S=\sum_{f=1}^{F} s_f(p)$ and s_f is upper hemicontinuous and compact-valued, from Theorem 3.4(vi) S is upper hemicontinuous and compact-valued. Convexity of $s_f(p)$ follows from the fact that the set of maximizing points for a linear function on a convex set is convex. $S(p)$ is convex as the sum of convex sets is convex. Thus, (R3) is established.

The following are standard assumptions regarding consumption sets. For all h:

A4. X_h is closed and convex,

A5. X_h is lowerbounded,

A6. $\precsim_h$ is a total preordering such that:

(i) for all $x_h^\circ \in X_h$, $\{x_h \in X_h : x_h^\circ \precsim_h x_h\}$ and $\{x_h \in X_h : x_h^\circ \succsim_h x_h\}$ are closed in X_h (continuity),

(ii) for all $x_h^\circ \in X_h$, $\{x_h : x_h^\circ \precsim_h x_h\}$ is convex (convexity),

(iii) X_h has no maximal element for $\precsim_h$ (non-satiation).

Given A4–A5 it can be shown that X is closed and convex and lowerbounded. It can also be shown (Debreu, 1959) that if A4 and A6 hold then a continuous real-valued utility function (order-preserving mapping) $U_h:X_h \to \mathbb{R}$ exists.

Define $B_h=\{(p,w) \in \mathbb{R}^{H+N}$: there is x_h in X_h such that $p'x_h \leqslant w_h\}$. Then the mapping $\psi_h:B_h \to X_h$ defined by $\psi_h(p,w)=\{x_h \in X_h : p'x_h \leqslant w_h)$ is the budget correspondence for consumer h.

It can be shown (Debreu, 1959, p. 64) that if X_h is compact, convex, and if (p°, w°) is a point of B_h such that $w_h^\circ \neq \min p^\circ X_h$, then ψ_h is continuous at (p°, w°). Henceforth we assume that X_h is compact and not just closed and lowerbounded. The requirement that $w_h^\circ \neq \min pX_h$ will require some additional assumption. We assume that there is $x_h^\circ \in X_h$ such that $x_h^\circ \ll \bar{x}_h$. This is a particularly strong assumption which guarantees the continuity of the mapping ψ_h at (p, w).

Given the price-wealth pair (p, w) in B_h, consumer h is assumed to choose in the set $\psi_h(p, w)$ a greatest element for his preference ordering $\lesssim_h$, or, equivalently, he maximizes his utility function U_h on the set ψ_h. Since X_h is compact, the set $\psi_h(p, w)$ is non-empty and compact, and a utility maximizing consumption plan therefore exists for each $(p, w) \in B_h$. A demand correspondence for consumer h is a mapping $d_h: B_h \rightarrow X_h$ defined by $d_h(p, w) = \{x_h \in \psi_h(p, w): x_h$ is a maximal element of $\psi_h(p, w)$ for $\lesssim_h\}$.

The maximum utility, when the price-wealth pair is (p, w) in B_h, is denoted $V_h(p, w)$. The function $V_h: B_h \rightarrow R$ is the *indirect utility function* for consumer h: $V_h(p, w) = \max U_h[\psi_h(p, w)]$.

For the economy as a whole an aggregate demand correspondence can be defined by $D(p, w) = \sum_{h=1}^{H} d_h(p, w)$. The wealth of consumer h is $w_h = p'\bar{x}_h + \sum_{f=1}^{F} \alpha_{hf}\pi_f(p)$, so that the wealth distribution is the H-tuple $w = (w_1, \ldots, w_h)$. The demand correspondence for consumer h is thus a point-set mapping of p only; $d_h(p) = d_h\{p, [p'\bar{x}_h + \sum_{f=1}^{F} \alpha_{hf}\pi_f(p)]\}$. The aggregate demand correspondence is thus also a mapping of p only.

Given the above assumptions on consumption sets we have:

R4. For all h, d_h is upper hemicontinuous and V_h is continuous,

R5. D is upper hemicontinuous,

R6. $d_h(p)$ and $D(p)$ are convex sets.

Since $x_h^\circ \ll \bar{x}_h$ for every $p \in P$ we have $p'x_h^\circ < p'\bar{x}_h$. Furthermore, since $\theta \in Y_f$ for every $p \in P$, we have $\pi_f(p) \geqslant 0$. Hence, for every $p \in P$, the inequality $p'x_h^\circ < p'\bar{x}_h + \sum_{f=1}^{F} \alpha_{hf}\pi_f(p)$ holds, and thus the correspondence ψ_h is continuous at the point $(p, [p'\bar{x}_h + \sum^{F}{}_{f=1} \alpha_{hf}\pi_f(p)])$. ψ_h is also compact-valued. Since $\pi_f(p)$ is continuous on P, so is the function $p \rightarrow \{w_h(p)\}$ which assigns to each $p \in P$ the H-tuple $w = (w_1, \ldots, w_h)$. As ψ_h is continuous at $(p, [p'\bar{x}_h + \sum_{f=1}^{F} \alpha_{hf}\pi_f(p)])$ and V_h is continuous, the Maximum Theorem can be applied to obtain R4. R5 and R6 are established along the same lines as R2 and R3.

3.3.4 Equilibrium

DEFINITION 3.8 A **competitive equilibrium** for the private ownership economy E is a triplet $[(x_h*),(y_f*),p*]$ such that:

(i) For all h, x_h* is a maximal element of $\{x_h \in X_h : p*'x_h \leqslant p*'\bar{x}_h + \sum_{f=1}^{F} \alpha_{hf} p*'y_f*\}$ for $\lesssim_h$,
(ii) For all f, y_f* is a maximizer of $p*Y_f$ on Y_f,
(iii) $x* - y* = \bar{x}$.

Consider a state of the economy characterized by a pair $[(x_h),(y_f)]$ and total resources $\bar{x}$. If $x_h \in X_h$ for all h and $y_f \in Y_f$ for all f, we call $x - y - \bar{x} = z$ an excess demand. Clearly, $z \in Z = X - Y - \{\bar{x}\}$. Given the aggregate demand correspondence, $D(p)$, and the aggregate supply correspondence, $S(p)$, the point-set mapping $z: P \to Z$, where $z(p) = D(p) - S(p) - \{\bar{x}\}$, is the aggregate excess demand correspondence.

Since for a given p the action of an economic agent has to satisfy the wealth constraint, it must be that $p'x_h \leqslant p'\bar{x}_h + \sum_{f=1}^{F} \alpha_{hf} p'y_f$ for all h. Summing over all h, we have $\sum_{h=1}^{H} p'x_h \leqslant \sum_{h=1}^{H} p'\bar{x}_h + \sum_{h=1}^{H} \sum_{f=1}^{F} \alpha_{hf} p'y_f$; that is, $p'x \leqslant p'\bar{x} + p'y$, or $p'z \leqslant 0$. The condition $p'z \leqslant 0$ for all $p \in P$ is known as *Walras' Law*. Since $p* \in P$ we must necessarily have $z(p*) \leqslant \theta$. The essential piece in the proof of the existence of an equilibrium is thus the proof that there exists a $p* \in P$ such that $z(p*) \leqslant \theta$, or, what is the same, there exists a $p* \in P$ such that $z(p*) \cap (-\mathbb{R}^n_+) \neq \phi$. This is done by means of the so-called Excess Demand Theorem. The second stage in the existence proof is done by showing that the described economy satisfies the conditions for the application of the Excess Demand Theorem.

THEOREM 3.10 (Excess Demand Theorem) Let $Z \subset \mathbb{R}^n$ be compact. If the mapping $z: P \to Z$ is an upper hemicontinuous correspondence such that for all $p \in P$ the image set $z(p)$ is non-empty, convex and satisfies Walras' Law $[p'z(p) \leqslant 0]$ then there exists a $p* \in P$ such that $z(p*) \cap (-\mathbb{R}^n_+) \neq \phi$.

Proof. P is non-empty, compact and convex. Z can be replaced by any compact subset Z' of $\mathbb{R}^n$ containing it; Z' is chosen to be convex. As P is non-empty, so clearly is Z, and hence Z'. Let $z \in Z'$. Call $\mu(z)$ the set of all $p \in P$ which maximize $p'z$ on P. Since P is non-empty and compact, $\mu(z)$ is non-empty. The point-set mapping $z \rightarrow \mu(z)$ from Z' to P is upper hemicontinuous on Z' (this follows in exactly the same way as the upper hemicontinuity of firms' supply correspondences). Since P is convex, $\mu(z)$ is also, for either (i) $z = \theta$ – and then $\mu(z)$ is P itself, or (ii) $z \neq \theta$ – and then $\mu(z)$ is the intersection of P and $\{p : p'z = \max Pz)$. Consider now the correspondence $\psi : P \otimes Z' \rightarrow P \otimes Z'$ defined by $\psi(p, z) = \mu(z) \otimes z(p)$. The set $P \otimes Z'$ is non-empty, compact and convex since P and Z' have these properties. The correspondence ψ is upper hemicontinuous as μ and z are upper hemicontinuous (Theorem 3.4(v)). Finally, for all $(p, z) \in P \otimes Z'$, the set $\psi(p, z)$ is non-empty and convex, as $\mu(z)$ and $z(p)$ have these properties (Theorem 1.2). All the conditions of Kakutani's Fixed-Point Theorem are satisfied, so that ψ has a fixed-point $(p*, z*)$. Thus $(p*, z*) \in \mu(z*) \otimes z(p*)$, or $p* \in \mu(z*)$ and $z* \in z(p*)$. From the definition of $\mu(z)$ it follows that $p* \in \mu(z*)$ implies $p'z* \leqslant p*'z*$. But $z* \in z(p)$ implies from Walras' Law that $p*'z* \leqslant 0$. Hence, for every $p \in P$, we have $p'z* \leqslant 0$. Therefore it must be that $z* \leqslant \theta$, and consequently $z* \in -\mathbb{R}^n_+$. This with $z* \in z(p*)$ proves that $p*$ has the desired property.

Returning to the private ownership economy described above we have already established that the aggregate supply correspondence S is upper hemicontinuous and $S(p)$ is compact and convex; furthermore, the aggregate demand correspondence D is upper hemicontinuous and $D(p)$ is compact and convex. The excess demand correspondence $z(p) = D(p) - S(p) - \{\bar{x}\}$ is therefore upper hemicontinuous and the set $z(p)$ is convex and compact. Hence, the Excess Demand Theorem is applicable, and an equilibrium exists for the private ownership economy.

3.4 EQUILIBRIUM IN n-PERSON NON-COOPERATIVE GAMES

A *non-cooperative game* is a situation where a number of players, n, must each make independently a choice of an action and then, based on all the choices, some outcome occurs. The set of actions under player i's control is his **strategy set** S^i. The elements of S^i are player i's **strategies** (actions). When certain aspects of the game are random as in, say, poker, then it is conventional to treat *Nature* as a player with Nature choosing the random action to be taken. A player's strategy itself might involve a random variable. Such a strategy is called a **mixed-strategy**. For instance, if there are a finite number m of 'pure' strategies, then we can identify a 'mixed-strategy' with a vector in $\mathbb{R}^m$, the components of which indicate the probability of taking the corresponding 'pure' action. A **strategy vector** is a vector $s=(s_1,\ldots,s_n)$ with $s_i \in S^i, i=1,\ldots,n$; a strategy vector consists of a list of the choices of strategy for each player. Each strategy vector determines an outcome of the game. Although the outcome may be a random variable, its distribution is determined by the strategy vector. Players have preferences over outcomes and this induces preferences over strategy vectors. We assume that each player i has a pattern of preferences over outcomes (strategy vectors), so that for each player i there is a real-valued utility function P^i, defined over all possible strategy vectors and unique up to a positive affine transformation, known as his **payoff function**. Given the rules of the game, it is possible to tabulate $P(s)=[P^1(s),\ldots,P^n(s)]$, with $s=(s_1,\ldots,s_n)$ for all possible values of s_i, $s_i \in S^i, i=1,\ldots,n$, by setting up an n-dimensional array of n-vectors. This n-dimensional array is called the **normal** or **strategic form** of the game. A game in normal form is specified by a list of strategy sets and payoffs.

In a non-cooperative game there is absence of trust among the players and in such a situation a player i who expects the other players to choose strategies $(s_1,\ldots,s_{i-1},s_{i+1},\ldots,s_n)$ can himself be expected to select a strategy s_i°, maximizing his payoff given his anticipation of the other players' actions. For the strategy vector $s=(s_1,\ldots,s_n)$ let $s\backslash s_i^\circ \equiv (s_1,\ldots,s_{i-1},s_i^\circ,s_{i+1},\ldots,s_n)$, so that $s\backslash s_i^\circ$ is the strategy vector s in which s_i° has replaced the strategy s_i. We define a

best reply s_i° for player i by $P^i(s\backslash s_i^\circ) \geqslant P^i(s)$ for all $s_i \in S^i$. However, if he ascribes the same degree of rationality to the other players, player i will expect them to follow the same decision rule. This is likely to lead into an expectational circle of the kind 'If they know that I know that they know. . . .', which is typical of the decision problem any individual has to face when he recognizes the interdependence of actions in a game. Only if a strategy vector $s*$ exists for which $P^i(s*) \geqslant P^i(s*\backslash s_i^\circ)$ for all $s_i^\circ \in S^i$ and all $i, i=1,\ldots,n$ is this expectational circle broken, since in $s*$ each player's strategy $s*_i$ is a best reply to the other players' strategies.

DEFINITION 3.9 A strategy vector $s*$ is an **equilibrium point** or **Nash equilibrium** if, and only if, for each player i, $i=1,\ldots,n$, and any $s_i^\circ \in S^i$, $P^i(s*) \geqslant P^i(s*\backslash s_i^\circ)$.

The Nash equilibrium is the basic solution concept for n-person non-cooperative games. The definition of a Nash equilibrium implies that no single player has an incentive to deviate from it.

THEOREM 3.11 (Nash) In the n-person non-cooperative game G, if, for every player i, $i=1,\ldots,n$, the set S^i is a non-empty, compact, convex subset of a Euclidean space and P^i is a continuous real-valued function on $S \equiv S^1 \otimes \cdots \otimes S^n$ that is linear in its ith variable, then G has an equilibrium point.

Proof. For each i, define the correspondence μ_i from S to S^i as follows. Given an element s of S, let $\mu_i(s) = \{x_i \in S^i : P^i(s\backslash x_i) = \max_{y \in S} P^i(s\backslash y_i)\}$; that is $\mu_i(s)$ is the set of strategies for player i that maximize his payoff given the strategies of the other players prescribed by s. Note that $\mu_i(s)$ is non-empty and compact since P^i is continuous and S^i is non-empty and compact. Note also that $\mu_i(s)$ is convex since P^i is linear in s_i and the maximization is carried out over a convex set. We now show that the correspondence μ_i is upper hemi-continuous. Let $\{s^q\}$ be a sequence in S converging to s°, and $\{x^q\}$ be a sequence in S^i converging to x_i° and such that for every q one has $x_i^q \in \mu_i(s^q)$. Consider an arbitrary point y_i in S^i. For every q one has $P^i(s^q\backslash x_i^q) \geqslant P^i(s^q\backslash y_i)$. In the limit, $P^i(s^\circ\backslash x_i^\circ) \geqslant P^i(s^\circ\backslash y_i)$. Since this inequality holds for any y_i in S^{ι}, we have established that $x_i^{\circ} \in \mu_i(s^{\circ})$. Hence, from Theorem 3.2, μ_i is upper hemicontinuous. Now define the correspondence $\mu: S \to S$ by $\mu(s) = \mu_i(s) \otimes \ldots \otimes \mu_n(s)$. A point $s* \in S$ is an equilibrium point if, and only if, for every i, $s_i* \in \mu_i(s*)$; hence, if, and only if, $s* \in \mu(s*)$. An equilibrium point of G is therefore equivalent to a fixed-point of the correspondence μ. Since each μ_i is upper hemicontinuous and compact-valued from Theorem 3.4(v), we have that μ is upper hemicontinuous and compact-valued. Moreover, since $\mu_i(s)$ is convex, $\mu(s)$ is convex, (Theorem 1.2(iv)). The conditions of Kakutani's Fixed-Point Theorem are thus satisfied and we can conclude that μ has a fixed-point and hence that G has an equilibrium-point.

Note that, in the statement of Theorem 3.11, the condition that P^i is linear in its ith variable can be replaced by the assumption that P^i is quasiconcave with respect to its ith variable, since the quasiconcavity of P^i with respect to its ith variable also ensures that $\mu_i(s)$ is convex (Theorem 1.42).

EXAMPLE 3.3 The classic example of a Nash equilibrium in economics is that of the Cournot equilibrium in oligopoly theory. In the Cournot model there is a fixed number of firms producing a homogenous good in a single period. Price competition is absent in the model, being ruled out by the assumption that each firm chooses an output level and that the market price is determined from the demand function as the price that equates the amount demanded to the total amount produced by the firms. Let q_i denote the output level of firm i, $Q = \sum_{i=1} q_i$ the total output of the industry and p the market price. The inverse demand function is $p = f(Q)$, where (i) $f(0) > 0$, (ii) if $Q > 0$ and $f(Q) > 0$ then $f'(Q) < 0$, (iii) if $Q > 0$ then $Qf(Q) \leqslant M$, where M is some finite number. Let $C_i(q_i)$ be firm i.'s total cost function, with $C_i(q_i) > 0$ for $q_i > 0$, $C_i'(q_i) > 0$ and $C_i(0) \geqslant 0$. Let $q = (q_1, \ldots, q_n)$. Then firm i's profit function is $\pi_i(q) = q_i f(Q) - C_i(q_i)$. The Cournot equilibrium is a vector of outputs $q^c = (q_1^c, \ldots, q_n^c)$ such that no single firm i can increase its profit by choosing an output level different from q_i^c, given that the other firms are choosing q_j^c, $j \neq i$.

The Cournot model is easily interpreted as an n-person non-cooperative game. The firms are the players. For each player i the strategy set S^i is the interval $[0, \bar{Q}_i]$ of output levels from which firm i must choose its output level, and the profit function $\pi^i(q)$ is the payoff function of firm i. The firm's output level is its strategy. The Cournot equilibrium is precisely the equilibrium point of the game. For extensive discussion of the application of game theory to oligopoly problems the reader is referred to Friedman (1977).

CHAPTER 4

COMPARATIVE STATICS AND DUALITY

4.1 INTRODUCTION

In any static optimization problem, the objective function and the constraint functions will contain certain parameters and the optimal solution will depend on the values taken by these parameters. Thus, if any particular parameter value is altered, then we should expect the optimal choice of control variables and the maximum value of the objective function to change. The determination of the effects of parameter variations on the optimal choice of control variables and the maximum value of the objective function is referred to in the economics literature as *comparative statics analysis*. Section 4.2 is devoted to comparative statics analysis.

Closely related to comparative statics analysis is the *theory of duality*. At the heart of duality theory in economics is the notion of 'equivalent representations'. Following Epstein (1981) we may say that:

THE THEORY OF DUALITY describes alternative equivalent representations of consumers' preferences (direct or indirect utility function, expenditure function), or of a competitive producer's technology (production, profit or cost function).

Thus, in economics, duality refers to the existence of 'dual functions' which, under appropriate regularity conditions, embody the same information on preferences or technology as the more familiar 'primal functions' such as the utility or production function. Dual functions describe the results of optimizing responses to input and output prices and constraints rather than global responses to input and output quantities as in the corresponding primal functions. Dual functions are typically maximum value functions. Consider a familiar pair of primal and dual functions – the single-product firm's product and cost functions

$$q = f(x_1, \ldots, x_n), \tag{4.1}$$

$$c = g(q, r_1, \ldots, r_n), \tag{4.2}$$

where q denotes output, c is total cost, and the x_i's and r_i's are input quantities and prices, respectively. The production function (4.1), referred to as the 'primal', describes output response globally to all possible combinations of input quantities. The cost function (4.2), which is a 'dual' of the production function, describes the minimum cost of producing any level of output, given a set of input prices and the firm's production technology. An example from consumer theory would be the indirect utility function (dual) which shows the maximum value of utility associated with given commodity prices and level of income. The familiar utility function (primal), on the other hand, describes the level of utility associated with all possible combinations of commodity quantities. In Section 4.3, where the theory of duality is introduced, we shall see that dual functions contain information about both optimal behaviour and the structure of the underlying technology or preferences, whereas the primal function describes only the latter.

4.2 COMPARATIVE STATICS AND MAXIMUM VALUE FUNCTIONS

The following result provides the key to the determination of the manner in which parameter variations affect the optimal choice of control variables in an unconstrained optimization problem.

THEOREM 4.1 For the problem maximize$_x$ $f(x{:}\alpha)$ subject to $x \in S$, where $\alpha \in A \subset \mathbb{R}^p$ is a vector of parameters and S is an open subset of $\mathbb{R}^n$, if f is twice-continuously differentiable on S, if x° is a local maximum point of f for given $\alpha^\circ \in A$ and if $|\partial^2 f(x^\circ{:}\alpha^\circ)/\partial x_i \partial x_j| \neq 0$ (this will be so if the sufficient conditions for a strict local maximum hold), then

$$\begin{bmatrix} \partial^2 f/\partial x_1^2 & \cdots & \partial^2 f/\partial x_n \partial x_1 \\ \vdots & & \vdots \\ \partial^2 f/\partial x_1 \partial x_n & \cdots & \partial^2 f/\partial x_n^2 \end{bmatrix} \begin{bmatrix} \partial x_1/\partial \alpha_k \\ \vdots \\ \partial x_n/\partial \alpha_k \end{bmatrix} = \begin{bmatrix} -\partial^2 f/\partial \alpha_k \partial x_1 \\ \vdots \\ -\partial^2 f/\partial \alpha_k \partial x_n \end{bmatrix} \tag{4.3}$$

for $k = 1, \ldots, p$, where all derivatives are understood to be evaluated at $(x^\circ{:}\alpha^\circ)$.

Proof. Since x° is a local maximum point,

$$f_x(x^\circ{:}\alpha^\circ) = \theta. \tag{4.4}$$

Furthermore, as $|\partial^2 f(x^\circ{:}\alpha^\circ)/\partial x_i \partial x_j| \neq 0$, the rank of the Jacobian matrix of (4.4) is n and we can apply the Implicit Function Theorem to conclude that there exist continuously differentiable functions

$$x_j^\circ = x_j(\alpha), \qquad j = 1, \ldots, n \tag{4.5}$$

such that

$$\partial f(x_1(\alpha), \ldots, x_n(\alpha))/\partial x_j \equiv 0, \qquad j = 1, \ldots, n \tag{4.6}$$

for all $\alpha \in A$ in some neighbourhood of α°. Differentiating the identities in (4.6) partially with respect to α_k, $k = 1, \ldots, p$, we have

$$\sum_{i=1}^{n} (\partial^2 f/\partial x_i \partial x_j)(\partial x_i/\partial \alpha_k) + \partial^2 f/\partial \alpha_k \partial x_j = 0, \qquad k = 1, \ldots, p, \tag{4.7}$$

where all derivatives are understood to be evaluated at $(x^\circ{:}\alpha^\circ)$. Expressing (4.7) in matrix form gives (4.3).

Note that, since $|\partial^2 f(x^\circ:\alpha^\circ)/\partial x_i \partial x_j| \neq 0$, (4.3) can be uniquely solved to obtain explicit expressions for $\partial x_i(x^\circ:\alpha^\circ)/\partial\alpha_k$, $i=1,\ldots,n$; $k=1,\ldots,p$. The solutions are frequently obtained by applying Cramer's Rule. Of particular importance in economic applications are the signs of these derivatives. From the assumptions imposed on the objective function in the formulation of the problem it *may* be possible to determine their signs.

THEOREM 4.2 For the problem maximize$_x$ $f(x:\alpha)$ subject to $g^i(x:\alpha)=0$, $i=1,\ldots,m$, $m<n$, where $x \in X \subset \mathbb{R}^n$ and $\alpha \in A \subset \mathbb{R}^p$ is a vector of parameters, let f and each of the g^i's be continuously differentiable over the open set X, and let x° be a constrained local maximum point for given $\alpha^\circ \in A$. Assume that the Jacobian matrix $[\partial g^i(x^\circ:\alpha^\circ)/\partial x_j]$ has rank m and that the matrix

$$B \equiv \begin{bmatrix} L_{xx}(x^\circ,\lambda^\circ:\alpha^\circ) & [\partial g^i(x^\circ:\alpha^\circ)/\partial x_j]' \\ [\partial g^i(x^\circ:\alpha^\circ)/\partial x_j] & \theta \end{bmatrix}$$

is of full rank (this will be so if the sufficient conditions for a strict constrained local maximum hold). Then

$$\begin{bmatrix} L_{xx} & [\partial g^i/\partial x_j]' \\ & \\ [\partial g^i/\partial x_j] & \theta \\ & \end{bmatrix} \begin{bmatrix} \partial x_1/\partial\alpha_k \\ \vdots \\ \partial x_n/\partial\alpha_k \\ \partial\lambda_1/\partial\alpha_k \\ \vdots \\ \partial\lambda_m/\partial\alpha_k \end{bmatrix} = \begin{bmatrix} -\partial^2 L/\partial\alpha_k\partial x_1 \\ \vdots \\ -\partial^2 L/\partial\alpha_k\partial x_n \\ -\partial g^1/\partial\alpha_k \\ \vdots \\ -\partial g^m/\partial\alpha_k \end{bmatrix} \tag{4.8}$$

for $k=1,\ldots,p$, where it is understood that all derivatives are evaluated at $(x^\circ,\lambda^\circ:\alpha^\circ)$.

Proof. Since the Jacobian matrix $[\partial g^i(x^\circ:\alpha^\circ)/\partial x_j]$ has rank m and x° is a constrained local maximum point the following conditions necessarily hold:

$$L_x(x^\circ,\lambda^\circ:\alpha^\circ)=\theta, \tag{4.9}$$

$$g^i(x^\circ:\alpha^\circ)=0, \qquad i=1,\ldots,m.$$

Since $|B| \neq 0$, the Jacobian matrix of the system of equations (4.9) has rank $m+n$ and the Implicit Function Theorem can be applied to conclude that there exist continuously differentiable functions

$$x_j^\circ = x_j(\alpha), \qquad j=1,\ldots,n,$$
$$\lambda_i^\circ = \lambda_i(\alpha), \qquad i=1,\ldots,m$$

such that

$$L_x(x_1(\alpha),\ldots,x_n(\alpha),\lambda_1(\alpha),\ldots,\lambda_m(\alpha):\alpha) \equiv \theta \tag{4.10}$$
$$g^i(x_1(\alpha),\ldots,x_n(\alpha):\alpha) \equiv 0, \qquad i=1,\ldots,m$$

in some neighbourhood of α°. Differentiating the identities in (4.10) partially with respect to α_k, $k=1,\ldots,p$, gives (4.8).

Note that since $|B| \neq 0$, (4.8) can be solved uniquely for $\partial x_j/\partial \alpha_k$ and $\partial \lambda_i/\partial \alpha_k$.

EXAMPLE 4.1 (Taxation and labour supply) An individual's utility depends on consumption, c, and leisure, l, and he has a fixed amount of time, k, to devote to leisure or labour, L. A wage rate, w, is paid for each unit of labour provided. The individual pays income tax at the rate t on income above the level A. The individual's labour-leisure choice involves solving the problem:

$$\underset{c,l}{\text{maximize}}\ U(c,l) \tag{i}$$

subject to

$$k = l + L, \quad c = wL - t(wL - A) = (1-t)wL + tA, \quad k>0 \text{ a constant,} \tag{ii}$$

where it is assumed that $U(.)$ is twice-continuously differentiable with $U_c > 0$, $U_l > 0$, and $U(.)$ strictly concave. Maximization of (i) subject to (ii) is equivalent to maximizing

$$U[(1-t)w(k-l) + tA, l] \tag{iii}$$

with respect to l. A necessary and sufficient condition for utility to be maximized at l° is

$$dU(l^\circ)/dl = -w(1-t)U_c(c^\circ, l^\circ) + U_l(c^\circ, l^\circ) = 0, \tag{iv}$$

where $c^\circ = (1-t)w(k-l^\circ) + tA$.

Let $B \equiv d^2U(l^\circ)/dl^2 = w^2(1-t)^2U_{cc}(c^\circ, l^\circ) + U_{ll}(c^\circ, l^\circ) - 2U_{cl}(c^\circ, l^\circ)$ $(1-t)w$. Strict concavity of U implies $B \leqslant 0$. Assume $B < 0$, so the Implicit Function Theorem can be applied to (iv) to obtain the individual's demand for leisure function $l^\circ = l(k, w, t, A)$, where l is continuously differentiable, and the identity

$$-w(1-t)U_c(c^\circ, l^\circ) + U_l(c^\circ, l^\circ) \equiv 0, \tag{v}$$

where

$$c^\circ = w(1-t)(k-l^\circ) + tA \qquad \text{and} \quad l^\circ = l(k, w, t, A).$$

Partially differentiating (v) with respect to A gives

$$-w(1-t)\{U_{cc}[-(1-t)w\,\partial l^\circ/\partial A + t] + U_{cc}\partial l^\circ/\partial A\}$$
$$+ U_{cl}[-(1-t)w\,\partial l^\circ/\partial A + t] + U_{ll}\partial l^\circ/\partial A = 0;$$

that is,

$$\partial l^\circ/\partial A = t[w(1-t)U_{cc} - U_{cl}]/B. \tag{vi}$$

Since $U_{cc} < 0$, from the strict concavity of U, we see that if $U_{cl} > 0$ then $\partial l^\circ/\partial A > 0$ and thus

$$\partial L^\circ/\partial A = -\partial l^\circ/\partial A < 0.$$

Partially differentiating (v) with respect to t gives

$$wU_c - w(1-t)\{U_{cc}[-w(k-l^\circ) - t)w\,\partial l^\circ/\partial A + A] + U_{cl}\partial l^\circ/\partial t\}$$
$$+ U_{lc}[-w(k-l^\circ) - (1-t)w\,\partial l^\circ/\partial t + A] + U_{ll}\partial l^\circ/\partial t = 0;$$

that is,

$$\partial l^\circ/\partial t = -\{wU_c + [w(k-l^\circ) - A][w(1-t)U_{cc} - U_{lc}]\}/B. \tag{vii}$$

From (vi) and the time constraint in (ii), (vii) may be expressed as

$$\partial l^\circ/\partial t = -wU_c/B - (wL^\circ - A)t^{-1}\partial l^\circ/\partial A. \tag{viii}$$

In general, the sign of $\partial l^\circ/\partial t$ is indeterminate, and thus the effect of income taxation on the individual's labour supply cannot be determined *a priori*.

For the problem maximize$_x$ $f(x:\alpha)$ subject to $x \in S$, where $\alpha \in A \subset \mathbb{R}^p$ is a vector of parameters and S is an open subset of $\mathbb{R}^n$, let x° be a maximizing choice of x given $\alpha \in A$: $x_j^\circ = x_j(\alpha), j = 1, \ldots, n$. Define the function $M: A \to \mathbb{R}$ by

$$M(\alpha) = f(x^\circ : \alpha) = f(x_1(\alpha), \ldots, x_n(\alpha) : \alpha), \tag{4.11}$$

so that $M(\alpha)$ is the maximum value of the objective function given $\alpha \in A$. $M(\alpha)$ is always a function as there can only be one maximal value of $f(x:\alpha)$ for each choice of $\alpha \in A$. $M(\alpha)$ is the **maximum value function**. Note, however, that $x_j(\alpha)$, $j = 1, \ldots, n$, need not be functions, since the optimal choice of control variables need not be unique for a given $\alpha \in A$.

THEOREM 4.3 (Envelope Theorem) For the problem maximize$_x$ $f(x:\alpha)$ subject to $x \in S$, where $\alpha \in A \subset \mathbb{R}^p$ is a vector of parameters and S is an open subset of $\mathbb{R}^n$, let x° be a maximizing choice of x given $\alpha \in A$. If f is twice-continuously differentiable and $|\partial^2 f(x^\circ : \alpha)/\partial x_i \partial x_j| \neq 0$, then

$$\partial M/\partial \alpha_k = \partial f/\partial \alpha_k, \qquad k = 1, \ldots, p. \tag{4.12}$$

Proof. Since f is twice-continuously differentiable and x° is a maximizing choice of x for a given α and $|\partial^2 f(x^\circ : \alpha)/\partial x_i \partial x_j| \neq 0$, then from the Implicit Function Theorem $x_j^\circ = x_j(\alpha), j = 1, \ldots, n$, where the x_j's are continuously differentiable functions in some neighbourhood of α. Hence, from (4.11), we have

$$\partial M/\partial \alpha_k = \sum_{j=1}^{n} [\partial f(x^\circ : \alpha)/\partial x_j][\partial x_j/\partial \alpha_k] + \partial f/\partial \alpha_k, \qquad k = 1, \ldots, p.$$

But since x° is optimal for given $\alpha \in A$, $\partial f(x^\circ : \alpha)/\partial x_j = 0, j = 1, \ldots, n$, so that $\partial M/\partial \alpha_k = \partial f/\partial \alpha_k$, $k = 1, \ldots, p$.

A similar Envelope Theorem is available for the equality constrained optimization problem:

THEOREM 4.4 (Envelope Theorem) For the problem maximize$_x$ $f(x:\alpha)$ subject to $g^i(x:\alpha) = 0$, $i = 1, \ldots, m < n$, where

$x \in X \subset \mathbb{R}^n$ and $\alpha \in A \subset \mathbb{R}^p$ is a vector of parameters, let f and each of the g^i's be twice-continuously differentiable over the open set X and let x° be a constrained maximizing point for given $\alpha \in A$. Assume that the Jacobian matrix $[\partial g^i(x^\circ : \alpha)/\partial x_j]$ has rank m and that the matrix

$$B \equiv \begin{bmatrix} L_{xx}(x^\circ, \lambda^\circ : \alpha) & [g^i(x^\circ : \alpha)/\partial x_j]' \\ [\partial g^i(x^\circ : \alpha)/\partial x_j] & \theta \end{bmatrix}$$

is of full rank. Then,

$$\partial M/\partial \alpha_k = \partial L(x^\circ, \lambda^\circ : \alpha)/\partial \alpha_k, \qquad k = 1, \ldots, p. \tag{4.13}$$

Proof. Since $|B| \neq 0$, from the Implicit Function Theorem $x_j^\circ = x_j(\alpha)$, $j = 1, \ldots, n$, where the x_j's are continuously differentiable functions in an appropriate neighbourhood. Then from (4.8)

$$\partial M/\partial \alpha_k = \sum_{j=1}^{n} [\partial f/\partial x_j][\partial x_j/\partial \alpha_k] + \partial f/\partial \alpha_k, \quad k = 1, \ldots, p, \tag{4.14}$$

and from the constraint equations

$$\sum_{j=1}^{n} [\partial g^i/\partial x_j][\partial x_j/\partial \alpha_k] + \partial g^i/\partial \alpha_k = 0, \qquad k = 1, \ldots, p; \; i = 1, \ldots, m, \tag{4.15}$$

where all derivatives are evaluated at $(x^\circ, \lambda^\circ : \alpha)$. Multiplying each equation in (4.15) by λ_i° and adding the result obtained to (4.14) gives for $k = 1, \ldots, p$,

$$\begin{aligned} \partial M/\partial \alpha_k = {} & \sum_{j=1}^{n} [\partial f/\partial x_j][\partial x_j/\partial \alpha_k] + \partial f/\partial \alpha_k + \\ & \sum_{i=1}^{m} \lambda_i^\circ \sum_{j=1}^{n} [\partial g^i/\partial x_j][\partial x_j/\partial \alpha_k] + \sum_{i=1}^{m} \lambda_i^\circ [\partial g^i/\partial \alpha_k] \\ = {} & \sum_{j=1}^{n} \{\partial f/\partial x_j + \sum_{i=1}^{m} \lambda_i^\circ [\partial g^i/\partial x_j]\}[\partial x_j/\partial \alpha_k] \\ & + \partial f/\partial \alpha_k + \sum_{i=1}^{m} \lambda_i^\circ [\partial g^i/\partial \alpha_k]. \end{aligned} \tag{4.16}$$

But since x° is a constrained local maximum point $\{\partial f/\partial x_j + \sum_{i=1}^m \lambda_i^\circ[\partial g^i/\partial x_j]\} = 0$, $j = 1, \ldots, n$, so that $\partial M/\partial \alpha_k = \partial f/\partial \alpha_k + \sum_{i=1}^m \lambda_i^\circ[\partial g^i/\partial \alpha_k] = \partial L(x^\circ, \lambda^\circ : \alpha)/\partial \alpha_k$.

Note that, in the special case where α_k does not enter any constraint function, we have $\partial M/\partial \alpha_k = \partial f/\partial \alpha_k$, as in the unconstrained case.

Using Theorem 4.4 we can obtain an important result concerning the interpretation of the Lagrange multipliers. Consider the problem maximize$_x$ $f(x)$ subject to $g^i(x:\alpha) \equiv \alpha_i - h^i(x) = 0$, $i = 1, \ldots, m$, $m < n$. The Lagrangean for this problem is $L(x, \lambda : \alpha) = f(x) + \sum_{i=1}^m \lambda_i(\alpha_i - h^i(x))$. If the hypotheses of Theorem 4.4 are satisfied then from (4.13) we have

$$\partial M/\partial \alpha_k = \partial L(x^\circ, \lambda^\circ : \alpha)/\partial \alpha_k = \lambda_k^\circ, \qquad k = 1, \ldots, p. \tag{4.17}$$

The Lagrange multiplier λ_i, at its optimal value, measures the sensitivity of the maximized value of the objective function to variations in the parameter α_i. In economic problems in which f has the dimensions of a 'value' (price $\times$ quantity), such as profit or revenue, and the α_i have the dimensions of a 'quantity', such as output or input, then the multipliers λ_i° have the interpretation of a 'price', and are frequently referred to as 'shadow prices'.

The Envelope Theorem can be used to obtain some important results in economics, as Example 4.2 below illustrates.

EXAMPLE 4.2 (Roy's Identity) Consider the utility maximization problem: maximize$_x$ $U(x)$ subject to $p'x = y$, where $x \gg \theta$ is a vector of commodities, $p \gg \theta$ a vector of commodity prices and y is the consumer's income. If the sufficient conditions for a strict constrained local maximum hold at the point x° then the first-order necessary conditions can be uniquely solved to obtain $x_j^\circ = x_j(p, y)$, $j = 1, \ldots, n$, where the x_j's, the ordinary demand functions, are continuously differentiable. The maximum value of the objective function is then $W(p, y) \equiv U[x_1(p, y), \ldots, x_n(p, y)]$. $W(p, y)$ is the **indirect utility function**. The Lagrangean function for this utility maximization problem is $L(x, \lambda) = U(x) + \lambda(y - p'x)$.

From the Envelope Theorem (Theorem 4.4) we have

$$\partial W/\partial y = \partial L(x^\circ, \lambda^\circ, :p, y)/\partial y = \lambda^\circ,$$

and

$$\partial W/\partial p_j = \partial L(x^\circ, \lambda^\circ :p, y)/\partial p_j = -\lambda^\circ x_j^\circ, \qquad j=1,\ldots,n.$$

Hence,

$$x_j^\circ = -[\partial W/\partial p_j]/[\partial W/\partial y], \qquad j=1,\ldots,n, \tag{i}$$

Equation (i) is **Roy's Identity**. Given an indirect utility function, it is thus straightforward to derive a system of ordinary (Marshallian) demand functions using Roy's Identity.

To this point, we have required the optimal choice of control variables to be continuously differentiable functions of the parameters. We now present some useful properties of maximum value functions which do not depend on this assumption.

THEOREM 4.5 Consider the problem maximize$_x$ $f(x:\alpha)$ subject to $g^i(x:\alpha) \geqslant 0$, $i=1,\ldots,m$, where $\alpha \in A \subset \mathbb{R}^p$ is a vector of parameters and f and each g^i are defined on an open set $X \subset \mathbb{R}^n$. If f and the g^i's are concave, jointly in x and α, then $M(\alpha)$ is concave.

Proof. Let x', x'' be optimal for the parameters α', $\alpha'' \in A$, respectively, and let $\hat{x}$ be optimal for the parameters $\lambda\alpha' + (1-\lambda)\alpha''$, $0 \leqslant \lambda \leqslant 1$, for given λ. Since each g^i is concave jointly in x and α we have $g^i(\lambda x' + (1-\lambda)x'' : \lambda\alpha' + (1-\lambda)\alpha'') \geqslant \lambda g^i(x':\alpha') + (1-\lambda)g^i(x'':\alpha'')$, for $i=1,\ldots,m$; $0 \leqslant \lambda \leqslant 1$. But $g^i(x':\alpha') \geqslant 0$ and $g^i(x'':\alpha'') \geqslant 0$, $i=1,\ldots,m$, since x', x'' are optimal for the parameters α', α'', respectively. Hence, $g^i(\lambda x' + (1-\lambda)x'' : \lambda\alpha' + (1-\lambda)\alpha'') \geqslant 0$, $i=1,\ldots,m$, so that $\lambda x' + (1-\lambda)x''$ is feasible for the problem with parameters $\lambda\alpha' + (1-\lambda)\alpha''$, $0 \leqslant \lambda \leqslant 1$, for given λ. However, $\lambda x' + (1-\lambda)x''$ is not necessarily optimal when the parameters are $\lambda\alpha' + (1-\lambda)\alpha''$. Hence, for $0 \leqslant \lambda \leqslant 1$, we have $M(\lambda\alpha' + (1-\lambda)\alpha'') = f(\hat{x} : \lambda\alpha' + (1-\lambda)\alpha'') \geqslant f(\lambda x' + (1-\lambda)x'' : \lambda\alpha' + (1-\lambda)\alpha'')$. But for

$0 \leqslant \lambda \leqslant 1$, $f(\lambda x' + (1-\lambda)x'' : \lambda\alpha' + (1-\lambda)\alpha'') \geqslant \lambda f(x' : \alpha') + (1-\lambda) f(x'' : \alpha'')$, as f is jointly concave in x and α. Therefore, for $0 \leqslant \lambda \leqslant 1$, $M(\lambda\alpha' + (1-\lambda)\alpha'') \geqslant \lambda f(x', \alpha') + (1-\lambda) f(x'' : \alpha'') = \lambda M(\alpha') + (1-\lambda)M(\alpha'')$, since $M(\alpha') = f(x' : \alpha')$ and $M(\alpha'') = f(x'' : \alpha'')$. Thus, $M(\alpha)$ is concave.

Note that in the above theorem the parameters α enter into both the objective and constraint functions, whereas in the following theorem the parameters appear only in the objective function.

THEOREM 4.6 Consider the problem maximize$_x$ $f(x : \alpha)$ subject to $g^i(x) \geqslant 0, i = 1, \ldots, m$, where $\alpha \in A \subset \mathbb{R}^p$ is a vector of parameters and f and g^i, $i = 1, \ldots, m$, are defined on an open set $X \subset \mathbb{R}^n$. If f is convex in α then $M(\alpha)$ is convex.

Proof. Let x', x'' be optimal for the parameters α', $\alpha'' \in A$, respectively, and let $\hat{x}$ be optimal for the parameters $\lambda\alpha' + (1-\lambda)\alpha''$, $0 \leqslant \lambda \leqslant 1$, for given λ. Then we have $g^i(x') \geqslant 0$, $g^i(x'') \geqslant 0$, and $g^i(\hat{x}) \geqslant 0$, $i = 1, \ldots, m$. Now, for $0 \leqslant \lambda \leqslant 1$, $M(\lambda\alpha' + (1-\lambda)\alpha'') = f(\hat{x} : \lambda\alpha' + (1-\lambda)\alpha'') \leqslant \lambda f(\hat{x} : \alpha') + (1-\lambda) f(\hat{x} : \alpha'')$, since f is convex in α, for given x. But $\hat{x}$, for given λ, is feasible for the parameters α', $\alpha'' \in A$, but not necessarily optimal, so that $M(\alpha') = f(x' : \alpha') \geqslant f(\hat{x} : \alpha')$ and $M(\alpha'') = f(x'' : \alpha'') \geqslant f(\hat{x} : \alpha'')$. Thus, for $0 \leqslant \lambda \leqslant 1$, $M(\lambda\alpha' + (1-\lambda)\alpha'') \leqslant \lambda M(\alpha') + (1-\lambda)M(\alpha'')$, and hence $M(\alpha)$ is convex.

EXAMPLE 4.3 (Expenditure minimization) Consider a consumer minimizing the expenditure required to attain a target utility level. The consumer's problem is

$$\underset{x}{\text{minimize}}\ p'x \text{ subject to } U(x) = u,$$

or equivalently,

$$\underset{x}{\text{maximize}}\ -p'x \text{ subject to } U(x) = u, \tag{i}$$

where x is a vector of commodities, p the vector of corresponding commodity prices, $U(.)$ the consumer's utility function and u the target level of utility.

Suppose prices are the only parameters of interest. Then the parameters of interest appear only in the objective function, $-p'x$. The objective function is linear in p and therefore convex in p, so that the maximum value function associated with (i), $M(p) = -p'x(p)$, is convex, from Theorem 4.6.

Define $E(p) \equiv -M(p) = p'x(p)$, so that $E(p)$ is the minimum level of expenditure required to attain the target level of utility at prices p. $E(p)$ is the consumer's **cost (expenditure) function**. $E(p)$ is clearly concave. Cost functions are considered in detail in Section 4.3, where the concavity of $E(p)$ plays a vital role.

Theorem 4.6 was concerned with problems in which the parameters of interest appeared only in the objective function. Theorems 4.7–4.9 below are concerned with problems in which the parameters of interest appear only in the constraint functions.

THEOREM 4.7 For the problem maximize$_x$ $f(x)$ subject to $g^i(x) \geqslant \alpha_i$, $i = 1, \ldots, m$, where f and each g^i are defined on an open set $X \subset \mathbb{R}^n$ and $\alpha_i \in \mathbb{R}$, $i = 1, \ldots, m$:

(i) $M(\alpha)$ is non-increasing in each α_i.

(ii) If f and each g^i are concave then $M(\alpha)$ is concave.

Proof.

(i) Let $A \equiv \{\alpha \in \mathbb{R}^m$: there exists an $x \in X$ for which $g^i(x) \geqslant \alpha_i$, $i = 1, \ldots, m\}$. For each $\alpha \in A$ it is clear that the set of points which satisfies the constraints is reduced when α_k increases. Consequently, the maximum value of the objective function cannot increase when α_k is increased.

(ii) Let α', α'' be two vectors of parameters admitting optimal solutions x', x'', respectively, and let $\hat{x}$ be the optimal solution for the parameter vector $\lambda\alpha' + (1-\lambda)\alpha''$, $0 \leqslant \lambda \leqslant 1$, so that $M(\alpha') = f(x')$, $M(\alpha'') = f(x'')$ and $M[\lambda\alpha' + (1-\lambda)\alpha''] = f(\hat{x})$. Since each g^i is concave, for $i = 1, \ldots, m$ and $0 \leqslant \lambda \leqslant 1$, we have $g^i[\lambda x' + (1-\lambda)x''] \geqslant \lambda g^i(x') + (1-\lambda)g^i(x'') \geqslant \lambda\alpha_i' + (1-\lambda)\alpha_i''$, the last inequality following as x', x'' are optimal for the parameters α', α'', respectively. Thus, $\lambda x' + (1-\lambda)x''$ is feasible for the problem with parameters $\lambda\alpha' + (1-\lambda)\alpha''$, but not necessarily optimal. Hence, for $0 \leqslant \lambda \leqslant 1$, we have $M(\lambda\alpha' + (1-\lambda)\alpha'') = f(\hat{x}) \geqslant f(\lambda x' + (1-\lambda)x'') \geqslant \lambda f(x') + (1-\lambda)f(x'')$, since f is concave. But $f(x') = M(\alpha')$ and $f(x'') = M(\alpha'')$, so that for $0 \leqslant \lambda \leqslant 1$ we have $M(\lambda\alpha' + (1-\lambda)\alpha'') \geqslant \lambda M(\alpha') + (1-\lambda)M(\alpha'')$, and thus $M(\alpha)$ is concave.

THEOREM 4.8 Let f and g^i, $i = 1, \ldots, m$ be real-valued concave functions on an open set $X \subset \mathbb{R}^n$. Assume that there exists some $x \in X$ for which $g^i(x) > 0$, $i = 1, \ldots, m$, and that for given $\alpha \in \mathbb{R}^m$ there exists some $x \in X$ for which $g^i(x) > \alpha_i$, $i = 1, \ldots, m$. Suppose x° solves the problem

$$\underset{x}{\text{maximize}}\ f(x) \text{ subject to } g^i(x) \geqslant 0, \qquad i = 1, \ldots, m, \tag{4.18}$$

and suppose $\tilde{x}$ solves the problem

$$\underset{x}{\text{maximize}}\ f(x) \text{ subject to } g^i(x) \geqslant \alpha_i, \qquad i = 1, \ldots, m. \tag{4.19}$$

Let $\lambda^\circ, \tilde{\lambda} \in \mathbb{R}^m$, $\lambda^\circ \geqslant \theta$, $\tilde{\lambda} \geqslant \theta$ be the Lagrange multipliers associated with the problems (4.18) and (4.19), respectively. Then

$$\sum_{i=1}^{m} \lambda_i^\circ \alpha_i \leqslant f(x^\circ) - f(\tilde{x}) \leqslant \sum_{i=1}^{m} \tilde{\lambda}_i \alpha_i. \tag{4.20}$$

Proof. From Theorems 2.10 and 2.11 x° solves problem (4.18) if, and only if, there exists a $\lambda^\circ \geqslant \theta$ such that $L(x, \lambda^\circ) \leqslant L(x^\circ, \lambda^\circ) \leqslant L(x^\circ, \lambda)$ for all $x \in X$, $\lambda \geqslant \theta$, that is, if, and only if,

$$f(x) + \sum_{i=1}^{m} \lambda_i^\circ g^i(x) \leqslant f(x^\circ) + \sum_{i=1}^{m} \lambda_i^\circ g^i(x^\circ) \leqslant f(x^\circ) + \sum_{i=1}^{m} \lambda_i g^i(x^\circ) \tag{4.21}$$

$$\text{for all } x \in X, \qquad \lambda_i \geqslant \theta, \qquad i = 1, \ldots, m.$$

From (4.21), and the complementary slackness condition, $\lambda_i^\circ g^i(x^\circ) = 0$, $i = 1, \ldots, m$, we have

$$f(x) + \sum_{i=1}^{m} \lambda_i^\circ g(x) \leqslant f(x^\circ), \text{ for all } x \in X. \tag{4.22}$$

Setting $x=\tilde{x}$ in (4.22), we have

$$f(\tilde{x})-f(x^\circ)\leqslant -\sum_{i=1}^{m} \lambda_i^\circ g^i(\tilde{x}). \tag{4.23}$$

But $g^i(\tilde{x})\geqslant \alpha_i$, $i=1,\ldots,m$, since $\tilde{x}$ solves (4.19), and since $\lambda_i^\circ \geqslant 0$, $i=1,\ldots,m$, (4.23) implies

$$f(\tilde{x})-f(x^\circ)\leqslant -\sum_{i=1}^{m} \lambda_i^\circ \alpha_i. \tag{4.24}$$

Similarly, $\tilde{x}$ solves problem (4.19) if, and only if, there exists a $\tilde{\lambda}\geqslant \theta$ such that

$$f(x)+\sum_{i=1}^{m} \tilde{\lambda}_i[g^i(x)-\alpha_i]\leqslant f(\tilde{x})+\sum_{i=1}^{m} \tilde{\lambda}_i[g^i(\tilde{x})-\alpha_i]\leqslant f(\tilde{x})$$
$$+\sum_{i=1}^{m} \lambda_i[g^i(\tilde{x})-\alpha_i], \text{ for all } x\epsilon X$$
$$\lambda_i\geqslant 0, \qquad i=1,\ldots,m. \tag{4.25}$$

From (4.25) and the complementary slackness condition, $\sum_{i=1}^{m} \tilde{\lambda}_i[g^i(\tilde{x})-\alpha_i]=0$, we have, setting $x=x^\circ$

$$f(\tilde{x})-f(x^\circ)\geqslant \sum_{i=1}^{n} \tilde{\lambda}_i g^i(x^\circ)-\sum_{i=1}^{m} \tilde{\lambda}_i\alpha_i. \tag{4.26}$$

But $\sum_{i=1}^{m} \tilde{\lambda}_i g^i(x^\circ)\geqslant 0$, since $\tilde{\lambda}_i\geqslant 0$ and $g^i(x^\circ)\geqslant 0$, the latter holding since x° solves problem (4.19). Hence

$$f(\tilde{x})-f(x^\circ)\geqslant -\sum_{i=1}^{m} \tilde{\lambda}_i\alpha_i. \tag{4.27}$$

The result follows from (4.24) and (4.27).

Note that from (4.20) we have $(\lambda^\circ-\tilde{\lambda})'\alpha\leqslant 0$. This result is often referred to as an example of the *Le Chatelier Principle*. The Le Chatelier Principle, originally formulated by the French chemist Le Chatelier in 1884, can be phrased as follows (Eichhorn and Oettli, 1972) 'if a system is in stable equilibrium, and one of the conditions is changed, then the equilibrium will shift in such a way as to tend to annul the applied change in the conditions'. We have already noted that, in economic resource allocation problems, the Lagrange multipliers (at their optimal values) can be interpreted as prices for available resources. If we suppose that the quantity of one of the available resources changes, we would expect, by the Le Chatelier Principle, (4.20), that its price changes in the opposite direction.

Theorem 4.8 together with Theorem 4.7(ii) provide an interpretation of the Lagrange multipliers in the concave programming problem.

THEOREM 4.9 (Interpretation of multipliers) Consider the concave programming problem maximize$_x$ $f(x)$ subject to $g^i(x) \geqslant \alpha_i$, $i=1,\ldots,m$, where $\alpha_i \in \mathbb{R}$ and f and each g^i are real-valued concave functions defined on an open convex set $X \subset \mathbb{R}^n$. Suppose there is some $\hat{x} \in X$ for which $g^i(\hat{x}) > \alpha_i$, $i=1,\ldots,m$. Let $M(\alpha)$ be the maximized value of the objective function for the vector of parameters α. Then,

$$\frac{\partial M(\alpha^+)}{\partial \alpha_k} \leqslant \lambda_k^\circ \leqslant \frac{\partial M(\alpha^-)}{\partial \alpha_k} \tag{4.28}$$

Proof. Let α, α' be two parameter vectors admitting optimal solutions x°, x', respectively, so that $M(\alpha) \equiv f(x^\circ)$ and $M(\alpha') \equiv f(x')$. Then from Theorem 4.8 we have

$$\Sigma \lambda_i^\circ (\alpha_i' - \alpha_i) \leqslant M(\alpha') - M(\alpha). \tag{4.29}$$

Setting $\alpha_i' = \alpha_i$ for all $i=1,\ldots,m$, $i \neq k$, in (4.29), we have for each k

$$\begin{aligned} \frac{M(\alpha')-M(\alpha)}{\alpha_k' - \alpha_k} &\geqslant \lambda_k^\circ, \qquad \text{if} \quad \alpha_k > \alpha_k', \\ \frac{M(\alpha')-M(\alpha)}{\alpha_k' - \alpha_k} &\leqslant \lambda_k^\circ, \qquad \text{if} \quad \alpha_k < \alpha_k', \end{aligned} \tag{4.30}$$

Taking the limit as $\alpha_k' \to \alpha_k$ in (4.30), we have

$$\begin{aligned} \lim_{\alpha_k' \to \alpha_k} \frac{M(\alpha')-M(\alpha)}{\alpha_k' - \alpha_k} &= \frac{\partial M(\alpha^+)}{\partial \alpha_k} \geqslant \lambda_k^\circ, \\ \lim_{\alpha_k' \to \alpha_k} \frac{M(\alpha')-M(\alpha)}{\alpha_k' - \alpha_k} &= \frac{\partial M(\alpha^-)}{\partial \alpha_k} \leqslant \lambda_k^\circ. \end{aligned} \tag{4.31}$$

Both limits exist since $M(\alpha)$ is concave from Theorem 4.7(ii), and the result is established.

Note that in Theorems 4.8–4.9 the functions involved do *not* have to be differentiable. In particular, the existence of differentiable functions $x_j(\alpha), j=1,\ldots,n$, is not required. If $M(\alpha)$ is continuously differentiable in α then (4.28) becomes $\partial M(\alpha)/\partial\alpha_k=\lambda_k^\circ$, and the multipliers at their optimal values measure the sensitivity of the maximized value of the objective function to changes in the constraint constants.

We now consider how the maximum value function is affected by the addition of constraints. Given the problem maximize$_x$ $f(x{:}\alpha)$ subject to $g^i(x{:}\alpha)\geqslant 0, i=1,\ldots,m$, where $\alpha\in A\subset\mathbb{R}^p$ is a vector of parameters, if additional constraints are added that happen to be satisfied at the existing optimum $x_j^\circ=x_j(\alpha), j=1,\ldots,n$, then the maximum value of the objective function will not be affected, but the additional constraints will affect the maximum value function $M(\alpha)$. A general result of this nature has been established by Silberberg (1971) for a problem with equality constraints. His result is often described as a 'Generalized Envelope Theorem'. We give below Silberberg's result without proof.

THEOREM 4.10 (Generalized Envelope Theorem) Let $f(x{:}\alpha)$ be twice-continuously differentiable and strictly concave in x with $\alpha\in A\subset\mathbb{R}^p$ a vector of parameters. For $\alpha\in A$, let x° be a unique local maximum point of f. Suppose that r constraints $g^i(x)=0$ are imposed, such that $g^i(x^\circ)=0$, $i=1,\ldots,r$, $r\leqslant n$. Furthermore, assume that $[\partial g^i(x^\circ{:}\alpha)/\partial x_j]$ has rank r, and that for all r the sufficient conditions for a strict local maximum hold at x° given $\alpha\in A$. Then $x_j^\circ=x_j(\alpha)$, $j=1,\ldots,n$, and letting $M^r(\alpha)$ denote the maximum value function in the presence of the r constraints, $M^r(\alpha)=f(x_1(\alpha),\ldots,x_n(\alpha))$, $r\leqslant n$, we have

$$M^r(\alpha)=M^{r-1}(\alpha), \tag{4.32}$$

$$M_\alpha^r(\alpha)=M_\alpha^{r-1}(\alpha), \tag{4.33}$$

$$M^r_{\alpha\alpha}(\alpha) < M^{r-1}_{\alpha\alpha}(\alpha). \tag{4.34}$$

Consider the case where $\alpha \in A \subset \mathbb{R}$, i.e., there is but a single parameter. Let $\alpha^\circ \in A$ and suppose that x° is optimal for α°, so that $M(\alpha^\circ) = f(x^\circ{}_1(\alpha^\circ), \ldots, x^\circ_n(\alpha^\circ))$. Suppose that r additional constraints are added to the problem and that these constraints are satisfied at x°, the optimal solution to the problem with parameter α°. Let $M^r(\alpha)$ denote the maximum value function in the presence of the r additional constraints. Assuming $M^r(\alpha)$ is differentiable, $M^r(\alpha^\circ)$ will be tangential to $M(\alpha^\circ)$ and $M^r(\alpha) < M(\alpha)$ for $\alpha \in A$, $\alpha \neq \alpha^\circ$. Moreover, if $M(\alpha)$ is concave, then $M^r(\alpha)$ will be 'more concave' than $M(\alpha)$, having greater curvature at $\alpha = \alpha^\circ$, and the higher r is (the greater the number of additional constraints) the 'more concave' $M^r(\alpha)$ will become (see Figure 4.1).

A particular instance of the Generalized Envelope Theorem occurs in the problem of optimizing with a restricted subset of control variables with the remaining control variables held fixed. Let $\alpha \in A \subset \mathbb{R}^m$ be a vector of parameters and consider the problem

$$\underset{x}{\text{maximize}}\, f(x) \text{ subject to } g^i(x) = \alpha_i,\ i = 1, \ldots, m,\ m < n. \tag{4.35}$$

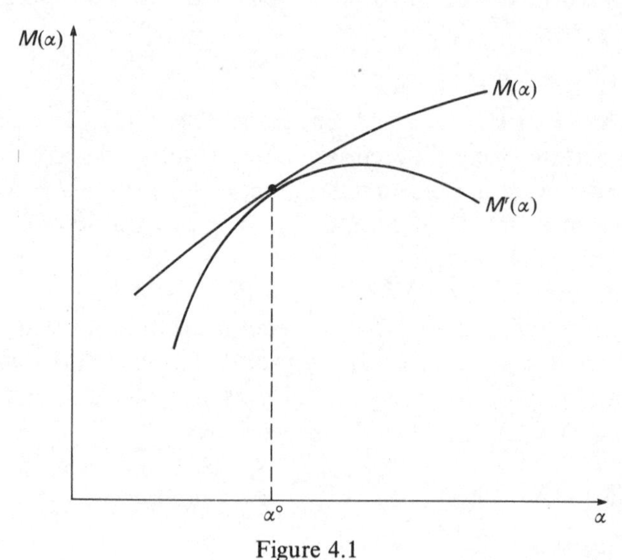

Figure 4.1

Partition the vector x as $x=(y,z)'$, where $y\in\mathbb{R}^k$, $z\in\mathbb{R}^p$, $k+p=n$, and let $x^\circ=(y^\circ,z^\circ)$ be a solution to (4.35) for $\alpha\in A$ when all control variables are allowed to vary. Then $y^\circ=y(\alpha)$, $z^\circ=z(\alpha)$ and the maximum value function is $M(\alpha)=f(y_1(\alpha),\ldots,y_k(\alpha), z_1(\alpha),\ldots,z_p(\alpha))$. Suppose now that the control variables y are held fixed. Then y is a vector of parameters which appear in the objective and constraint functions. For $\alpha\in A$, let $z^\circ=z(y,\alpha)$ be a solution to (4.35) when the control variables y are held fixed. The maximum value function is then $M(y,\alpha)=f(y,z_1(y,\alpha), \ldots,z_p(y,\alpha))$. (In general, $M(y,\alpha)\leqslant M(\alpha)$, since we cannot do better with some control variables fixed than if they are free to vary.) But if y is held fixed at y° then $z^\circ=z(y^\circ,\alpha)$ and $M(y^\circ,\alpha)=M(\alpha)$. Now consider a change $d\alpha$ in α, with $\alpha+d\alpha\in A$. Although $y^\circ=y(\alpha)$ was optimal for (4.35), it need not remain so when the parameters are changed to $\alpha+d\alpha$. Thus, we have

$$M(y^\circ,\alpha)=M(\alpha), \tag{4.36}$$

$$M(y^\circ,\alpha+d\alpha)\leqslant M(\alpha+d\alpha), \tag{4.37}$$

so that

$$M(y^\circ,\alpha+d\alpha)-M(y^\circ,\alpha)\leqslant M(\alpha+d\alpha)-M(\alpha). \tag{4.38}$$

Dividing both sides of (4.38) by $d\alpha$ the inequality is maintained if $d\alpha\gg\theta$ and is reversed if $d\alpha\ll\theta$, and taking the limit as $d\alpha\to\theta$ we have

$$M_\alpha(y^\circ,\alpha)=M_\alpha(\alpha) \tag{4.39}$$

provided both functions are differentiable. Taking a Taylor Series expansion of (4.37), we have

$$\begin{aligned} M(y^\circ,\alpha)+M_\alpha(y^\circ,\alpha)d\alpha+\tfrac{1}{2}(d\alpha)'M_{\alpha\alpha}(y^\circ,\alpha)(d\alpha)+\ldots \\ \leqslant M(\alpha)+M_\alpha(\alpha)d\alpha+\tfrac{1}{2}(d\alpha)'M_{\alpha\alpha}(\alpha)(d\alpha)+\ldots \end{aligned}$$

But from (4.36) $M(y^\circ,\alpha)=M(\alpha)$ and from (4.30) $M_\alpha(y^\circ,\alpha)=M_\alpha(\alpha)$, so that for $d\alpha$ sufficiently small $(d\alpha)'M_{\alpha\alpha}(y^\circ,\alpha)(d\alpha)\leqslant(d\alpha)'M_{\alpha\alpha}(\alpha)(d\alpha)$; that is

$$(d\alpha)'[M_{\alpha\alpha}(y^\circ,\alpha)-M_{\alpha\alpha}(\alpha)](d\alpha)\leqslant 0. \tag{4.40}$$

Letting $d\alpha=(0,\ldots,0,\ d\alpha_j,\ 0,\ldots,0)'$, (4.40) becomes

$$(d\alpha_j)^2[\partial^2 M(y^\circ,\alpha)/\partial\alpha_j^2-\partial^2 M(\alpha)/\partial\alpha_j^2]\leqslant 0,$$

or

$$\partial^2 M(y^\circ,\alpha)/\partial\alpha_j^2\leqslant\partial^2 M(\alpha)/\partial\alpha_j^2, \tag{4.41}$$

Inequalities (4.41) are often referred to as the *Le Chatelier effects.*

EXAMPLE 4.4 (Rationing constraints and consumer demand) Consider a consumer minimizing the expenditure required to attain a target utility level. The consumer's problem is

$$\underset{x}{\text{maximize}}-p'x \text{ subject to } U(x)=u, \tag{i}$$

where x is a vector of commodities, p the vector of corresponding commodity prices, $U(.)$ the consumer's utility function and u the target utility level. Assume that the conditions underlying the Generalized Envelope Theorem are satisfied, and let $x*=x(p,u)$ be the solution to the consumer's problem. $x(p,u)$ are the consumer's compensated demand functions. The maximum value function associated with the consumer's problem is

$$M(p,u)=-p'x(p,u). \tag{ii}$$

Suppose now that rationing constraints are imposed on the consumer, one by one, which permit him to continue to purchase his original optimum commodity bundle x^*, but which constrain his responses to price changes. From (4.34) of the Generalized Envelope Theorem we have

$$\frac{\partial^2 M^0(p,u)}{\partial p_k^2}\leqslant\frac{\partial^2 M^1(p,u)}{\partial p_k^2}\leqslant\frac{\partial^2 M^2(p,u)}{\partial p_k^2}\leqslant\ldots, \tag{iii}$$

where M^r is the maximum value function in the presence of the r additional constraints. From Shephard's Lemma I (see Theorem 4.13, below) we have

$$\partial M(p,u)/\partial p_k=-x_k(p,u), \tag{iv}$$

so that

$$\partial^2 M(p,u)/\partial p_k^2 = -\partial x_k(p,u)/\partial p_k, \tag{v}$$

and hence (iii) becomes

$$-\left(\frac{\partial x_k(p,u)}{\partial p_k}\right)^0 \leqslant -\left(\frac{\partial x_k(p,u)}{\partial p_k}\right)^1 \leqslant -\left(\frac{\partial x_k(p,u)}{\partial p_k}\right)^2 \leqslant \ldots, \tag{vi}$$

where the superscripts denote the number of rationing constraints. Multiplying (vi) through by p_k/x_k* we obtain

$$e_k^0 \leqslant e_k^1 \leqslant e_k^2 \leqslant \ldots, \tag{vii}$$

where e_k^r is the price-elasticity of the compensated demand for good k in the presence of r rationing constraints. Thus, from (vii), we see that the more rationing constraints are placed upon the consumer the more inelastic the compensated demands become.

4.3 AN INTRODUCTION TO DUALITY THEORY

The ideas underlying duality theory in economics have been around in the literature for some considerable time (see the historical notes in Diewert, 1974 and 1982). However, it is only in recent years that duality theory has been widely applied. Indeed, the use of duality concepts is now so widespread that we can do no more than provide the reader with an introduction to the theory and its applications. In this book we consider only duality relationships between production and cost functions. However, the results obtained have an immediate application to consumer theory. For surveys of duality theory, the reader is referred to Diewert (1974, 1982), and Blackorby, Primont and Russell (1978).

In the theory of production, the firm is assumed to seek to minimize the cost of producing any given level of output, subject to its technological constraints, where the latter are summarized by a production function. Specifically the firm solves the problem

$$\underset{x}{\text{minimize}}\, p'x \text{ subject to } f(x) \geqslant q, \qquad x \geqslant \theta, \qquad p \gg \theta,$$

where x is a vector of input quantities, p the vector of corresponding input prices, q the level of output and f is the production function.

DEFINITION 4.1 Given a set of input prices $P \equiv \{p:p \gg \theta\}$ and a production function $f:\mathbb{R}^n_+ \to \mathbb{R}$, with range $R(f)$, the **cost function**, C, is defined for all $q \in R(f)$ by

$$C(q,p) = \min_x \{p'x : f(x) \geqslant q,\ x \geqslant \theta\}, \tag{4.42}$$

where x is a vector of input quantities.

THEOREM 4.11 If f is real-valued and continuous then the cost function C has the following properties:

AI
- (i) $C(q,p)$ is real-valued and non-negative,
- (ii) $C(q,p)$ is positively homogeneous in p for every $q \in R(f)$; $q \in R(f)$, $\lambda > 0$ implies $C(q, \lambda p) = \lambda C(q,p)$,
- (iii) $C(q,p)$ is non-decreasing in q for every $p \in P$; $p \in P$, q_1, $q_2 \in R(f)$ with $q_1 \leqslant q_2$ implies $C(q_1,p) \leqslant C(q_2,p)$,
- (iv) $C(q,p)$ is non-decreasing in p for every $q \in R(f)$; $q \in R(f)$, $p_2 \geqslant p_1 \gg \theta$ implies $C(q,p_2) \geqslant C(q,p_1)$,
- (v) $C(q,p)$ is concave in p for every $q \in R(f)$,
- (vi) $C(q,p)$ is continuous in p for $p \in P$.

Proof. Note that the continuity of f can be relaxed to the requirement that f be uppersemicontinuous.

Since f is continuous, for every $q \in R(f)$ the constraint set $\{x:f(x) \geqslant q, x \geqslant \theta\}$ in the cost minimization problem is closed and bounded below, so that $C(q,p)$ exists, since it is the minimum of a linear function over a closed set that is bounded below. $C(q,p) \equiv \min_x\{p'x:f(x) \geqslant q, x \geqslant \theta\} = p'\hat{x}$, say, where $\hat{x} \geqslant \theta, f(\hat{x}) \geqslant q$. But $p'\hat{x} \geqslant 0$, since $p \gg \theta$ and $\hat{x} \geqslant \theta$. Hence $C(q,p)$ is non-negative.

To establish AI(ii), let $p \gg \theta$, $q \in R(f)$ and $\lambda > 0$. Then we have $C(q,\lambda p) \equiv \min_x\{\lambda p'x:f(x) \geqslant q, x \geqslant \theta\} = \lambda \min_x\{p'x:f(x) \geqslant q, x \geqslant \theta\} = \lambda C(q,p)$.

To establish AI(iii), let $q_1, q_2 \in R(f)$ with $q_1 \leqslant q_2$. Then $C(q_2,p) \equiv \min_x\{p'x:f(x) \geqslant q_2, x \geqslant \theta\} \geqslant \min_x\{p'x:f(x) \geqslant q_1, x \geqslant \theta\} \equiv C(q_1,p)$, since, if $q_1 \leqslant q_2$, $\{x:f(x) \geqslant q_1, x \geqslant \theta\} \supset \{x:f(x) \geqslant q_2, x \geqslant \theta\}$

and the minimum of $p'x$ over a larger set cannot increase. Hence, $C(q_1,p) \leqslant C(q_2,p)$ for $q_2 \geqslant q_1$.

To show $AI(iv)$, let $q \in R(f)$ and $p_2 > p_1 \gg \theta$. Then $C(q,p_2) \equiv \min_x\{p_2'x : f(x) \geqslant q, x \geqslant \theta\} = p_2'\hat{x}$, say, where $\hat{x} \geqslant \theta$, $f(\hat{x}) \geqq q$. Since $p_2 > p_1$ and $\hat{x} \geqslant \theta$, we have $p_2'\hat{x} \geqslant p_1'\hat{x}$. But $\hat{x}$ is feasible, but not necessarily optimal, for the problem $\min_x\{p_1'x : f(x) \geqslant q, x \geqslant \theta\}$. Hence, $p_1'\hat{x} \geqslant \min_x\{p_1'x : f(x) \geqslant q, x \geqslant \theta\} \equiv C(q,p_1)$. Hence, $C(q,p_2) \geqslant C(q,p_1)$.

To establish AI(v), the concavity of $C(q,p)$ in p, let $q \in R(f)$, $p_1 \gg \theta$, $p_2 \gg \theta$ and $0 \leqslant \lambda \leqslant 1$. Then we have $C(q,p_1) \equiv \min_x\{p_1'x : f(x) \geqslant q, x \geqslant \theta\} = p_1'\hat{x}$, say, and $C(q,p_2) \equiv \min_x \{p_2'x : f(x) \geqslant q, x \geqslant \theta\} = p_2'\tilde{x}$, say. Also, $C(q, \lambda p_1 + (1-\lambda)p_2) \equiv \min_x \{(\lambda p_1 + (1-\lambda)p_2)'x : f(x) \geqslant q, x \geqslant \theta\} = (\lambda p_1 + (1-\lambda)p_2)'x^*$, say. But $(\lambda p_1 + (1-\lambda)p_2)'x^* = \lambda p_1'x^* + (1-\lambda)p_2'x^*$. Now x^* is feasible for the cost minimization problems associated with the price vectors p_1, p_2, but x^* is not necessarily optimal for these problems. Hence, $\lambda p_1'x^* + (1-\lambda)p_2'x^* \geqslant \lambda p_1'\hat{x} + (1-\lambda)p_2'\tilde{x} = \lambda C(q,p_1) + (1-\lambda)C(q,p_2)$. Hence, for $0 \leqslant \lambda \leqslant 1$, $C(q, \lambda p_1 + (1-\lambda)p_2) \geqslant \lambda C(q,p_1) + (1-\lambda)C(q,p_2)$ and $C(q,p)$ is concave in p.

To establish AI(vi), simply note that continuity of $C(q,p)$ in p follows directly from the concavity of $C(q,p)$ in p, since $p \gg \theta$.

Typically, production functions are assumed to be more than just real-valued and continuous; in particular, it is often assumed that they are increasing; $x_1 > x \geqslant \theta$ implies $f(x_1) > f(x)$, and quasiconcave. With these additional restrictions on f, we should expect to obtain somewhat stronger conditions on the cost function.

THEOREM 4.12 If f is real-valued and continuous over the non-negative orthant of $\mathbb{R}^n$ and f is increasing and quasiconcave with $f(\theta)=0$, then the cost function has the following properties:

AII
- (i) $C(q,p)$ is real-valued and jointly continuous in (q,p),
- (ii) $C(q,p)$ is non-negative and $C(0,p)=0$ for every $p \in P$,
- (iii) $C(q,p)$ is positively homogeneous in p for every $q \in R(f)$,
- (iv) $C(q,p)$ is increasing in q for every $p \in P$; $p \in P, q_1, q_2 \in R(f)$ with $q_1 < q_2$ implies $C(q_1,p) < C(q_2,p)$,

(v) $C(q,p)$ is increasing in p for every $q \in R(f)$; $q \in R(f)$, $p_2 > p_1 \gg \theta$ implies $C(q,p_2) > C(q,p_1)$,

(vi) $C(q,p)$ is concave in p for every $q \in R(f)$.

The proof of Theorem 4.12, which we do not give, is relatively straightforward with the exception of AII(i), concerning the continuity properties of C.

THEOREM 4.13 (Shephard's Lemma I) Suppose the production function f is real-valued and continuous over the non-negative orthant of $\mathbb{R}^n$ and that the cost function is defined by (4.42). Let $q* \in R(f)$, $p* \gg \theta$ and suppose that $x*$ is a solution to the problem of minimizing the cost of producing the output level $q*$ at the prices $p*$; i.e.

$$C(q*,p*) \equiv \min_x \{p*'x : f(x) \geqslant q*, x \geqslant \theta\} = p*'x*. \tag{4.43}$$

If, in addition, C is differentiable with respect to input prices at $(q*,p*)$, then

$$x* = C_p(q*,p*). \tag{4.44}$$

Proof. Given any vector of prices $p \gg \theta$, $x*$ is feasible for the problem defined by $C(q*,p)$, but it is not necessarily optimal:

$$p'x* \geqslant C(q*,p) \text{ for all } p, \qquad p \gg \theta. \tag{4.45}$$

For $p \gg \theta$, define the function $h(p) \equiv p'x* - C(q*,p)$. From (4.45) $h(p) \geqslant 0$ for $p \gg \theta$, and from (4.43), $h(p*) = 0$. Thus, $h(p)$ attains a global maximum at $p = p*$. Since $h(p)$ is differentiable at $p*$, the first-order necessary conditions for a local minimum must be satisfied; $h_p(p*) = x* - C_p(q*,p*) = \theta$, which implies (4.44).

EXAMPLE 4.5 (Cost minimization and input demand functions) Suppose the firm's production function f is real-valued and continuous, increasing and quasiconcave with $f(\theta) = 0$, and that the firm takes input prices r, $r \gg \theta$, as given. The firm's cost function $C(q,r) \equiv \min_x \{r'x : f(x) \geqslant q, \ x \geqslant \theta\}$ then has properties AII. Suppose in addition that C is twice-continuously differentiable at $(\hat{q}, \hat{r})$, where $\hat{q} \in R(f)$, where $R(f)$ is the range of the production function, and $\hat{r} \gg \theta$. Then, from Shephard's Lemma I, we know that the solution to the cost minimization problem $\min_{x \geqslant \theta}\{r'x : f(x) \geqslant \hat{q}\}$ is given by $x* = C_r(\hat{q}, \hat{r})$. Thus

$$x*_i(\hat{q}, \hat{r}) = \partial C(\hat{q}, \hat{r})/\partial r_i, \qquad i = 1, \ldots, n \tag{i}$$

are the firm's cost minimizing input demand functions.

Let $[\partial x*_i/\partial r_j] \equiv [\partial x*_i(\hat{q}, \hat{r})/\partial r_j]$ denote the $n \times n$ matrix of derivatives of the n input demand functions $x*_i(\hat{q}, \hat{r})$ with respect to the n input prices r_j. Then we have

$$[\partial x*_i/\partial r_j] = C_{rr}(\hat{q}, \hat{r}), \tag{ii}$$

where $C_{rr}(\hat{q}, \hat{r})$ is the Hessian matrix of the cost function with respect to input prices, evaluated at $(\hat{q}, \hat{r})$. Since C is twice-continuously differentiable, it follows that

$$\frac{\partial x*_i(\hat{q}, \hat{r})}{\partial r_j} = \frac{\partial^2 C(\hat{q}, \hat{r})}{\partial r_j \partial r_i} = \frac{\partial^2 C(\hat{q}, \hat{r})}{\partial r_i \partial r_j} = \frac{\partial x*_j(\hat{q}, \hat{r})}{\partial r_i}, \qquad \text{for all } i, j. \tag{iii}$$

Furthermore, since C is concave in r, $C_{rr}(\hat{q}, \hat{r})$ is negative semidefinite, so

$$z'[\partial x*_i/\partial r_j]z \leqslant 0, \quad \text{for all} \quad z \in \mathbb{R}^n, \tag{iv}$$

and letting $z = e_i$, the ith unit vector, we have

$$\partial x*_i(\hat{q}, \hat{r})/\partial r_i \leqslant 0, \qquad i = 1, \ldots, n. \tag{v}$$

Thus, the ith cost minimizing input demand function cannot be upward sloping with respect to the ith input price.

C is also positively homogeneous in r, so that

$$C(\hat{q}, \lambda\hat{r}) \equiv \lambda C(\hat{q}, \hat{r}), \text{ for all } \lambda > 0. \tag{vi}$$

Differentiating (vi) with respect to r_i yields

$$\lambda \partial C(\hat{q}, \lambda\hat{r})/\partial r_i = \lambda \partial C(\hat{q}, \hat{r})/\partial r_i, \tag{vii}$$

and differentiating (vii) with respect to λ yields

$$\partial C(\hat{q}, \lambda\hat{r})/\partial r_i + \lambda \sum_{j=1}^{n} \hat{r}_j(\partial^2 C(\hat{q}, \lambda\hat{r})/\partial(\lambda r_j)\partial r_i) = \partial C(\hat{q}, \hat{r})/\partial r_i,$$

and letting $\lambda = 1$, we have

$$\sum_{j=1}^{n} \hat{r}_j(\partial^2 C(\hat{q}, \hat{r})/\partial r_j \partial r_i) = 0,$$

or equivalently

$$\sum_{j=1}^{n} \hat{r}_j(\partial x*_i(\hat{q}, \hat{r})/\partial r_j) = 0. \tag{viii}$$

There is a further general restriction on the input demand functions, which can be obtained in a similar manner to the derivation of (viii). Differentiate (vi) with respect to q to obtain

$$\partial C(\hat{q}, \lambda\hat{r})/\partial q = \lambda\, \partial C(\hat{q}, \hat{r})/\partial q, \tag{ix}$$

and then differentiate (ix) with respect to λ to get

$$\sum_{j=1}^{n} \hat{r}_j(\partial^2 C(\hat{q}, \lambda\hat{r})/\partial(\lambda r_j)\partial q) = \partial C(\hat{q}, \hat{r})/\partial q.$$

Letting $\lambda = 1$, we obtain

$$\sum_{j=1}^{n} \hat{r}_j(\partial^2 C(\hat{q}, \hat{r})/\partial r_j \partial q) = \partial C(\hat{q}, \hat{r})/\partial q. \tag{x}$$

Now $\partial^2 C(\hat{q}, \hat{r})/\partial r_j \partial q = \partial^2 C(\hat{q}, \hat{r})/\partial q\, \partial r_j = \partial/\partial q[\partial C(\hat{q}, \hat{r})/\partial r_j] = \partial x*_j(\hat{q}, \hat{r})/\partial r_j$, so that (x) can be written

$$\sum_{j=1}^{n} \hat{r}_j(\partial x*_j(\hat{q}, \hat{r})/\partial r_j) = \partial C(\hat{q}, \hat{r})/\partial q > 0, \tag{xi}$$

the inequality following from the fact that C is increasing in q. From (xi) we see that the changes in the cost minimizing input demands brought about by an increase in output cannot all be negative; in other words, not all inputs can be inferior.

EXAMPLE 4.6 Suppose the firm's technology can be characterized by a real-valued continuous production function f, defined over the non-negative orthant of $\mathbb{R}^n$, which is in addition positive, positively homogeneous and concave. Then the cost function which corresponds to f has the form

$$C(q,p)=qC(1,p), \qquad \text{for} \quad q>0, \qquad p \gg \theta.$$

By definition

$$C(q,p) \equiv \min_x \{p'x: f(x) \geqslant q, x \geqslant \theta\},$$

$$= \min_x \{p'x: q^{-1} f(x) \geqslant 1, x \geqslant \theta\}, \text{ since } q>0,$$

$$= \min_x \{p'x: f(q^{-1}x) \geqslant 1, x \geqslant \theta\}, \quad \text{since } f \text{ is positively}$$

homogeneous,

$$= \min_x \{qp'(q^{-1}x): f(q^{-1}x) \geqslant 1, (q^{-1}x) \geqslant \theta\},$$

$$= q.\min_x \{p'z: f(z) \geqslant 1, z \geqslant \theta\}, \text{ letting } z \equiv (q^{-1}x),$$

$$= qC(1,p),$$

by the definition of $C(q,p)$, and the minimum exists since, as f is positive and concave, there will exist a $\hat{z} \geqslant \theta$ such that $f(\hat{z}) \geqslant 1$ and the minimum of the continuous function f over the non-empty compact set $\{z: f(z) \geqslant 1, p'z \leqslant p'\hat{z}\}$ will exist.

EXAMPLE 4.7 (Slutsky conditions of consumer demand) Suppose a consumer has a utility function U which is real-valued continuous, increasing and quasiconcave, so that his expenditure function C, defined by $C(u,p) \equiv \min_x \{p'x: U(x) \geqslant u, x \geqslant \theta\}$ has properties AII. If the consumer faces prices $\hat{p} \gg \theta$ and has income $\hat{y}$ to spend on commodities, he will choose the largest $u \in R(U)$ such that his expenditure on commodities does not exceed his income. Thus, the consumer's equilibrium utility level $\hat{u}$ is defined by

$$\hat{u} \equiv \max_x \{u \in R(U): C(u,\hat{p}) \leqslant \hat{y}, x \geqslant \theta\}. \qquad \text{(i)}$$

Suppose that C is twice-continuously differentiable in its arguments at $(\hat{u},\hat{p})$. Then, since C is increasing in u, we have

$$\partial C(\hat{u},\hat{p})/\partial u > 0, \qquad \text{(ii)}$$

and it follows that the consumer will spend all his income on purchasing commodities

$$C(\hat{u}, \hat{p}) = \hat{y}. \tag{iii}$$

Furthermore, since C is positively homogeneous in p, for $\lambda > 0$,

$$C(\hat{u}, \lambda\hat{p}) = \lambda C(\hat{u}, \hat{p}), \tag{iv}$$

so that, letting $\lambda = \hat{y}^{-1}$, we have from (iii) and (iv)

$$C(\hat{u}, \hat{p}/\hat{y}) = 1. \tag{v}$$

As C is twice-continuously differentiable at $(\hat{u}, \hat{p})$, and $\partial C(\hat{u}, \hat{p})/\partial u > 0$, we can apply the Implicit Function Theorem to (v) and solve for u as a differentiable function V of p/y in some neighbourhood of $\hat{p}/\hat{y}$; i.e.,

$$u = V(p/y) \text{ in some neighbourhood of } \hat{p}/\hat{y}. \tag{vi}$$

V is the consumer's indirect utility function.

The consumer's system of compensated demand functions

$$f_i(u, p), \qquad i = 1, \ldots, n \tag{vii}$$

are defined as the solution to the expenditure minimization problem, $\min_x\{p'x : U(x) \geqslant u,\ x \geqslant \theta\}$, so that from Shephard's Lemma I we have

$$f_i(\hat{u}, \hat{p}) = \partial C(\hat{u}, \hat{p})/\partial p_i, \qquad i = 1, \ldots, n. \tag{viii}$$

The consumer's system of ordinary demand functions

$$x_i(y, p) \qquad i = 1, \ldots, n \tag{ix}$$

can be obtained from (vii) if we replace u in (vii) by $V(p/y)$;

$$x_i(\hat{y}, \hat{p}) \equiv f_i(V(\hat{p}/\hat{y}), \hat{p}), \qquad i = 1, \ldots, n. \tag{x}$$

If we replace y in (ix) by $C(u, p)$ then we have the consumer's system of compensated demands

$$f_i(\hat{u}, \hat{p}) \equiv x_i(C(\hat{u}, \hat{p}), \hat{p}), \qquad i = 1, \ldots, n. \tag{xi}$$

Differentiating both sides of (xi) gives, for $i, j = 1, \ldots, n$,

$$\partial f_i(\hat{u},\hat{p})/\partial p_j = \partial x_i(\hat{y},\hat{p})/\partial p_j + (\partial x_i(\hat{y},\hat{p})/\partial y)(\partial C(\hat{u},\hat{p})/\partial p_j),$$

using (iii),

$$= \partial x_i(\hat{y},\hat{p})/\partial p_j + f_i(\hat{u},\hat{p})(\partial x_i(\hat{y},\hat{p})/\partial y), \quad \text{using (viii)},$$

$$= \partial x_i(\hat{y},\hat{p})/\partial p_j + x_i(\hat{y},\hat{p})(\partial x_i(\hat{y},\hat{p})/\partial y), \qquad \text{(xii)}$$

using (x) and the fact that $\hat{u} = V(\hat{p}/\hat{y})$.

Equations (xii) are the *Slutsky equations*. The $n \times n$ matrix

$$\hat{K} \equiv [\partial f_i(\hat{u},\hat{p})/\partial p_j]$$

is the matrix of *Slutsky coefficients*. From Shephard's Lemma I,

$$\partial f_i(\hat{u},\hat{p})/\partial p_j = \partial^2 C(\hat{u},\hat{p})/\partial p_j \partial p_i,$$

so that from (iii), (iv), (viii) in Example 4.5 we have the *Slutsky conditions*

$$\hat{K} = \hat{K}'$$

$$z'\hat{K}z \leqslant 0, \text{ for every } z \in \mathbb{R}^n$$

$$\hat{K}\hat{p} = \theta.$$

Thus, the Slutsky matrix is symmetric and negative semidefinite.

In our discussion above we assumed the existence of a production function f from which we derived the firm's cost function. What is not so obvious is that given a cost function $C(q,p)$ we can, under certain regularity conditions, derive (construct) the firm's production function f which generates the cost function C. This two-way relationship between production and cost functions is referred to in the economics literature as a duality relationship between production and cost functions. We outline below how, in principle, the firm's technology can be derived from knowledge of its cost function.

If starting from the firm's cost function we can derive the firm's input requirement set $S(q) \equiv \{x : f(x) \geqslant q, \ x \geqslant \theta\}$ then we will have obtained the firm's technology. For any $q \in R(f)$, the cost function C can be used to construct an outer approximation to the set $S(q)$ in the following manner. Pick some input price vector p_1,

$p_1 \gg \theta$, and graph the isocost surface $\{x: p_1'x = C(q, p_1)\}$. The set $S(q)$ must lie above and intersect this set, since $C(q, p_1) \equiv \min_x\{p_1'x : x \in S(q)\}$. Thus, $S(q) \subset \{x: p_1'x \geqslant C(q, p)\}$. Pick additional input price vectors $p_2 \gg \theta, p_3 \gg \theta, \ldots, p_k \gg \theta, \ldots$, and graph the isocost surfaces $\{x: p_k'x = C(q, p_k)\}$. It is evident that $S(q)$ must be a subset of each of the sets $\{x: p_k'x \geqslant C(q, p_k)\}$. Thus,

$$S(q) \subset \bigcap_{p \gg \theta} \{x: p'x \geqslant C(q, p)\} \equiv S^c(q). \tag{4.46}$$

That is, the true input requirement set $S(q)$ must be contained in the outer approximation input requirement set $S^c(q)$ which is obtained as the intersection of all of the supporting total cost half-spaces to the true input requirement set $S(q)$. In Figure 4.2, $S^c(q)$ is the shaded area; the unbroken straight lines correspond to various isocost surfaces. The boundary of the set $S^c(q)$ forms an approximation to the true q-isoquant. Note that $S^c(q)$ must by construction be a convex set and its boundary cannot contain any backward bending sections.

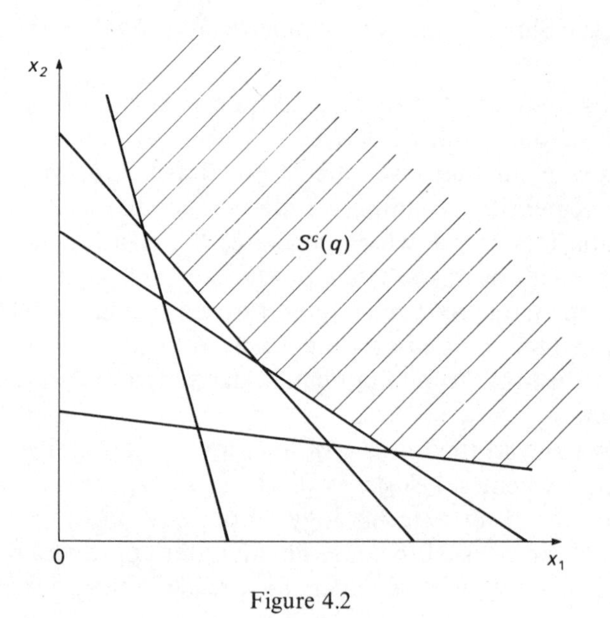

Figure 4.2

Once the family of approximating input requirement sets $S^c(q)$ has been constructed, the approximating production function f^c can be defined as,

$$\begin{aligned} f^c &\equiv \max_q \{q : x \in S^c(q)\} \\ &= \max_q \{q : p'x \geqslant C(q,p) \text{ for every } p \gg \theta\}. \end{aligned} \tag{4.47}$$

for $x \geqslant \theta$.

In general, the approximating production function f^c will not coincide with the true production function f. Suppose, for example, that the firm's true input requirement set $S(q)$ is of the form depicted in Figure 4.3. Then the approximating input requirement set $S^c(q)$ will not coincide exactly with the true input requirement set; in particular, any input combinations in the shaded areas will belong to $S^c(q)$ but they do not belong to $S(q)$. Thus, in general, $S(q) \subset S^c(q)$ but $S(q) \neq S^c(q)$. The constructed q-isoquant will coincide in part

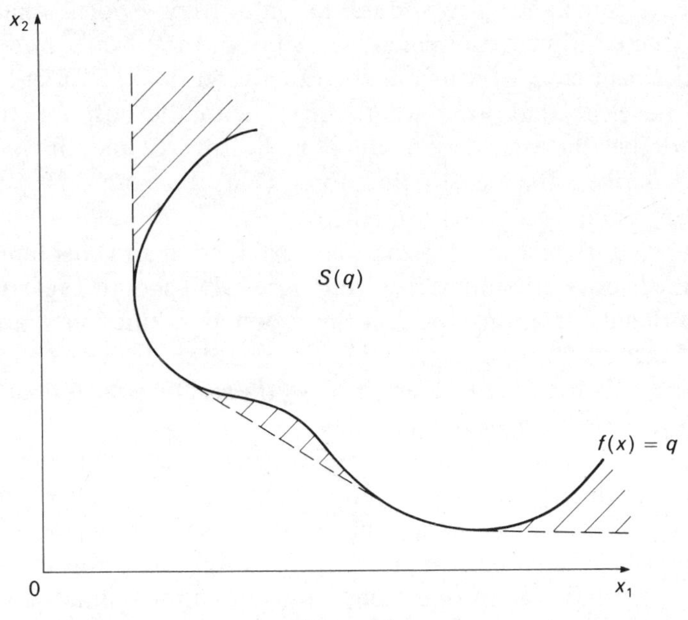

Figure 4.3

with the true q-isoquant but it will not have the backward bending and non-convex portions of the latter. Nevertheless, it is clear that, from the viewpoint of observable market behaviour, if the firm is cost minimizing then it does not matter whether the firm is minimizing cost subject to the production function constraint f or f^c; observable market data will never allow us to determine whether the firm has the production function f or the production function f^c.

If we want the production function f^c to coincide exactly with the true production function f, then we need to impose some requirements in addition to those of real-valuedness and continuity on the function f. Suppose we assume that the production function f has the following properties:

(i) f is real-valued and continuous,
AIII (ii) f is increasing: $x' > x$ implies $f(x') > f(x)$,
(iii) f is quasiconcave: for every $q \in R(f)$, $S(q) \equiv \{x : f(x) \geqslant q\}$ is a convex set.

Then backward bending and non-convex isoquants are ruled out. In these circumstances, it should be intuitively evident that our constructed input requirement set $S^c(q)$ and the firm's true input requirement set $S(q)$ will coincide exactly, so that $f^c = f$. It should also be clear that were we to begin with the cost function C and derive the production function f^c from C and then go on and derive the cost function $C^c(q,p)$ from f^c, where $C^c(q,p) \equiv \min_x\{p'x : f^c(x) \geqslant q\}$, then $C^c = C$.

The duality relationships between production and cost functions outlined above are summarized in Figure 4.4. It should be clear from these duality relationships that the production function f and the cost function C are equivalent descriptions of the firm's technology; to use McFadden's (1966) terminology *the cost function is a sufficient statistic for the production function.*

The above duality relationships have an immediate application in consumer theory; all we need do is interpret f as a utility function and C as the consumer's expenditure function. Similar duality relationships have been established for utility and indirect utility functions and for production (transformation) functions and maximized profit functions. For the statement of such results the

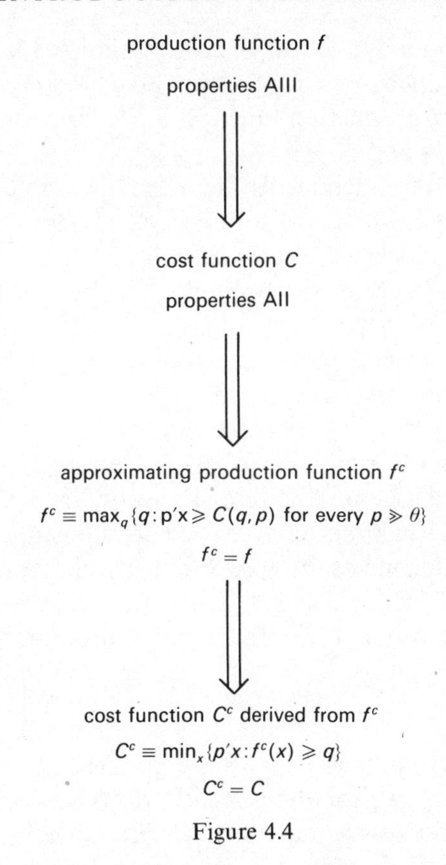

Figure 4.4

reader is referred to Diewert (1974) and (1982), McFadden (1966) and Blackorby, Primont and Russell (1978). These works contain rigorous proofs of the duality results discussed above.

The fact that the economically relevant aspects of a firm's technology are contained within its cost function is especially important in view of the following result, first derived by Shephard.

THEOREM 4.14 (Shephard's Lemma II) If the cost function has properties AII and, in addition, is differentiable at $(q*, p*)$ then

$$x(q*, p*) = C_p(q*, p*),$$

where $x(q*, p*)$ is the vector of cost minimizing input quantities needed to produce $q*$ units of output given input prices $p*$, where the underlying production function is the function f^c defined by (4.47), $q* \in R(f)$ and $p* \gg \theta$.

Before proving Shephard's Lemma II, we note the difference between it and Shephard's Lemma I. Shephard's Lemma I assumes the existence of the production function f and does not specify anything about the properties of the cost function other than differentiability. Shephard's Lemma II assumes only the existence of a cost function satisfying the appropriate regularity conditions and the corresponding production function f^c is defined using the given cost function. From an applied (econometric) viewpoint, Shephard's Lemma II is more useful than Shephard's Lemma I; in order to obtain a valid system of input demand functions, all one need do is postulate a functional form for C which meets the appropriate regularity conditions and differentiate C partially with respect to input prices. It is not necessary to compute the corresponding production function f^c.

(*Proof of Shephard's Lemma II.*) Our proof is basically in two parts. We first show (i) that, if C is concave in p and positively homogeneous in p, then $x*$ is a supergradient of C at $(\hat{q}, \hat{p})$ if, and only if, $C(\hat{q}, p) \leqslant x*'p$ for all $p \gg \theta$ and $C(\hat{q}, \hat{p}) = x*'\hat{p}$. We then show (ii) that the set of supergradients to C with respect to p at the point $(\hat{q}, \hat{p})$, where $q \in R(f)$ and $p \gg \theta$, $\partial C(\hat{q}, \hat{p})$, is the solution to the cost minimization problem $\min_x \{\hat{p}'x : f^c(x) \geqslant \hat{q}, x \geqslant \theta\}$, where f^c is the production function defined by (4.47) which corresponds to the given cost function, C, which satisfies AII. Shephard's Lemma II then follows immediately, since if C is differentiable the set of supergradients with respect to p at the point $(\hat{q}, \hat{p})$ is the single point $C_p(\hat{q}, \hat{p})$.

(i) From Definition 1.15, x^* is a supergradient of the function $C(\hat{q}, p)$, concave in p, at the point $p \in P \equiv \{p : p \gg \theta\}$ if, and only if, $C(\hat{q}, p) \leqslant C(\hat{q}, \hat{p}) + (p - \hat{p})'x*$ for every $p \in P$. We now show that if, in addition to being concave in p, C is positively homogeneous in p (i.e., $C(\hat{q}, \lambda p) = \lambda C(\hat{q}, p)$ for all $\lambda > 0$, where $p \in P$ and $\lambda p \in P$), then $x*$ is a supergradient of C at $(\hat{q}, \hat{p})$ if, and only if, $C(\hat{q}, p) \leqslant x*'p$

for all $p \in P$ and $C(\hat{q}, \hat{p}) = x*'\hat{p}$. Suppose $x*$ is a supergradient of C at $\hat{p} \in P$ and that C is positively homogeneous in p. Any $p \in P$ can be expressed as $p = \lambda \hat{p}$, where λ is some strictly positive number. Therefore, we have $C(\hat{q}, \lambda\hat{p}) \leqslant C(\hat{q}, \hat{p}) + (\lambda\hat{p} - \hat{p})'x*$, for all $\lambda > 0$. But since C is positively homogeneous in p the above inequality becomes $(\lambda - 1)C(\hat{q}, \hat{p}) \leqslant (\lambda - 1)\hat{p}'x*$, for all $\lambda > 0$, and hence $C(\hat{q}, \hat{p}) = p'x*$. Since $x*$ satisfies $C(\hat{q}, p) \leqslant C(\hat{q}, \hat{p}) + (p - \hat{p})'x*$ and $C(\hat{q}, \hat{p}) = \hat{p}'x*$, it follows that $C(\hat{q}, p) \leqslant p'x*$. Now let $C(\hat{q}, p) \leqslant p'x*$ and $C(\hat{q}, \hat{p}) = \hat{p}'x*$. Then we have $C(\hat{q}, \hat{p}) + (p - \hat{p})'x* = C(\hat{q}, \hat{p}) + p'x* - \hat{p}'x* = p'x* \geqslant C(\hat{q}, p)$, so that $x*$ is a supergradient of C at $\hat{p} \in P$.

(ii) Let $p \gg \theta$. Then

$$C^c(\hat{q}, \hat{p}) \equiv \min_x \{\hat{p}'x : f^c(x) \geqslant \hat{q}, x \geqslant \theta\},$$

$$= \min_x \{\hat{p}'x : f^c(x) = \hat{q}, x \geqslant \theta\},$$

since a linear function takes its minimum on the boundary of the feasible set,

$$= \min_x \{\hat{p}'x : \max_q \{q : C(q, p) \leqslant p'x, p \gg \theta\} = \hat{q}, x \geqslant \theta\},$$

from (4.47)

$$= \min_x \{\hat{p}'x : C(\hat{q}, p) \leqslant p'x \text{ for every } p \gg \theta \text{ with equality for at least one } p, x \geqslant \theta\}.$$

Since C is increasing, concave and positively homogeneous in p it follows that the solution set to the above minimization problem is the set of all supergradients to C with respect to p at the point $(\hat{q}, \hat{p})$, as any solution $x*$ must satisfy $C(\hat{q}, p) \leqslant \hat{p}'x$ for every $p \gg \theta$ and $\hat{p}'x* = C(\hat{q}, \hat{p})$. Thus, $C^c(\hat{q}, \hat{p}) = \hat{p}'x*$, where $x*$ is any supergradient to C at the point $(\hat{q}, \hat{p})$, and hence $C^c(\hat{q}, \hat{p}) = C(\hat{q}, \hat{p})$.

Since C is differentiable with respect to p at $(\hat{q}, \hat{p})$, the set of supergradients to C at $(\hat{q}, \hat{p})$ consists of the single point $x* = C_p(\hat{q}, \hat{p})$.

Given a cost function with appropriate properties, Shephard's Lemma II allows us to derive systems of input demand functions which are consistent with optimizing behaviour on the part of economic agents, by the simple process of differentiation. But how do we select a particular functional form for the dual function? Should we require that a particular functional form that is chosen satisfy the appropriate regularity conditions globally (for all $\mathbb{R}^n$) or only locally? For most practical purposes only satisfaction of the regularity conditions locally need be insisted upon, since only variables with values with a certain domain will be relevant. Moreover, in cases where the function has to be estimated from actual data, it does not make much sense to extrapolate the results too far away from the sample data.

Given that we are satisfied with obtaining only a local representation of the dual function, a wide variety of functional forms is available. How do we choose between those that are available? Here two principles of choice can be discerned in the literature. The first is that the functional form must be capable of approximating an arbitrary function to a desired order of accuracy. A functional form f is said to be **flexible** if it can provide a second-order (differential) approximation to an arbitrary twice-continuously differentiable function F at a point $\tilde{x}$. f 'differentially approximates' F at $\tilde{x}$ if, and only if, (i) $f(\tilde{x})=F(\tilde{x})$, (ii) $f_x(\tilde{x})=F_x(\tilde{x})$, and (iii) $f_{xx}(\tilde{x})=F_{xx}(\tilde{x})$, where both f and F are assumed twice-continuously differentiable at $\tilde{x}$ (so that the two Hessian matrices in (iii) will be symmetric). A general flexible functional form f must therefore have at least $1+n+\frac{1}{2}(n+1)n$ free parameters. If, however, both f and F are positively homogeneous, then the Euler relations imply $f_x(\tilde{x})'\tilde{x}=f(\tilde{x})$ and $f_{xx}(\tilde{x})\tilde{x}=\theta$, so that a flexible, positively homogeneous functional form need only have $\frac{1}{2}n(n+1)$ free parameters. The second principle of choice is that the functional form must result in systems that are relatively straightforward to estimate by known econometric techniques.

EXAMPLE 4.8 Consider the **translog cost function**

$$\ln C(q,p)\equiv\alpha_0+\sum_{i=1}^{n}\alpha_i\ln p_i+\tfrac{1}{2}\sum_{i=1}^{n}\sum_{j=1}^{n}\gamma_{ij}\ln p_i\ln p_j$$
$$+\delta_o\ln q+\sum_{i=1}^{n}\delta_i\ln p_i\ln q+\tfrac{1}{2}\varepsilon_0(\ln q)^2,$$

where q denotes output and p is a vector of input prices, and where the parameters satisfy the following restrictions;

$$\sum_{i=1}^{n} \alpha_i = 1, \qquad \gamma_{ij} = \gamma_{ji} \text{ for all } i,j \text{ with } \sum_{j=1}^{n} \gamma_{ij} = 0,$$

$$i = 1,\ldots,n, \text{ and } \sum_{i=1}^{n} \delta_i = 0.$$

These restrictions ensure that C is positively homogeneous in p. If we impose the additional restrictions $\delta_0 = 1$, $\delta_i = 0$, $i = 1,\ldots,n$, and $\varepsilon_0 = 0$, then $C(q,p) = qC(1,p)$, and the corresponding production function is positively homogeneous (see Example 4.6).

In general, C as defined above will not satisfy the appropriate regularity conditions globally; however, it is flexible: It can provide a good local approximation to an arbitrary twice-continuously differentiable, positively homogeneous in p cost function.

The cost minimizing input demand functions $x_i(q,p)$ which can be obtained from C as defined above, by Shephard's Lemma II, are non-linear in the unknown parameters. However, as the reader may verify, the factor share functions

$$\begin{aligned} s_i(q,p) &\equiv p_i x_i(q,p) \Big/ \left[\sum_{k=1}^{n} p_k x_k(q,p) \right] = p_i x_i(q,p)/[C(q,p)] \\ &= \partial \ln C(q,p)/\partial \ln p_i \end{aligned}$$

are linear in the unknown parameters

$$s_i(q,p) = \alpha_i + \sum_{j=1}^{n} \gamma_{ij} \ln p_j + \delta_i \ln q, \qquad i = 1,\ldots,n. \tag{i}$$

But, since the shares sum to unity, only $n-1$ of the n equations in (i) can be statistically independent. Furthermore, the parameters α_0, δ_0 and ε_0 do not appear in (i). Nevertheless, all of the parameters can be statistically estimated, given data on input prices, output and input levels, by appending the equation defining C, which is also linear in the unknown parameters, to $n-1$ of the n equations in (i).

CHAPTER 5

DYNAMICS AND STABILITY

5.1 INTRODUCTION

Differential and difference equations have been used extensively to model the way economic systems change over time. Differential equations may be used to model processes which change continuously over time whilst difference equations are more appropriate when adjustment is best viewed as a discrete period by period process. There are great similarities in the properties of differential and difference equations, but the differences are also important and may give rise to different kinds of dynamic behaviour. Essentially, a judgement must be made as to which technique is the more appropriate for a particular application (if some variables are best thought of as continuous whilst others are discrete, mixed difference/differential equation systems, not discussed here, may be considered; see Bellman and Cooke, 1963). Sections 5.2–5.7 are concerned with the dynamics and stability of differential equations, whilst Section 5.8 deals with difference equations.

5.2 DIFFERENTIAL EQUATIONS: BASIC CONCEPTS

5.2.1 Existence and uniqueness theorems

The equation

$$x^{(n)}(t)=F(t,x(t),x^{(1)}(t),\ldots,x^{(n-1)}(t)), \tag{5.1}$$

where $x^{(k)}(t)\equiv \mathrm{d}^k x(t)/\mathrm{d}t^k$ is termed a **non-linear**, ***n*th order**, **ordinary differential equation**. The order is given by the order of the highest derivative. Equation (5.1) is also **non-autonomous** since the independent variable t features explicitly; without t featuring explicitly, (5.1) would be **autonomous**. Equation (5.1) would also be linear if it were linear in the variables $x^{(k)}$; the equation

$$\sum_{k=0}^{n} a_k(t)x^{(k)}(t)+g(t)=0$$

is both linear and non-autonomous. Speaking informally, a differential equation is an equation involving derivatives of an unknown function x; as we shall see, the problem of solving the equation is one of finding a function that satisfies the equation.

Non-autonomous systems may be converted into autonomous systems by a suitable change of variable, although this approach is not adopted here (see Chapter 7, Section 7.3 for an application of this technique). We shall deal exclusively with first order systems of equations; that is, equations of the form

$$\dot{x}_i(t)=f^i(x_1(t),\ldots,x_n(t),t) \qquad i=1,\ldots,n \tag{5.2}$$

or, more compactly

$$\dot{x}(t)=f(x(t),t), \tag{5.3}$$

where $x(t)=(x_1(t),\ldots,x_n(t))'$, $f=(f^1,\ldots,f^n)'$ and $\dot{x}(t)=\mathrm{d}x(t)/\mathrm{d}t$. It is theoretically unnecessary to deal with higher order equations or systems of equations since they can be transformed to first order. Thus consider (5.1); the substitutions $z_i(t)=x^{(i)}(t)$, $i=0,\ldots,n-1$

convert this equation to the equivalent first order equations

$$\dot{z}_i(t)=z_{i+1}(t) \qquad i=0,\ldots,n-2$$
$$\dot{z}_{n-1}(t)=F(t,z_0(t),\ldots,z_{n-1}(t)). \tag{5.4}$$

An application of this procedure is given in Example 5.4.

Many economic systems can be characterized by first order autonomous differential equations of the form

$$\dot{x}(t)=f(x(t)), \qquad x(t)\in\mathbb{R}^n,$$

and such systems are often more tractable for being autonomous, as we shall see.

In Chapters 6 and 7, the following type of differential equation system will be encountered:

$$\dot{x}(t)=g(x(t),u(t),t), \tag{5.5}$$

where $g=(g^1,\ldots,g^n)'$, $x(t)\in\mathbb{R}^n$, $u(t)\in R^m$. u is a given, piecewise continuous, time varying function (called a control function). For a given function u, we may define $f(x(t),t)=g(x(t)\,u(t),t)$. Since $u(t)$ is piecewise continuous, $f(x,t)$ may feature discontinuities with respect to t (for example, if $g\in C_0$, f may be discontinuous at points at which u is discontinuous). Hence, the existence and uniqueness of solutions will be examined for the case where f may have a finite number of points at which it is discontinuous with respect to t.

DEFINITION 5.1 The system of differential equations $\dot{x}(t)=f(x(t),t)$, where $x(t)\in X$, an open connected subset of $\mathbb{R}^n$ and $t\in T=(T^1,T^2)\subset\mathbb{R}$, has a **solution** $\theta:\mathbb{R}\rightarrow\mathbb{R}^n$ on a subinterval $(t_1,t_2)\subset T$ if:

(1) $\theta(t)$ is continuous on (t_1,t_2),
(2) $\theta(t)\in X$ on (t_1,t_2),
(2) $\dot{\theta}(t)=f(\theta(t),t)$ on (t_1,t_2) almost everywhere (except possibly on a countable subset of (t_1,t_2)).

The problem is one of finding a function $\theta(t)$ which satisfies these properties. It is often termed an initial value problem (since, when t denotes time, (x_0,t_0) often describes the initial situation, the solution describing what happens later). Note that, since $f(x(t),t)$ may be discontinuous, $\dot{x}(t)$ does not necessarily exist everywhere. If two functions $\alpha(t)$, $\beta(t)$ satisfy $\alpha(t)=\beta(t)$ on an interval T except possibly at a countable number of points in T, we say $\alpha(t)=\beta(t)$ almost everywhere. It follows that $\theta(t)$ is required to be continuous, but may have kinks.

It is often notationally convenient in applied work not to distinguish the solution $\theta(t)$ from the variables $x(t)$ of the system $\dot{x}(t)=f(x(t),t)$ (i.e., to write $x(t)$ as the solution); the context usually makes clear which interpretation is correct.

Prior to studying their behaviour, it is important to establish whether solutions exist and, if so, whether they are unique. Theorem 5.1 below gives sufficient conditions for local, global existence and uniqueness of solutions.

Let t^i denote the ith point at which $f(x,t)$ is discontinuous and $W=\{t^i\}$, the set of such points. The following three assumptions are used in the statement of the theorem and are worth stating separately:

[A1] $f(x,t)$ is defined on $X\otimes T$ and is continuous on the set $A=X\otimes(T\backslash W)$ and the limits $f(x,\tau^+)$, $f(x,\tau^-)$ exist for all $\tau\in W$.

[A2] There exist piecewise continuous functions $\alpha(t)$, $\beta(t)$ such that $\|f(x,t)\|\leqslant\alpha(t)\|x\|+\beta(t)$ for all $(x,t)\in A$.

[A3] **Lipschitz continuity** in x: for all $(x,t)\in A$, there exists an $r>0$, an $L>0$ and an open interval (a,b) containing t such that

$B_r(x)\otimes(a,b)\in A$ and $\|f(z,\tau)-f(y,\tau)\|\leqslant L.\|z-y\|$
for all $z,y\in B_r(x)$ and $\tau\in(a,b)$.

THEOREM 5.1 For the system of equations $\dot{x}=f(x,t)$, $x\in X\subset\mathbb{R}^n$, $t\in T\subset\mathbb{R}$, with initial condition (x_0,t_0):

(i) **Local Existence:** if [A1] holds, then for $x\in X$, $t\in T$, there exists at least one solution satisfying the initial condition. That is, there exists an interval $(t_1,t_2)\subset T$ including t_0 and a solution $\theta(t)$ defined on (t_1,t_2) such that $\theta(t_0)=x_0$.

(ii) **Global Existence**: if [A1] holds and [A2] holds with $A = \mathbb{R}^n \otimes (t_1, t_2)$ (where t_1, t_2 could be $-\infty$, $+\infty$ respectively), then there exists a solution $\theta(t)$ on (t_1, t_2) with $\theta(t_0) = x_0$.

(iii) **Uniqueness**: if [A1] and [A3] hold and, for $(x_0, t_0) \in A$, if $\theta(t)$, $\psi(t)$ are two solutions defined on intervals T_1, T_2 respectively such that $t_0 \in T_1 \cap T_2$ and $\theta(t_0) = \psi(t_0) = x_0$, then $\theta(t) = \psi(t)$ on $T_1 \cap T_2$.

Proof. See Seierstad and Sydsaeter (1987), Hale (1969) and Pontryagin (1962).

Assumption [A1] will usually be satisfied; it will often be the case that $f \in C_0$, or, as in Chapters 6 and 7, that equations of type (5.5) feature $g \in C_q$, $q \geqslant 0$, with $u(t)$ piecewise continuous. In practical applications, global existence on some prescribed interval is usually required. Often this is simply assumed, although it is clearly preferable to establish that the sufficient conditions outlined in Theorem 5.1(ii) are in fact satisfied. Establishing the global existence of solutions is slightly more complex if there are restrictions on the possible values of x (i.e. $x(t) \in X$), simply because there is the additional requirement that, given any $(x_0, t_0) \in A$, the solution must also satisfy $\theta(t|x_0, t_0) \in X$ (see Seierstad and Sydsaeter, 1987, appendix).

Functions which are Lipschitz continuous are continuous but not necessarily differentiable; for example, the function $|x|$ is Lipschitz continuous but is not differentiable at $x = 0$. A function which is discontinuous is, of course, not Lipschitz continuous. In [A3], the requirement is that f be Lipschitz continuous in x. A sufficient, but not necessary, condition is that $f_x(x, t)$ has the same properties as f in [A1]. This condition is often satisfied in economic applications; however there are several important applications where f fails to be Lipschitz continuous (see Section 5.7).

EXAMPLE 5.1 (Existence and uniqueness)

(i) The equation $\dot{x} = f(x, t)$, where $f(x, t) = |x|$, defined on

$A=\mathbb{R}^2$, satisfies [A1]–[A3] and, applying Theorem 5.1, a unique solution $\theta(t)$ exists on the interval $(-\infty, +\infty)$ for any initial condition $(x_0, t_0) \in \mathbb{R}^2$. The solution is in fact given by $\theta(t)=x_0\exp(t-t_0)$ for $t \geqslant t_0$, and $\theta(t)=x_0\exp(-t-t_0)$ for $t \leqslant t_0$.

(ii) The equation $\dot{x}=x^\alpha$, $\alpha>1$, does not satisfy condition (A2). For example, $\dot{x}=x^2$, with initial condition $(x_0, t_0)=(1/c, 0)$, may be integrated to obtain the solution $\theta(t)=1/(c-t)$, which explodes as $t \to c$. Hence a global solution does not exist on an interval including c.

(iii) Consider the equation $\dot{x}=f(x,t)$, where $f(x,t)=3x^{2/3}$, defined for $t \geqslant 0$, $x \geqslant 0$ with initial condition $(x_0, t_0)=(0,0)$. Here, f is not Lipschitz continuous in x at $x=0$. Thus, for example, $|f(x,t)-f(0,t)|/|x-0|=3x^{-1/3}$ increases without bound as $x \to 0$. It is easy to verify that the solutions $\theta(t)=t^3$, $\psi(t)=0$, $t \geqslant 0$ both constitute solutions to this initial value problem.

For linear systems, global existence and uniqueness follow simply from assumptions of continuity.

THEOREM 5.2 Let $A(t)=\{a_{ij}(t)\}$ be a continuous function on an interval T and let u be a piecewise continuous function on T with $u(t) \in \mathbb{R}^n$. Given $x_0 \in \mathbb{R}^n$ and $t_0 \in T$, there exists a function θ from T into $\mathbb{R}^n$ such that:

(i) $\theta(t)$ is continuous on T,

(ii) $\theta(t_0)=x_0$,

(iii) $\theta(t)$ is a unique solution to the linear system $\dot{x}(t)=A(t)x(t)+u(t)$. That is, $\dot{\theta}(t)=A(t)\theta(t)+u(t)$ for $t \in T$.

Proof. See Pontryagin (1962), Hale (1969).

Given a unique solution exists, its dependence upon the initial condition (x_0, t_0) is of interest. We shall denote this dependence by writing $\theta(t; x_0, t_0)$. Parts (i)–(iii) of the following theorem establish that, under certain (often satisfied) assumptions, θ will be continuous and indeed differentiable with respect to these variables. Finally, part (iv) of the theorem considers the equation $\dot{x}=f(x,t,\lambda)$, where λ is a vector of parameters. It turns out that similar assumptions establish that θ (and $\dot{\theta}$) will also be differentiable with respect to such parameters.

THEOREM 5.3 (Dependence upon initial conditions and parameters) Consider the system of differential equations, $\dot{x}=f(x,t)$, satisfying [A1] and [A3] for which $\psi(t)$ denotes a unique solution on some interval $[t^1,t^2]\subset T$ such that $\psi(\bar{t})=\bar{x}$ for some $(\bar{x},\bar{t})\in X\otimes[t^1,t^2]$:

(i) Then there exists a neighbourhood $N=B_r(\bar{x})\otimes[\bar{t}-\mu,\bar{t}+\mu]$ (with $r,\mu>0$) of the point $(\bar{x},\bar{t})$ such that, if $(x_0,t_0)\in N$, there exists a unique solution $\theta(t;x_0,t_0)$ defined for $t\in T'=[t^1-\varepsilon,t^2+\varepsilon]$, for some $\varepsilon>0$ such that $\theta\in C_0$ on $T'\otimes N$.

(ii) If in addition f_x satisfies [A1] and if $t_0\notin W$, then θ is also differentiable with respect to x_0,t_0 on $T'\otimes N$.

(iii) Further, if $f\in C_q, q\geqslant 1$, then both $\theta(t;x_0,t_0)$ and $\dot{\theta}(t;x_0,t_0)\in C_q$ on $T'\otimes N$.

(iv) If f depends on a vector of parameters, $\lambda=(\lambda_1,\ldots,\lambda_m)'$, $\lambda\in\Lambda\subset\mathbb{R}^m$, where Λ is an open set such that $\dot{x}=f(x,t,\lambda)$, and suppose $\psi(t)$ denotes the unique solution, as above, for some given $\lambda_0\in\Lambda$, then, if $f\in C_q, q\geqslant 1$, both $\theta(t;x_0,t_0,\lambda)$ and $\dot{\theta}(t;x_0,t_0,\lambda)\in C_q$ on $T'\otimes N\otimes B_s(\lambda_0)$, where $s>0$.

Proof. See Hestenes (1966).

The above results can prove useful in comparative statics analysis. An extended application may be found in Section 6.11.

An important feature of the autonomous system $\dot{x}=f(x)$ is that, given existence of unique solutions,

$$\theta(\tau+t_0;x_0,t_0)=\theta(\tau+t_1;x_0,t_1) \text{ for any } t_0,t_1.$$

Thus the value of θ depends only upon the elapsed time since x_0 was reached. In view of this, the solution for the autonomous system will often be denoted by $\theta(t; x_0)$, where t denotes elapsed time and x_0 the initial condition such that $\theta(0; x_0) = x_0$. Clearly, we then have $\theta(t; \theta(\tau; x_0)) = \theta(t + \tau; x_0)$.

5.2.2 Equilibrium and stability concepts

In Example 5.1, $\psi(t) = 0$ was termed a singular solution. The trajectory is a single point, the origin. Such points are generally referred to as equilibrium or critical points.

DEFINITION 5.2 A point $\hat{x} \in X$ is an **equilibrium point**, or critical point, for the differential equation system $\dot{x} = f(x, t)$, $x \in X \subset \mathbb{R}^n$ if $f(\hat{x}, t) = \theta$ for all t.

The solution $\theta(t) = \hat{x}$ is termed an equilibrium solution; note that $\dot{\theta}(t) = \theta$ and the trajectory in $\mathbb{R}^n$ is a single point. An equilibrium may be unique but there may also be multiple equilibria, in which case it is useful to distinguish between 'isolated' and 'non-isolated' equilibria.

DEFINITION 5.3 An equilibrium $\hat{x}_j$ is said to be **isolated** if there exists an open neighbourhood $N(\hat{x}_j)$ such that $N(\hat{x}_j) \in X$ and such that, for any other equilibrium point, $\hat{x}_i \neq \hat{x}_j$, then $\hat{x}_i \notin N(\hat{x}_j)$.

The above definition is intuitive; in what follows, it is assumed that there is a unique equilibrium point. The case where there may be multiple equilibria is examined in Sections 5.6.4 and 5.7.

Stability is concerned with whether or not solutions converge on each other. In economics, equilibrium solutions are of particular interest and the concern is whether other solutions converge upon such solutions. From now on, unless otherwise stated, it is assumed that, for any admissible initial condition (x_0, t_0), a unique solution exists for all $t \geqslant t_0$.

DEFINITION 5.4 (Liapunov stability) An equilibrium $\hat{x}$ is said to be **Liapunov stable** if, given any $\varepsilon > 0$ and t_0, there exists a $\delta(\varepsilon, t_0) > 0$ such that, if $\|x_0 - \hat{x}\| \leqslant \delta$, then $\|\theta(t|x_0, t_0) - \hat{x}\| \leqslant \varepsilon$ for all $t \geqslant t_0$.

DEFINITION 5.5 (Asymptotic stability) An equilibrium $\hat{x}$ is said to be **asymptotically stable** (sometimes, **locally asymptotically stable**) if:

(i) it is Liapunov stable, and

(ii) there exists an $r > 0$ (generally depending on t_0) such that $\|x_0 - \hat{x}\| \leqslant r$ implies $\|\theta(t|x_0, t_0) - \hat{x}\| \to \theta$ as $t \to \infty$. That is, for any $\mu > 0$, there exists some $r(t_0) > 0$ and $t_1 > 0$ (generally depending on μ, x_0, t_0) such that $\|\theta(t|x_0, t_0) - \hat{x}\| \leqslant \mu$ for all $t \geqslant t_1$ whenever $\|x_0 - \hat{x}\| \leqslant r$.

DEFINITION 5.6 (Global asymptotic stability) An equilibrium is said to be **globally asymptotically stable** *if:*

(i) it is Liapunov stable, and

(ii) for any $r > 0$ (arbitrarily large) and any $\mu > 0$ (however small), there is a $t_1 > 0$ such that $\|\theta(t; x_0, t_0) - \hat{x}\| \leqslant \mu$ for all $t \geqslant t_1$ whenever $\|x - \hat{x}\| \leqslant r$.

For a more extensive discussion of stability concepts, see Kalman and Bertram (1960). Notice that, if there are multiple equilibria, none of these equilibria can be globally asymptotically stable in the sense of Definition 5.6. Given two equilibria $\hat{x}_i$, $\hat{x}_j$ such that $\hat{x}_i \neq \hat{x}_j$, then a solution with initial condition $x_0 = \hat{x}_i$ (respectively $\hat{x}_j$) does not converge on $\hat{x}_j$ (respectively $\hat{x}_i$). However, solutions may always converge on one or other of these equilibria; there is a sense in which the system may be globally 'stable'; this point is discussed further in Section 5.6.4. Of course, if the equilibria are isolated, the concepts of Liapunov and asymptotic stability may still prove useful.

Liapunov stability merely requires that a trajectory which starts sufficiently 'close' to the equilibrium point will stay 'close' for all $t \geqslant t_0$. The size of the neighbourhood defined by δ will generally depend upon ε (and t_0 in the non-autonomous case). Clearly $\delta \leqslant \varepsilon$ (at least). Asymptotic stability additionally requires that trajectories converge upon the equilibrium point (see Figure 5.1). Both are

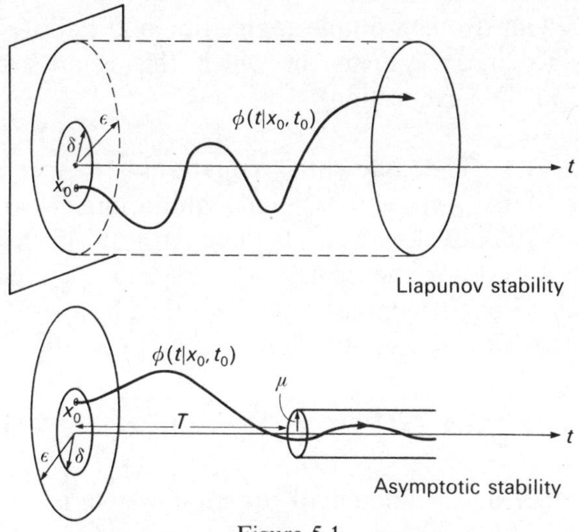

Figure 5.1

concepts of local stability. Global asymptotic stability requires that convergence occurs from all points in the domain.

More generally, the concept of stability need not be concerned solely with equilibrium solutions; if some particular solution $\bar{x}(t)$ is of interest, then $\bar{x}(t)$ simply replaces $\hat{x}$ in the above definitions (as in the following example).

EXAMPLE 5.2 Consider the differential equation $\dot{x}=\theta$, $x\in\mathbb{R}^n$. There is no unique equilibrium solution, but we can still enquire into the stability of solutions. For initial condition (x_0, t_0), the solution is clearly $\theta(t|x_0, t_0)=x_0$. To show Liapunov stability, for any $\varepsilon>0$, there must be a $\delta>0$ such that for any other solution $\theta(t|x_a, t_0)$ satisfying $\|x_0-x_a\|\leqslant\delta$, then $\|\theta(t|x_a, t_a)-\theta(t|x_0, t_0)\| = \|x_a-x_0\|\leqslant\delta$. Clearly setting $\delta\leqslant\varepsilon$ does this. Since δ depends upon ε but not t_0, we have demonstrated uniform Liapunov stability. However, solutions are not asymptotically stable since

$$\lim_{t\to\infty}\|\theta(t|x_a, t_0)-\theta(t|x_0, t_0)\| = \|x_a-x_0\| \neq 0$$

if $x_0\neq x_a$.

Example 5.3 illustrates a simple application of the above definitions (solutions to linear systems, of which this is an example, are developed in the next section).

EXAMPLE 5.3 The differential equations $\dot{x} = -x$, $x \in \mathbb{R}^n$ have solutions $\theta(t|x_0, t_0) = e^{(t_0 - t)}x_0$ and equilibrium $\hat{x} = \theta$. To show Liapunov stability, for any $\varepsilon > 0$, there must be a $\delta > 0$ such that, if $\|x_0 - \hat{x}\| = \|x_0\| \leqslant \delta$, then $\|\theta(t) - \hat{x}\| = e^{(t_0 - t)}\|x_0\| \leqslant \varepsilon$. Setting $\delta \leqslant \varepsilon$ does this. Global asymptotic stability then follows since, for all $x_0 \in \mathbb{R}^n$ with $\|x_0\| \leqslant r$, some r, $\lim_{t \to \infty} e^{(t_0 - t)}\|x_0\| = 0$.

5.3 LINEAR DIFFERENTIAL EQUATIONS

The first order linear differential equation system is

$$\dot{x}(t) = A(t)x(t) + g(t), \tag{5.6}$$

where $x(t) \in \mathbb{R}^n$, $A(t) = \{a_{ij}(t)\}$ and $g(t) = (g_1(t), \ldots, g_n(t))'$. The functions $a_{ij}(t)$, $g_i(t)$ are assumed continuous on some interval (t_1, t_2). Economic systems can often be modelled using linear systems, but they are also important because the local properties of non-linear systems can often be established through the study of associated linear approximation systems.

By Theorem 5.2, unique solutions to (5.6) exist on the whole interval (t_1, t_2). In fact, this is also the case if $A(t)$ and $g(t)$ are merely piecewise continuous (see Hestenes, 1966). If $g(t) = \theta$, we have the linear autonomous or homogeneous (of degree one in $x, \dot{x}$) system

$$\dot{x}(t) = A(t)x(t), \qquad x(t) \in \mathbb{R}^n \tag{5.7}$$

for which $\psi(t) = \hat{x} = \theta$ denotes the equilibrium solution. Now, any linear combination of solutions to (5.7) will also be a solution.

THEOREM 5.4 If $\theta_1(t), \ldots, \theta_n(t)$ are solutions to (5.7) on (t_1, t_2), then so is $\theta(t) = \sum_{i=1}^{n} c_i \theta_i(t)$, where $c_1, \ldots, c_n$ are arbitrary constants.

Proof. $\dot{\theta}(t) = \sum_{i=1}^{n} c_i \dot{\theta}_i(t) = \sum_{i=1}^{n} c_i A\theta_i(t) = A\theta(t)$.

Now, recall that vectors $y_1,\ldots,y_n \in \mathbb{R}^n$ are linearly dependent if, and only if, $|y_1,\ldots,y_n|=0$. The concept of linear dependence may be extended to functions as follows.

DEFINITION 5.7 Functions $\beta_1(t),\ldots,\beta_m(t)$ (where $\beta_i(t) \in \mathbb{R}^m$) are linearly dependent on (t_1,t_2) if there exists a set of constants, not all zero, such that $\sum_{i=1}^m c_i\beta_i(t)=\theta$ on (t_1,t_2). Otherwise, they are said to be linearly independent.

Note that if $\beta_1(t),\ldots,\beta_m(t)$ are linearly dependent on (t_1,t_2), they are linearly dependent at each point in the interval. However, if they are linearly independent, they may or may not be linearly independent at each point (e.g. they may be linearly dependent at each point but with different sets of constants at different points). Also, clearly, if $\beta_1(t),\ldots,\beta_m(t)$ are linearly dependent on (t_1,t_2), $|\beta_1(t),\ldots,\beta_m(t)|=0$ on that interval.

A collection of solutions $\theta_1,\ldots,\theta_n$ to (5.7) which are linearly independent is called a fundamental system of solutions to (5.7) and the matrix $\Phi(t)=[\theta_1(t),\ldots,\theta_n(t)]$ is termed a fundamental matrix. (Note that $\theta_i(t)$ is a column vector with n components.)

THEOREM 5.5 If $\Phi(t)$ is any matrix of solutions to (5.7) (not necessarily linearly independent), then either $|\Phi(t)|=0$ on (t_1,t_2) or $|\Phi(t)|\neq 0$ on (t_1,t_2). Furthermore, if the components of $\Phi(t)$ are linearly independent, then $|\Phi(t)|\neq 0$ on (t_1,t_2). Finally, if $\Phi(t)$ is a fundamental matrix, then $|\Phi(t)|\neq 0$ on (t_1,t_2).

Proof. Suppose there exists a $\tau \in (t_1,t_2)$ such that $|\Phi(\tau)|=0$. Then there exists a column vector of constants $c \in R^n$ with $c\neq\theta$ s.t. $\Phi(\tau).c=\theta$. But $\Phi(t).c$ is also a solution to (5.7) and so too is the singular solution $\psi(t)=\theta$. Hence uniqueness requires $\Phi(t).c=\theta$ for all $t\in(t_1,t_2)$ – and hence $|\Phi(t)|=0$ on (t_1,t_2). Alternatively, if $|\Phi(\tau)|\neq 0$ for some τ, then it follows that $|\Phi(t)|\neq 0$ for all $t\in(t_1,t_2)$. Finally, if $\Phi(t)$ is a fundamental matrix, $\theta_1(t),\ldots,\theta_n(t)$ are linearly independent on (t_1,t_2) so, for some $\tau\in(t_1,t_2)$, $|\Phi(\tau)|\neq 0$. Then, from the above, $|\Phi(t)|\neq 0$ for all $t\in(t_1,t_2)$.

Now, suppose n linearly independent solutions have been found for (5.7). The following theorem indicates that any other solution can always be expressed as a suitable unique combination of these solutions. Indeed, if $\theta_1,\ldots,\theta_n$ are linearly independent solutions to (5.7), then

$$\theta(t)=\sum_{i=1}^{n} c_i\theta_i(t), \tag{5.8}$$

where c_i, $i=1,\ldots,n$ are thought of as arbitrary constants, is customarily called a general solution to (5.7) and $\theta_1,\ldots,\theta_n$, a fundamental set of solutions.

THEOREM 5.6 If $\theta_1(t),\ldots,\theta_n(t)$ are linearly independent solutions to (5.7) for $t\in(t_1,t_2)$, then each and every solution $\theta(t)$ of (5.7) can be expressed as a unique linear combination: $\theta(t)=\sum_{i=1}^{n} c_i\theta_i(t)$.

Proof. Suppose the solution $\theta(t)$ passes through some point (x_a,t_a). We wish to establish that $\theta(t)$ may be written as $\sum_{i=1}^{n} c_i\theta_i(t)$, where θ_i, $i=1,\ldots,n$ are given and $c_1,\ldots,c_n$ are uniquely determined by

$$\sum_{i=1}^{n} c_i\theta_i(t_a)=x_a.$$

A necessary and sufficient condition for this equation to have a unique solution for $c_1,\ldots,c_n$ is that $|\Phi(t_a)|\neq 0$. This holds by Theorem 5.5. It then follows from uniqueness that $\theta(t)=\sum_{i=1}^{n} c_i\theta_i(t)$ for all $t\in(t_1,t_2)$.

Fundamental matrices are not unique; that is, if θ_1 is such a matrix, then so is θ_2 defined by $\theta_2=\theta_1 B$, where B is some non-singular constant matrix (although $\theta_1 B$ is not a fundamental matrix for *any* non-singular matrix B). In Chapter 7, we shall be concerned with the fundamental matrix defined as follows.

DEFINITION 5.8 Define $\Phi(t,t_0)$ to be the $n\times n$ matrix whose jth column is the solution $\theta_j(t)$ which satisfies $\theta_j(t_0)=e_j$, where e_j is a column vector with jth coordinate equal to unity, otherwise zero.

Thus, for the system $\dot{x}(t)=A(t)x(t)$, we have $\dot{\Phi}(t,t_0)=A(t)\Phi(t,t_0)$ and a solution with initial condition x_0 is given by

$$x(t)=\Phi(t,t_0)x_0.$$

The matrix $\Phi(t,t_0)$ has the following properties: first $\Phi(t,t)=I$ for all t (including t_0) since $x(\tau)=\Phi(\tau,t)x(t)$. Secondly, $\Phi(t+\tau,t_0)=\Phi(t+\tau,t)\Phi(t,t_0)$ as may be easily checked. These properties prove useful in Chapter 7.

Returning to (5.6), the non-homogeneous case, suppose a particular solution $\psi(t)$ has been found. Then, if $\theta_1(t),\ldots,\theta_n(t)$ denote a fundamental set of solutions to the associated homogeneous system (5.7), a general solution to (5.6) is given by $\psi+\sum_{i=1}^n\theta_i$, as may be easily verified.

Consider now the special case where $A(t)=A$, an $n\times n$ matrix of real constant coefficients; in this case, an explicit general solution is relatively straightforward to derive. The system is

$$\dot{x}=Ax, \tag{5.9}$$

where $x\in\mathbb{R}^n$ and A is an $n\times n$ matrix of real constant coefficients.

THEOREM 5.7 If λ_j is an eigenvalue of A, and v_j, its associated (column) eigenvector, then the function

$$\theta_j(t)=e^{\lambda_j t}v_j$$

is a solution to the differential equation (5.9).

Proof. We have $(A-\lambda_j I)v_j=\theta$. So

$$\dot{\theta}_j(t)=e^{\lambda_j t}\lambda_j v_j=e^{\lambda_j t}Av_j=A\theta_j(t).$$

THEOREM 5.8 If A is a real $n\times n$ matrix with n distinct eigenvalues $\lambda_1,\ldots,\lambda_n$ and $v_1,\ldots,v_n$ denote the corresponding eigenvectors, then a general solution to (5.9) is given by

$$\theta(t)=\sum_{i=1}^{n} e^{\lambda_i t}c_i v_i, \tag{5.10}$$

where $c_1,\ldots,c_n$ are arbitrary, possibly complex, constants.

Proof. If the eigenvalues are distinct, the eigenvectors are linearly independent. Hence $(e^{\lambda_1 t}v_1, \ldots, e^{\lambda_n t}v_n)$ constitutes a fundamental set of solutions. The result then follows from Theorem 5.6.

If $\lambda_1, \ldots, \lambda_n$ are not all distinct, in which case there are repeated roots, Theorem 5.8 does not apply, and a more sophisticated approach is required (see Murata, 1977, Chapter 3). It can be argued that the case of repeated roots is of little practical importance since they are a 'non-robust' property of the matrix A; that is, any small perturbation in the coefficients a_{ij} will yield a matrix with distinct roots. This may be stated in the form of a theorem.

THEOREM 5.9 Given a real constant matrix A, a matrix B can be found with distinct eigenvalues such that $\|A-B\| \leqslant \mu$, where $\mu>0$ may be chosen arbitrarily small.

Proof. See Bellman, 1960, pp. 199–200.

In view of this result, the repeated root case is not examined here.

Let $V=(v_1, \ldots, v_n)$ and let $e^{\Lambda t}$ denote the diagonal matrix

$$\begin{pmatrix} e^{\lambda_1 t} & 0 & \ldots & 0 \\ 0 & e^{\lambda_2 t} & \ldots & 0 \\ \vdots & & & \vdots \\ 0 & 0 & \ldots & e^{\lambda_n t} \end{pmatrix}$$

The general solution may now be written as

$$\theta(t) = Ve^{\Lambda t}c, \tag{5.11}$$

where c is a column vector of arbitrary constants.

If an initial condition (x_0, t_0) is specified, then $x_0 = Ve^{\Lambda t}c$, so $c = (e^{\Lambda t_0})^{-1}V^{-1}x_0$. Now, $(e^{\Lambda t_0})^{-1} = e^{-\Lambda t_0}$ since $e^{\Lambda t}$ is a diagonal matrix, so the particular solution becomes

$$\theta(t|x_0, t_0) = Ve^{\Lambda(t-t_0)}V^{-1}x_0. \tag{5.12}$$

A is a real matrix, hence eigenvalues, if complex, will occur in conjugate pairs. Since the solutions must be real, it follows that some of the constants $c_i, i=1,\ldots,n$ may be complex. Thus, if λ_i, λ_j, and v_i, v_j are complex conjugates, c_i, c_j can also be chosen as complex conjugates. This point is illustrated in Example 5.4 which also deals with the reduction of a second order equation to a pair of first order equations.

The non-homogeneous system $\dot{x}(t)=Ax(t)+g(t)$ has a general solution $\theta(t)+\psi(t)$, where $\theta(t)$ is the general solution to the associated homogeneous system $\dot{x}=Ax$ and $\psi(t)$ is a particular solution to the non-homogeneous system. When $g(t)=B$, a vector of constants, a particular solution can be obtained by setting $\dot{x}=\theta$ and solving $Ax+B=\theta$; assuming $|A|\neq 0$, $\psi(t)=-A^{-1}B$. The following example illustrates this along with the point about complex conjugates and the technique for reducing higher order differential equations to systems of first order equations (discussed in Section 5.2).

EXAMPLE 5.4 Consider the second order non-homogeneous linear differential equation $\ddot{x}+2\dot{x}+2x+4=0$ with initial condition $x(0)=\dot{x}(0)=1$. Let $y_1=x$ and $y_z=\dot{x}=\dot{y}_1$. With these substitutions, the equivalent first order system is

$$\begin{pmatrix}\dot{y}_1\\ \dot{y}_2\end{pmatrix}=\begin{pmatrix}0 & 1\\ -2 & -2\end{pmatrix}\begin{pmatrix}y_1\\ y_2\end{pmatrix}+\begin{pmatrix}0\\ -4\end{pmatrix} \tag{i}$$

or $\dot{y}=Ay+B$, where $A=\begin{pmatrix}0 & 1\\ -2 & -2\end{pmatrix}$ and $B=(0,-4)'$. A particular (equilibrium) solution θ_p is given by $\theta_p(t)=-A^{-1}B=-\begin{pmatrix}-1 & -\frac{1}{2}\\ 1 & 0\end{pmatrix}\begin{pmatrix}0\\ -4\end{pmatrix}=\begin{pmatrix}-2\\ 0\end{pmatrix}$. The eigenvalues for A are $\lambda_1=-1+i$, $\lambda_2=-1-i$. The associated eigenvectors are also complex. Thus writing $v_1=\begin{bmatrix}a_{11}+ib_{11}\\ a_{21}+ib_{21}\end{bmatrix}$, the characteristic equation $(A-\lambda_1 I)v_1=\theta$ gives the equations

$$a_{11}+ib_{11}-ia_{11}+b_{11}+a_{21}+ib_{21}=0,$$

$$-2a_{11}-2ib_{11}-a_{21}-ib_{21}-ia_{21}+b_{21}=0.$$

These equations must hold both for real and imaginary parts, so

$$a_{11}+b_{11}+a_{21}=0 \qquad b_{11}-a_{11}+b_{21}=0,$$

$$-2a_{11}-a_{21}+b_{21}=0, \; -2b_{11}-b_{21}-a_{21}=0.$$

The eigenvectors are determined up to a multiplicative constant hence we may set $a_{11}=b_{11}=1$, so that $a_{21}=-2$ and $b_{21}=0$. Thus $v_1=(1+i,\,-2)'.v_2$ is its complex conjugate $(1-i,\,-2)'$. This satisfies $(A-\lambda_2 I)v_2=\theta$. The general solution for the homogeneous system is given by

$$\theta(t)=k_1 e^{(-1+i)t}\begin{pmatrix}1+i\\-2\end{pmatrix}+k_2 e^{(-1-i)t}\begin{pmatrix}1-i\\-2\end{pmatrix} \qquad \text{(ii)}$$

and the general solution to (i) is

$$y(t)=k_1 e^{(-1+i)t}\begin{pmatrix}1+i\\-2\end{pmatrix}+k_2 e^{(-1-i)t}\begin{pmatrix}1-i\\-2\end{pmatrix}+\begin{pmatrix}-2\\0\end{pmatrix}. \qquad \text{(iii)}$$

The initial condition implies that $y(0)=(1,1)'$. The arbitrary constants may be determined from (iii):

$$1=k_1(1+i)+k_2(1-i)-2,$$

$$1=k_1(-2)+k_2(-2).$$

So $k_1=-(1+7i)/4$, $k_2=-(1-7i)/4$. Now, if r is a real constant, $e^{irt}=(\cos rt+i\sin rt)$, hence the solution may be finally simplified to give

$$\begin{pmatrix}y_1(t)\\y_2(t)\end{pmatrix}=e^{-t}\begin{pmatrix}3\cos t+4\sin t\\\cos t-7\sin t\end{pmatrix}+\begin{pmatrix}-2\\0\end{pmatrix}.$$

The solution may be checked by differentiation and substitution in (i). The solution spirals toward the equilibrium solution $(-2,0)'$.

The above example illustrates that, even for $x \in \mathbb{R}^2$, explicit solutions are quite tedious to calculate. In economic models, explicit solutions can rarely be calculated; the rest of this chapter is concerned with investigating the properties of trajectories where analytical solutions are not available.

5.4 STABILITY OF LINEAR DIFFERENTIAL EQUATIONS

In this section, the stability of the homogeneous system $\dot{x}=Ax$, where $x \in \mathbb{R}^n$, is considered. It is often possible to determine the stability of such systems even if the only information available for the coefficients of the matrix A is qualitative (i.e. $a_{ij}>,=,<0$).

5.4.1 Necessary and sufficient conditions for stability

THEOREM 5.10 An equilibrium point $\hat{x}$ for the differential equation system $\dot{x}=Ax$, $x \in \mathbb{R}^n$, is uniformly globally asymptotically stable if, and only if, the real parts of the eigenvalues of A are negative.

Proof. See Coddington and Levinson (1955).

For the case where the eigenvalues are distinct, the proof trivially follows from the general solution (5.10); writing $\lambda_j = p_j + iq_j$, where p_j, q_j are real constants, this becomes

$$\theta(t) = \sum_{j=1}^{n} e^{p_j t}(\cos q_j t + i \sin q_j t) c_j v_j.$$

Thus $\theta(t) \to \theta$ for all initial conditions if, and only if, $p_j < 0$ for all j.

DEFINITION 5.9 A real constant $n \times n$ matrix which has eigenvalues all with negative real parts is termed a **stability** or **stable matrix**.

Necessary and sufficient conditions for the eigenvalues to have negative real parts are given by the Routh–Hurwitz Theorem (for a statement and proof, see Gantmacher, 1959, pp. 190–6). The conditions are defined in terms of determinants involving the coefficients of the characteristic equation and are tedious to work with. If we write the characteristic equation as

$$|\lambda I - A| = \lambda^n + b_{n-1}\lambda^{n-1} + \ldots + b_1\lambda + b_0 = 0, \tag{5.13}$$

the following coefficients are of particular interest (see appendix):

$$b_0 = (-1)^n|A|, \tag{5.14}$$

$$b_{n-1} = \sum_{i=1}^{n} a_{ii} = -\mathrm{tr}A. \tag{5.15}$$

A statement of the Routh–Hurwitz Theorem is omitted in view of the fact that, more recently, certain computational simplifications have been made so the calculations can be based directly upon the elements of the matrix A itself. Unfortunately, the computational burden remains significant if A is larger than 3×3. The following exposition of the so-called 'Modified Routh–Hurwitz conditions' follows Murata (1977).

DEFINITION 5.10 Let A and B be $n \times n$ matrices. The bi-alternate product, denoted $C = A * B$, is a square matrix of order $p = n(n-1)/2$ such that if its rows are labelled gk $(g = 2, 3, \ldots, n;$ $k = 1, 2, \ldots, g-1)$ and its columns rs $(r = 2, 3, \ldots, n;$ $s = 1, 2, \ldots, r-1)$ then the element $c_{gk,rs}$ is defined by

$$c_{gk,rs} = \frac{1}{2}\left\{\begin{vmatrix} a_{gr} & a_{gs} \\ b_{kr} & b_{ks} \end{vmatrix} + \begin{vmatrix} b_{gr} & b_{gs} \\ a_{kr} & a_{ks} \end{vmatrix}\right\}. \tag{5.16}$$

Let $D = 2A * I$. Then it can be shown that

$$d_{gk,rs} = \begin{vmatrix} a_{gr} & a_{gs} \\ \delta_{kr} & \delta_{ks} \end{vmatrix} + \begin{vmatrix} \delta_{gr} & \delta_{gs} \\ a_{kr} & \alpha_{ks} \end{vmatrix} \text{ for } g > k,\ r > s, \text{ where}$$

$$\delta_{ij} = 1 \text{ if } i = j, \qquad \delta_{ij} = 0 \text{ if } i \neq j.$$

A useful reformulation of these conditions which aids computation is as follows:

$$d_{gk,rs}=\begin{cases} -a_{gs} & \text{if } r=k(s<k), \\ a_{gr} & \text{if } s=k \text{ and } r\neq g(r>k), \\ a_{ks} & \text{if } r=k \text{ and } s\neq k(s<g), \\ -a_{kr} & \text{if } s=g(r>g), \\ a_{gg}+a_{kk} & \text{if } r=g \text{ and } s=k, \\ 0 & \text{otherwise.} \end{cases} \tag{5.17}$$

THEOREM 5.11 (The modified Routh–Hurwitz (MR–H) conditions) Given A, B, C, D as defined above, then, for the eigenvalues of A to have negative real parts, it is necessary and sufficient that all the coefficients in the expansion form of $|\lambda I - A|$ and $|\mu I - D|$ are positive, including the constant terms.

Proof. Murata (1977), p. 190.

The following two examples apply the above theorem to deduce stability conditions for 2×2 and 3×3 A-matrices.

EXAMPLE 5.5 Suppose A is 2×2; $A=\begin{pmatrix} a_{11} & a_{12} \\ a_{21} & a_{22} \end{pmatrix}$.

Then $|\lambda I - A| = \lambda^2 - \operatorname{tr} A\lambda + |A| = 0$, so the MR–H conditions require that

$$\operatorname{tr} A<0 \quad \text{and} \quad |A|>0.$$

The bi-alternate product $D=2A*I$ is now calculated. Here $n=2$, so the definition implies that $gk=21$ and $rs=21$; the determinant has just one element. Noting that $g=r=2$ and $k=s=1$, then, from (5.17)

$$D=2(a_{22}+a_{11}).$$

The MR–H conditions require that $|\mu I–D|=\mu-2(a_{22}+a_{11})=0$ should have positive coefficients, which is so if $a_{11}+a_{22}<0$; nothing is added to the earlier conditions. Hence the necessary and sufficient conditions for stability are

$$\operatorname{tr} A<0, \quad \text{and} \quad |A|>0.$$

EXAMPLE 5.6 If A is 3×3, then the MR–H conditions may be simplified to obtain the following necessary and sufficient conditions for stability.

(i) $\operatorname{tr} A<0$,
(ii) $|A|<0$,
(iii) $|2A*I|<0$, which is equivalent to,

$$\begin{vmatrix} a_{22}+a_{11} & a_{23} & -a_{13} \\ a_{32} & a_{33}+a_{11} & a_{12} \\ -a_{31} & a_{21} & a_{33}+a_{22} \end{vmatrix} <0.$$

For economic applications of these conditions, see Example 5.8 and Section 5.7.

5.4.2 Necessary conditions for stability

The Modified Routh–Hurwitz conditions are necessary and sufficient for stability; some of these conditions are easier to check than others and these must, of course, be necessary conditions.

THEOREM 5.12 If A is a stable matrix, then $\operatorname{tr} A<0$.

THEOREM 5.13 If A is a stable matrix, then $(-1)^n|A|>0$.

These follow directly from the Modified Routh–Hurwitz conditions of Theorem 5.11 and equations (5.14) and (5.15). Of course, if A is 2×2, Theorems 5.12 and 5.13 together constitute necessary and sufficient conditions for stability. Note that necessary conditions for stability, if not satisfied, constitute sufficient conditions for instability.

5.4.3 Sufficient conditions for stability

THEOREM 5.14 A real $n\times n$ matrix A is a stable matrix:

(i) if, and only if, there exists a symmetric positive definite matrix B such that $BA+A'B$ is negative definite,
(ii) if A is quasinegative definite,
(iii) if A is symmetric and negative definite.

Proof. (i) will be proved as an application of Liapunov's second method (Example 5.11). (ii) follows immediately from (i); quasinegative definiteness requires that $A+A'$ be negative definite. Choosing $B=I$ in (i) gives this result. (iii) is just a special case of (ii).

That (ii) is sufficient but not necessary is demonstrated in the following example.

EXAMPLE 5.7 Let $A=\begin{pmatrix} a_{11} & a_{12} \\ a_{21} & a_{22} \end{pmatrix}$. The necessary and sufficient conditions for stability were (a) $\operatorname{tr} A<0$ and (b) $|A|>0$. Now, suppose A is quasinegative definite; that is $\begin{pmatrix} 2a_{11} & a_{12}+a_{21} \\ a_{21}+a_{12} & 2a_{22} \end{pmatrix}$ is negative definite. Hence the leading principal minor conditions must be satisfied; (c) $2a_{11}<0$ and (d) $4a_{11}a_{22}-(a_{12}+a_{21})^2>0$. Now, (c) and (d) imply that $a_{22}<0$. Requiring both a_{11} and $a_{22}<0$ is stronger than requiring $\operatorname{tr} A<0$. Thus (d) may be rewritten as

$$a_{11}a_{22}-a_{12}a_{21}-\left(\frac{a_{12}-a_{21}}{2}\right)^2>0$$

which clearly implies, but is not implied by, (b).

Some other types of sufficient condition arise in the context of general equilibrium models and are discussed in Section 5.7.

5.5 TRAJECTORIES, CONDITIONAL STABILITY AND SADDLEPOINT EQUILIBRIA

5.5.1 Trajectories

We observed in Section 5.2.1 that, for an autonomous system, we could write the solution as $\theta(t;x_0)$, where t denotes the elapsed time since the initial condition x_0 was reached; thus $\theta(0;x_0)=x_0$. For such a system, we may define a trajectory as follows.

DEFINITION 5.11 Given a solution $\theta(t; x_0)$ to an autonomous system of differential equations, the set of points $\{\theta(t; x_0); t\in\mathbb{R}\}$ is termed the **trajectory** passing through x_0.

Clearly, trajectories are not generally unique with respect to the initial condition x_0. If, after some elapsed time t_a, $\theta(t_a; x_0) = x_a$, then $\{\theta(t; x_a);\ t\in\mathbb{R}\} = \{\theta(t; x_0);\ t\in\mathbb{R}\}$ (assuming unique global solutions exist).

Apart from studying the stability of equilibrium, we may also be interested in the behaviour of trajectories around such an equilibrium point. For the system $\dot{x}(t) = Ax(t)$, $x(t)\in\mathbb{R}^2$, trajectories may be depicted in a diagram. The characteristic equation is

$$|\lambda I - A| = \lambda^2 - \operatorname{tr} A\,\lambda + |A| = 0, \tag{5.18}$$

so that, defining $\Delta = (\operatorname{tr} A)^2 - 4|A|$, the roots are

$$\lambda = (\operatorname{tr} A \pm \Delta^{1/2})/2. \tag{5.19}$$

Thus the eigenvalues are real if $\Delta > 0$ and complex if $\Delta < 0$. If $|A| < 0$, then $\Delta > (\operatorname{tr} A)^2$ and the roots are real and of opposite sign.

The structure of trajectories can be derived analytically if the explicit solution (5.10) is known, or qualitatively graphed using phase plane methods (see Section 5.6.5). Table 5.1 sets out the various conditions for the various types of trajectory map and these are illustrated in Figure 5.2. The structure of these trajectory maps should be intuitively clear. With equal roots, trajectories must be straight lines. If they are complex but have negative (positive) real parts, they tend to spiral into (away from) the equilibrium. With zero real parts, they orbit the equilibrium. Notice that small perturbations in the a_{ij} coefficients will usually destroy this latter property. The case where the roots are real and of opposite sign, the saddlepoint equilibrium, is considered in more detail in Sections 5.5.2 and 5.5.3 in view of its prominence in economic dynamics.

5.5.2 The stable manifold

Consider first the linear system $\dot{x} = Ax$, $x\in\mathbb{R}^2$, where $|A| < 0$. Here

Table 5.1

Case	Type of equilibrium	Eigenvalues	Trace, determinant, Δ
1	linear trajectory paths		
	(a) asym. stable.	$\lambda_1<0$, $\lambda_2<0$ $\lambda_1=\lambda_2$	tr $A<0$, $\|A\|>0$, $\Delta=0$
	(b) unstable	$\lambda_1>0$, $\lambda_2>0$ $\lambda_1=\lambda_2$	tr $A>0$, $\|A\|>0$, $\Delta=0$
2	generally non-linear trajectory paths		
	(a) asym. stable.	$\lambda_1<0$, $\lambda_2<0$ $\lambda_1\neq\lambda_2$	tr $A<0$, $\|A\|>0$, $\Delta>0$
	(b) unstable	$\lambda_1>0$, $\lambda_2>0$ $\lambda_1\neq\lambda_2$	tr $A>0$, $\|A\|>0$, $\Delta>0$
3	saddlepoints, unstable	$\lambda_1>0$, $\lambda_2<0$	$\|A\|<0$
4	spiral points		
	(a) asym. stable	λ_1,λ_2 complex $\mathrm{Re}(\lambda_i)<0$, $i=1,2$	tr $A<0$, $\Delta<0$
	(b) unstable	λ_1,λ_2 complex $\mathrm{Re}(\lambda_i)>0$, $i=1,2$	tr $A>0$, $\Delta<0$
5	centre Liapunov stable	λ_1,λ_2 complex $\mathrm{Re}(\lambda_i)=0$, $i=1,2$	tr $A=0$, $\Delta<0$

$\Delta>(\text{tr } A)^2>0$ and one root is positive and the other, negative; let $\lambda_1<0$, $\lambda_2>0$. The general solution is

$$\theta(t)=c_1 e^{\lambda_1 t}v_1+c_2 e^{\lambda_2 t}v_2, \tag{5.20}$$

where v_1, v_2 are the associated eigenvectors. With one root positive, the system is unstable. However, notice that, if $c_2 v_2=\theta$, then $\theta(t)\to\theta$ as $t\to\infty$, whilst, if $c_2 v_2 \neq\theta$, the solution may converge upon the equilibrium $\hat{x}=\theta$ for some time but eventually must diverge away. Thus, if the initial starting point x_0 is proportional to v_1 such that $x_0=\alpha v_1$, for some $\alpha\in\mathbb{R}$), then the solution will converge upon the equilibrium. All solutions of the type $\theta(t|\alpha v_1, t_0)$ converge on the equilibrium. The trajectory associated with such solutions is $\{x:x=\alpha v_1, \alpha\in\mathbb{R}\}$. This is termed the **stable arm** of the saddlepoint equilibrium. The trajectory $\{x:x=\alpha v_2, \alpha\in\mathbb{R}\}$ is often termed the

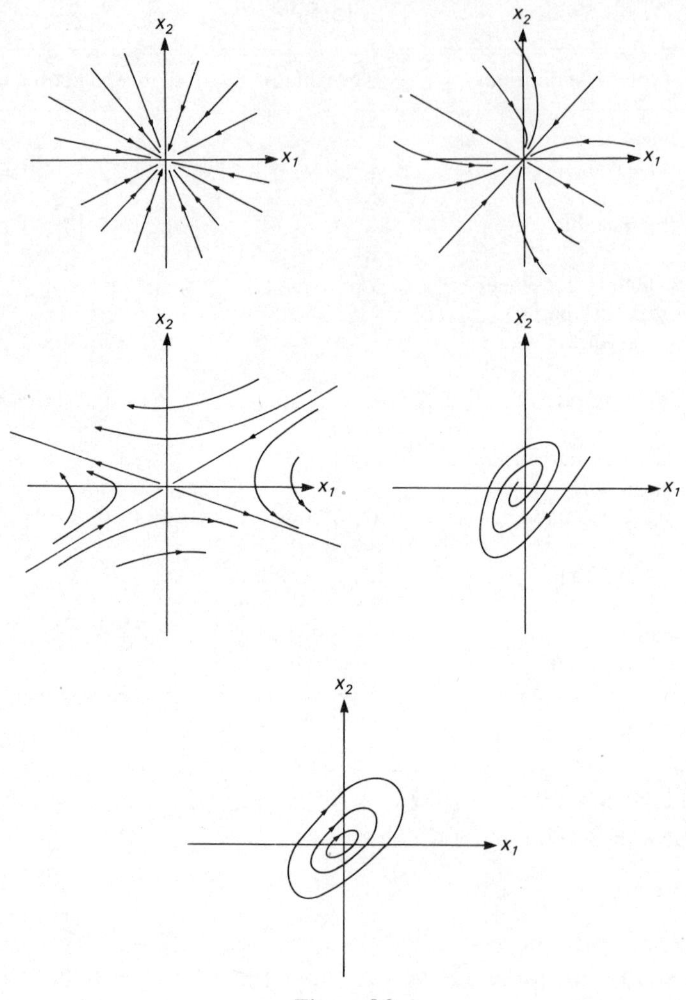

Figure 5.2

unstable arm. More formally, we have:

THEOREM 5.15 Given the general solution $\theta(t)=\sum_{j=1}^{n} c_j e^{\lambda_j t} v_j$ for the system $\dot{x}=Ax$, the behaviour of the solution is determined solely by one particular eigenvalue λ_i if, and only if, the initial condition vector x_0 (at $t=0$) is proportional to the eigenvector associated with λ_i (i.e., $x_0=\beta_i v_i$ for some $\beta_i \neq 0$).

Proof. Since $v_j, j=1,\ldots,n$ are linearly independent, $\sum_{j=1}^{n} \beta_j v_j=\theta$ if, and only if, $\beta_j=0$ for all j. Substituting the initial condition x_0 into the general solution yields $x_0=\sum_{j=1}^{n} c_j v_j$. Since $v_j, j=1,\ldots,n$ are linearly independent, any vector $x_0 \in \mathbb{R}^n$ may be written as a linear combination of these; let this be $x_0=\sum_{i=1}^{n} \beta_j v_j$. Hence $\sum_{j=1}^{n} (c_j-\beta_j)v_j=\theta$, so $c_j-\beta_j=0$ for all j. The solution depends on λ_i if, and only if, $c_i \neq 0$ and $c_j=0$ for all $j, j\neq i$. But this implies $\beta_i \neq 0$ and $\beta_j=0$ for all $j, j\neq i$, so $x_0=\beta_i v_i$ and the solution becomes

$$\theta(t)=\beta_i e^{\lambda_i t} v_i=e^{\lambda_i t} x_0.$$

This theorem implies that, in the case where there is just one root negative and the rest positive, there is a trajectory path defined by the eigenvector associated with the negative root such that solutions lying on this path converge on the equilibrium. The stable manifold is defined as the set of points for which $\theta(t)\rightarrow\hat{x}$ as $t\rightarrow\infty$. Where this has dimension one, it is referred to as the stable arm. More generally, the dimension of the stable manifold is equal to the number of roots with negative real part; in view of this, it is worth stating the generalization of Theorem 5.15.

THEOREM 5.16 Given the general solution $\theta(t)=\sum_{j=1}^{n} c_j e^{\lambda_j t} v_j$ for the system $\dot{x}=Ax$, the behaviour of the solution depends solely upon a subset of eigenvalues, $\{\lambda_k; k\in K\subset(1,\ldots,n)\}$ if, and only if, the initial condition vector is a linear combination of the associated eigenvectors $\{v_k; k\in K\subset(1,\ldots,n)\}$.

Proof. As for Theorem 5.15. Here the solution depends on $\lambda_k, k\in K$ if, and only if, $c_j \neq 0$ for all $j\in K$ and $c_j=0$ for all $j\notin K$. This implies $\beta_j\neq 0$ for all $j, j\in K$ and $\beta_j=0$ for all $j\notin K$. Hence $x_0=\sum_{k\in K}\beta_k v_k$ and the solution is $\theta(t)=\sum_{k\in K} e^{\lambda_k t}\beta_k v_k$.

5.5.3 Conditional stability

The saddlepoint equilibrium is unstable in the sense of Definitions

5.4–5.6 However, we have already noted that there is a subregion from which the system is convergent. Solutions may be said to be conditionally stable.

THEOREM 5.17 Given $\dot{x}=Ax$, where A has k eigenvalues with negative real parts and $n-k$ with positive real parts, then there exists a real k-dimensional manifold M containing the equilibrium point $\hat{x}=\theta$ such that, if $x_0 \in M$, then $\theta(t) \to \hat{x}$ as $t \to \infty$. Moreover, for any finite $\varepsilon > 0$ and any solution $\theta(t|x_0, t_0)$ with $x_0 \notin M$ at $t=t_0$, there is some t' for which the solution does not satisfy $\|\theta(t)-\hat{x}\| \leqslant \varepsilon$ for $t \geqslant t'$.

Proof. See Coddington and Levinson (1955).

Conditional stability is thus an intermediate concept lying between full stability (where M has dimension n) and instability (where M contains only the equilibrium point).

5.5.4 Identifying the stable manifold

The dimension of the stable manifold is determined by the number of positive and negative roots. Some of the difficulties and possibilities for qualitatively signing the eigenvalues are indicated below.

The eigenvalues are denoted as the (possibly complex) numbers $\lambda_1, \ldots, \lambda_n$ and it is assumed that $|A| \neq 0$. The characteristic equation is

$$\prod_{i=1}^{n} (\lambda - \lambda_i) = \lambda^n + b_{n-1}\lambda^{n-1} + \ldots + b_0 = 0, \tag{5.21}$$

where

$$b_0 = (-1)^n |A| = (-1)^n \lambda_1 \lambda_2 \ldots \lambda_n \Rightarrow \prod_{i=1}^{n} \lambda_i = |A|.$$

Hence we have:

THEOREM 5.18 A necessary condition for the matrix A to have $2k$

roots with negative real part ($0 \leqslant 2k \leqslant n$) is that $(-1)^n|A|<0$, and for A to have $2k+1$ roots with negative real part ($1 \leqslant 2k+1 \leqslant n$), that $(-1)^n|A|>0$.

For A, 2×2, the manifold M may be of dimension 0, 1, or 2. Which is the case may be determined by knowledge of the signs of tr A and $|A|$. Note that, in general, if the roots are complex, they occur in conjugate pairs of the form $a+ib$, $a-ib$. From (5.21), we have

$$\lambda_1\lambda_2 = |A|,$$
$$\text{and } \lambda_1+\lambda_2 = \operatorname{tr} A.$$

If $|A|>0$, then the real parts of the roots are either both positive or both negative. If tr $A>0$, they are positive [$\dim(M)=0$], whilst if tr $A<0$, they are both negative [$\dim(M)=2$]. If $|A|<0$, then one is positive and the other negative [$\dim(M)=1$]. Note that, in this latter case, the saddlepoint, the direction of the stable arm is given by the eigenvector associated with the negative root (see the example in Section 5.7.3, page 198).

For A, 3×3, we often, but not always, need additional information in order to fully identify the dimension of the stable manifold. From (5.21), we have

$$-b_0 = \lambda_1\lambda_2\lambda_3 = |A|, \tag{5.22}$$

$$-b_2 = \lambda_1+\lambda_2+\lambda_3 = \operatorname{tr} A, \tag{5.23}$$

and

$$b_1 = (\lambda_1\lambda_2+\lambda_2\lambda_3+\lambda_3\lambda_1)$$
$$= \begin{vmatrix} a_{11} & a_{12} \\ a_{21} & a_{22} \end{vmatrix} + \begin{vmatrix} a_{11} & a_{13} \\ a_{31} & a_{33} \end{vmatrix} + \begin{vmatrix} a_{22} & a_{23} \\ a_{32} & a_{33} \end{vmatrix}. \tag{5.24}$$

If the signs of b_0, b_1, b_2 are known, $\dim(M)$ may often be established. The conditions are examined sequentially. Note that, with three roots, either none or two will be complex. Accordingly, λ_3 is taken to be real in what follows.

Case 1: $|A|<0$
This implies that either the real parts of all roots are negative, or that

two are positive and one negative. The latter must be real, hence take $\lambda_3<0$. Now λ_1, λ_2 may be examined.
Case 1.1: tr $A>0$
Then at least one of λ_1, λ_2 has positive real part. Hence $\mathrm{Re}(\lambda_1)$, $\mathrm{Re}(\lambda_2)>0$ and $\lambda_3<0$ and $\dim(M)=1$.
Case 1.2: tr $A<0$
Here, additional information is required about b_1; either $b_1<0$ or $b_1>0$.
Case 1.2.1: $b_1<0$
Then $\lambda_1\lambda_2+\lambda_3(\lambda_1+\lambda_2)<0$. If $\mathrm{Re}(\lambda_1)$, $\mathrm{Re}(\lambda_2)<0$ then $\lambda_1\lambda_2>0$ and $\lambda_3(\lambda_1+\lambda_2)>0$, and there is a contradiction. Hence $\mathrm{Re}(\lambda_1)$, $\mathrm{Re}(\lambda_2)>0$, $\lambda_3<0$, and $\dim(M)=1$.
Case 1.2.2: $b_1>0$
Then the result is ambiguous and depends upon quantitative magnitudes (however, if the roots are real, then λ_1, λ_2, $\lambda_3<0$ and $\dim(M)=3$).

The same process of reasoning establishes the following results:
Case 2: $|A|>0$
Case 2.1: tr $A<0$
Then $\mathrm{Re}(\lambda_1)$, $\mathrm{Re}(\lambda_2)<0$, $\lambda_3>0$; $\dim(M)=2$.
Case 2.2: tr $A>0$
Case 2.2.1: $b_1<0$
Then $\mathrm{Re}(\lambda_1)$, $\mathrm{Re}(\lambda_2)<0$, $\lambda_3>0$; $\dim(M)=2$,
Case 2.2.2: $b_1>0$
Ambiguous (however, if the roots are real, λ_1, λ_2, $\lambda_3>0$ and $\dim(M)=0$).

5.6 STABILITY ANALYSIS FOR NON-LINEAR SYSTEMS

In Section 5.1, it was noted that analytic solutions for non-linear systems are not usually available, particularly for the kinds of equations which crop up in economic analysis. Fortunately qualitative information often suffices to establish results on stability and the nature of trajectories.

5.6.1 The linear approximation method

Consider the non-linear differential equations

$$\dot{x} = f(x), \qquad x \in \mathbb{R}^n. \tag{5.25}$$

It is assumed that $f \in C_2$ and $\hat{x}$ is an isolated equilibrium such that $f(\hat{x}) = \theta$. An exact Taylor's series expansion around $\hat{x}$ yields

$$\dot{x} = f_x(\hat{x})(x - \hat{x}) + (x - \hat{x})' f_{xx}(\hat{x} + \varepsilon(x - \hat{x}))(x - \hat{x}), \tag{5.26}$$

(where $0 < \varepsilon < 1$), since $f(\hat{x}) = \theta$. More compactly, we may write

$$\dot{x} = A(x - \hat{x}) + g(x), \tag{5.27}$$

where $A = [\partial f^i / \partial x_j]$, the Jacobian matrix evaluated at $\hat{x}$, and $g^i(x) = (x - \hat{x})' f_{xx}(\hat{x} + \varepsilon(x - \hat{x}))(x - \hat{x})$. Note that $g(\hat{x}) = \theta$. If $\|g(z)\| / \|z\| \to 0$ as $\|z\| \to 0$, we might expect the behaviour of solutions to (5.25) to be similar to that of

$$\dot{x} = A(x - \hat{x}) \tag{5.28}$$

in a sufficiently small neighbourhood of $\hat{x}$; this is usually, but not always, the case.

THEOREM 5.19. The system $\dot{x} = A(x - \hat{x}) + g(x, t)$, where A is a real constant $n \times n$ matrix whose eigenvalues all have negative real parts, and where g is a real and continuous function, with equilibrium $\hat{x}$ such that $g(\hat{x}, t) = \theta$ and

$$\|g(x, t)\| / \|x - \hat{x}\| \to 0 \text{ as } \|x - \hat{x}\| \to 0 \text{ uniformly in } t,$$

is asymptotically (locally) stable. However, if A has one or more eigenvalues with positive real part, the system is unstable.

Proof. See Coddington and Levinson (1955).

This theorem is slightly more general than is required. In our case, $g(x, t)[= g(x)]$ is independent of t and $f \in C_2$ hence $\|g(x)\| / \|x - \hat{x}\| \to 0$ as $\|x - \hat{x}\| \to 0$. The theorem will be proved using Liapunov's second method (Example 5.12). It follows that, if the

Jacobian matrix evaluated at the equilibrium point has negative real parts, the non-linear system is locally asymptotically stable. Theorem 5.17 on conditional stability has its parallel here too.

THEOREM 5.20 The system $\dot{x}=A(x-\hat{x})+g(x,t)$ where A has k eigenvalues with negative, and $n-k$ with positive, real parts and where $g \in C_0$ and $g(\hat{x},t)=\theta$, has a k-dimensional manifold M containing the equilibrium point $\hat{x}$ such that the solution $\theta(t) \to \hat{x}$ as $t \to \infty$ if $x_0 \in M$. If $x_0 \notin M$ at $t=t_0$, then, for any finite $\eta>0$, there is some t' for which the solution does not satisfy $\|\theta(t)-\hat{x}\| \leqslant \eta$ for $t \geqslant t'$.

Proof. See Coddington and Levinson (1955), pp. 329–33.

Equation (5.28) is termed the local approximation system for the non-linear system (5.25). If $\hat{x}$ is (globally) asymptotically stable (unstable) for the linear system, then $\hat{x}$ is locally asymptotically stable (unstable) for the non-linear system. However, if the linear approximation system is unstable, the trajectories for the non-linear system may still remain in the vicinity of the equilibrium due to the contribution of the non-linear terms (which may become significant as the distance from $\hat{x}$ increases).

In fact, the trajectory map in the neighbourhood of $\hat{x}$ for the linear system will have a similar structure to that of the non-linear system so long as the Jacobian matrix $f_x(\hat{x})$ does not have purely imaginary or zero roots. For a formalization and proof of this, see Hartman (1964). Thus, if the linear system has a saddle, spiral or node, so does the non-linear system in the neighbourhood of the equilibrium. However, if the linear system exhibits orbits, this implies little about the stability of the non-linear system. Orbits in the linear system arise when roots are purely imaginary, and such orbits are not structurally stable; a small perturbation in the coefficients of the matrix will destroy the orbits and leave some other form of behaviour.

If there are zero eigenvalues (this is so if, and only if, $|f_x(\hat{x})|=0$), the theorems no longer apply. For instance, the equation $\dot{x}=-x^3$ has a local approximation $\dot{x}=0$. The latter equation is not asymptotically stable (see Example 5.1) yet the former is (direct integration gives the solution $\theta(t)=[2(t+c)]^{-1/2}$ where c is an arbitrary constant). In fact, it is also the case that, if the Jacobian has zero roots, $|A|=0$ and the equilibrium of the linear approximation system is no longer unique; the point is that, for the linear approximation system

$$\dot{x}=\theta \Rightarrow A(x-\hat{x})=\theta.$$

This latter equation has a unique solution $x=\hat{x}$ only if $|A|\neq 0$. (Note that, as in Theorem 5.9, a small perturbation in the coefficients can restore the matrix to full rank.)

The principal limitation of the linear approximation method lies in the fact that, having established local stability for the non-linear system, the extent of the region of stability remains unknown. However, it may prove useful:

(a) where global methods prove to be analytically intractable,
(b) where the system is not globally stable – the equilibrium may still be locally stable,
(c) in characterizing more completely the nature of the trajectory map in the locality of the equilibrium.

EXAMPLE 5.8 A firm produces output (x) and some form of local environmental pollution (E) in a fixed relationship

$$E=f(x), \tag{i}$$

where $f\in C_1$ and $f'(x)>0$ for all x. The environmental control authority influences the firm's output (and hence pollution) by imposing a tax on output. The firm's optimal choice of x for a given tax rate τ is

$$x=a(\tau) \tag{ii}$$

with $a\in C_1$ and $a'(\tau)<0$ for all τ. The stock of pollutant(s) in the environment decays at a rate $\gamma(\gamma>0)$ so the rate of change of the stock is given by

$$\dot{s}=E-\gamma s. \tag{iii}$$

Social 'damage' costs (D) associated with the pollution are given by

$$D=h(s), \tag{iv}$$

where $h \in C_2$ and $h'(s), h''(s) > 0$. The authority is assumed to follow the tax adjustment rule

$$\dot{\tau} = \mathrm{d}D/\mathrm{d}s - \tau. \tag{v}$$

That is, the tax rate is adjusted in the direction of equating the tax rate with the marginal damage rate. The behaviour of the system may be described by a pair of differential equations in s and τ. From (ii), (iii) and (iv)

$$\dot{s} = f(x) - \gamma s = f(a(\tau)) - \gamma s. \tag{vi}$$

Differentiating (iv) and substituting into (v) yields

$$\dot{\tau} = h'(s) - \tau. \tag{vii}$$

The equilibrium point $(\hat{s}, \hat{\tau})$ may be found by setting $\dot{\tau} = 0$, $\dot{s} = 0$ in (vi), (vii). The linear approximation system $\dot{x} = A(x - \hat{x})$ is here given as

$$\begin{pmatrix} \dot{s} \\ \dot{\tau} \end{pmatrix} = \begin{pmatrix} -\gamma & f'(a(\hat{\tau}))a'(\hat{\tau}) \\ h''(\hat{s}) & -1 \end{pmatrix} \begin{pmatrix} s - \hat{s} \\ \tau - \hat{\tau} \end{pmatrix}. \tag{viii}$$

Necessary and sufficient conditions for stability are that $\operatorname{tr} A < 0$ and $|A| > 0$. $\operatorname{Tr} A = -\gamma - 1 < 0$ and $|A| = \gamma - f'(a(\hat{\tau}))a'(\hat{\tau})h''(\hat{s}) > 0$ given the signs of the derivatives involved. Hence the system is locally asymptotically stable.

It is also possible to determine the structure of the trajectory map in the neighbourhood of the equilibrium. Recall Table 5.1; the equilibrium will be of the type depicted in Figure 5.2 case 2(a) (a node), or case 4(a), (a spiral) as $\Delta = (\operatorname{tr} A)^2 - 4|A|$ is $>$ or < 0. Here, $\Delta = (\gamma + 1)^2 - 4(\gamma - f'a'h'') = (\gamma - 1)^2 + f'a'h''$, which is ambiguous and depends upon the magnitudes of γ, $f'(\alpha(\hat{\tau}))$, $a'(\hat{\tau})$ and $h''(\hat{s})$.

5.6.2 The Olech sufficient conditions for global stability

There are many special case results for particular kinds of non-linear system. Olech's Theorem is the most important of these for economic analysis. It is concerned with the stability of the 2-equation non-linear system.

THEOREM 5.21 (Olech) The differential equations $\dot{x}=f(x)$, $x \in \mathbb{R}^2$, where $f(x)=(f^1(x), f^2(x))'$, for which $\hat{x}=\theta$ is an equilibrium point, is uniformly globally asymptotically stable if:
(a) $\operatorname{tr} f_x(x)<0$ for all $x \in \mathbb{R}^2$,
(b) $|f_x(x)|>0$ for all $x \in \mathbb{R}^2$,
(c) $f_1^1(x)f_2^2(x) \neq 0$ for all $x \in \mathbb{R}^2$ or $f_2^1(x)f_1^2(x) \neq 0$ for all $x \in \mathbb{R}^2$.

Proof. Olech (1963). (See also Hartman, 1961, Hartman and Olech, 1962, for related results.)

EXAMPLE 5.9 Global stability may be proved for the pollution/taxation problem of Example 5.8 (if s and τ are unrestricted; for non-negativity restrictions, see Section 5.7). The equilibrium was $(\hat{s}, \hat{\tau})$, so the change of variables $s_1=s-\hat{s}$, $\tau_1=\tau-\hat{\tau}$ shifts the equilibrium to the origin of the new coordinate system. From equations (vi), (vii) of Example 5.8, we obtain

$$\dot{s}_1=f(a(\tau_1+\hat{\tau}))-\gamma(s_1+\hat{s}),$$
$$\dot{\tau}_i=h'(s_1+\hat{s})-(\tau_1+\hat{\tau}).$$

The Jacobian A matrix is then

$$\begin{pmatrix} -\gamma & f'(a(\tau_1+\hat{\tau}))a'(\tau_1+\hat{\tau}) \\ h''(s_1+\hat{s}) & -1 \end{pmatrix}$$

where, for all τ_1, s_1, the derivatives $f', h'', -a'>0$ and $\gamma>0$. Hence $\operatorname{tr} A<0$ and $|A|>0$ for all s_1, τ_1 and condition (c) holds since $f_{s_1}^1 f_{\tau_1}^2=\gamma \neq 0$ for all s_1, τ_1. Hence the system is uniformly globally asymptotically stable.

5.6.3 Liapunov's second method

For economic models, Liapunov's second (or 'direct') method is attractive since it does not require an explicit solution to be found for the differential equations. It may be used to establish various types of stability and may also sometimes be used to investigate regions of stability. However, although used extensively in general equilibrium analysis, the approach is not always easy to apply since a Liapunov function is not always easy to 'find'; there are certain basic principles and knowledge of previous applications may be helpful – but the rest remains essentially that of intelligent trial and error.

The first section deals with basic theorems, the second with the problem of finding a Liapunov function, and the third with applications, including proofs of several propositions and theorems from earlier sections, along with some economic examples.

Basic theorems

The following presentation is based on Kalman and Bertram (1960). It is assumed that unique solutions exist and that a unique equilibrium exists.

THEOREM 5.22 (Liapunov) Consider the system of differential equations $\dot{x}=f(x,t)$, $x\in\mathbb{R}^n$ which has $f(\hat{x},t)=\theta$ for all t, where $\hat{x}$ denotes the unique equilibrium point. Suppose a function $V(x,t)\in C_1$ exists such that:

[1] $V(\hat{x},t)=0$,

[2] there exist continuous non-decreasing real valued functions α, β such that $\alpha(0)=\beta(0)=0$ and such that:

(a) $0<\alpha(\|x-\hat{x}\|)\leqslant V(x,t)$ for all t if $x\neq\hat{x}$,

(b) $V(x,t)\leqslant\beta(\|x-\hat{x}\|)$ for all t if $x\neq\hat{x}$.

[3] $\alpha(\|x-\hat{x}\|)\rightarrow\infty$ as $\|x-\hat{x}\|\rightarrow\infty$,

[4] (a) $\dot{V}(x,t)=\mathrm{d}V(x,t)/\mathrm{d}t=V_x(x,t)\dot{x}+V_t(x,t)$

$$=V_x(x,t)f(x,t)+V_t(x,t)\leqslant 0$$

for all t if $x\neq\hat{x}$.

(b) There exists a continuous real valued function γ such that $\gamma(0)=0, \gamma(\tau)>0$ if $\tau>0$ and $\dot{V}\leqslant-\gamma(\|x-\hat{x}\|)<0$ for all t if $x\neq\hat{x}$.

(Note: this is slightly weaker than requiring V to be negative definite since γ is merely required to be continuous.)

Then, if [1]–[4] hold, $\hat{x}$ is uniformly globally asymptotically stable and $V(x,t)$ is a Liapunov function for the system $\dot{x}=f(x,t)$.

Proof. Denote the solution to $\dot{x}=f(x,t)$ as $\theta(t|x_0,t_0)$, or simply $\theta(t)$ or θ where no confusion will arise. From [4](b), $V(\theta(t),t)-V(x_0,t_0)=\int_{t_0}^{t}\dot{V}(\theta(\tau),\tau)\mathrm{d}\tau<0$ for all $t>t_0$. Hence

$$V(\theta(t),t)<V(x_0,t_0) \qquad \text{for all } t>t_0. \tag{5.29}$$

(a) Proving uniform Liapunov stability. Definition 5.4 requires that, given an $\varepsilon>0$, arbitrarily small, a $\delta>0$ can be found depending only on ε (and not on t_0) such that $\|x_0-\hat{x}\|\leqslant\delta\Rightarrow\|\theta-\hat{x}\|\leqslant\varepsilon$ for all $t\geqslant t_0$. Take $\varepsilon>0$ and $\delta(\varepsilon)>0$ such that $\beta(\delta)<\alpha(\varepsilon)$ (see Figure 5.3). This is always possible since β is continuous, non-decreasing and $\beta(0)=0$. Now, for

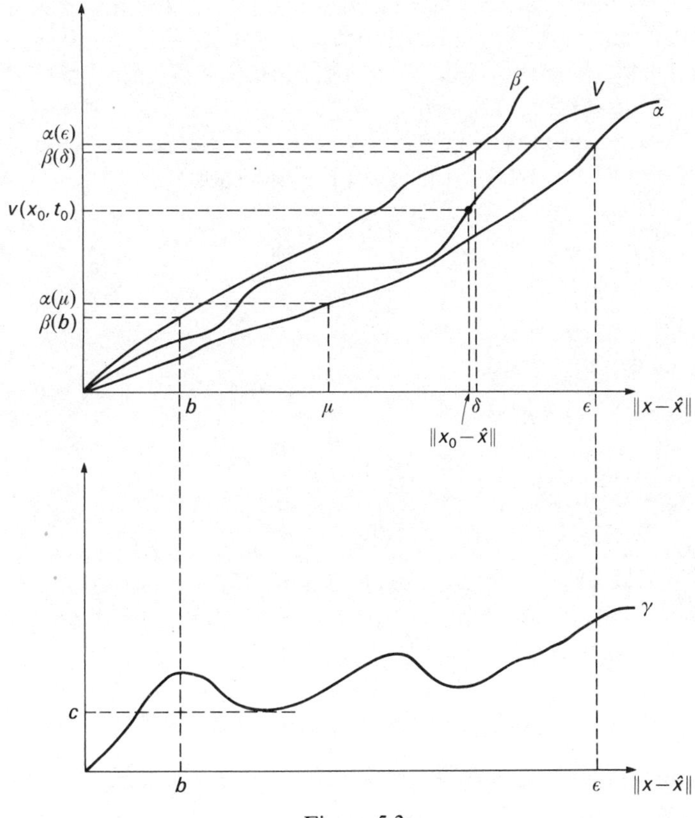

Figure 5.3

$\|x_0-\hat{x}\|<\delta$ (arbitrary t_0), clearly [2](a) and (b) imply

$V(x_0,t_0)\leqslant\beta(\delta)$. Thus, for $t\geqslant t_0$, $\alpha(\varepsilon)>\beta(\delta)\geqslant V(x_0,t_0)$

(refer to Figure 5.3). Since $V(x_0,t_0)>V(\theta,t)$ from (5.29) and $V(\theta,t)\geqslant\alpha(\|\theta-\hat{x}\|)$, then $\alpha(\varepsilon)>\alpha(\|\theta-\hat{x}\|)$. Now α is continuous and non-decreasing, hence this implies $\|\theta-\hat{x}\|<\varepsilon$, and hence we have uniform Liapunov stability. Note that [3] is not required and also that it is sufficient to have [4](a) without [4](b).

(b) Uniform asymptotic stability (Definition 5.5). As in (a), for any $\varepsilon>0$, find a $\delta>0$ such that $\beta(\delta)<\alpha(\varepsilon)$. Take any initial state x_0 such that $\|x_0-\hat{x}\|<\delta$. Hence, by part (a), $\|\theta(t)-\hat{x}\|<\varepsilon$ for all $t>t_0$.

Now take any $\mu>0$ such that $\mu<\|x_0-\hat{x}\|$ and find a constant $b>0$ such that $\beta(b)<\alpha(\mu)$ (thus $b<\mu$). Let c be the minimum for γ on $[b,\varepsilon]$. Clearly $c>0$ since $0\notin[b,\varepsilon]$ (see Figure 5.3). Define $T=\beta(\delta)/c$ and $t_1=t_0+T$.

Now, suppose $\|\theta-\hat{x}\|>b$ over $[t_0,t_1]$. Since $\|\theta-\hat{x}\|<\varepsilon$, then $\dot{V}\leqslant-\gamma(\|\theta-\hat{x}\|)\leqslant-c$ on $[t_0,t_1]$. Therefore

$$V(\theta(t_1),t_1)=V(x_0,t_0)+\int_{t_0}^{t_1}\dot{V}\,\mathrm{d}t\leqslant V(x_0,t_0)-(t_1-t_0)c.$$

Now, $t_1-t_0=T$ and $V(x_0,t_0)\leqslant\beta(\|x_0-\hat{x}\|)\leqslant\beta(\delta)$ since $\|x_0-\hat{x}\|\leqslant\delta$. Hence

$$V(\theta(t_1),\,t_1)\leqslant\beta(\delta)-Tc=0 \tag{5.30}$$

from the definition of T. However,

$$V(\theta(t_1),t_1)\geqslant\alpha(\|\theta(t)-\hat{x}\|)\geqslant\alpha(b)>0 \tag{5.31}$$

since $\|\theta(t)-\hat{x}\|>b>0$ by assumption. Equation (5.31) contradicts (5.30), hence $\|\theta(t)-\hat{x}\|>b$ cannot hold on $[t_0,t_1]$. Hence, by continuity, there exists a time $t_2\in[t_0,t_1]$ and $x_2=\theta(t_2|x_0,t_0)$ such that $\|x_2-\hat{x}\|=b$. Hence $\alpha(\|\theta(t|x_2,t_2)-\hat{x}\|)$ $\leqslant V(\theta(t|x_2,t_2),t)\leqslant V(x_2,t_2)\leqslant\beta(b)\leqslant\alpha(\mu)$ for all $t\geqslant t_2$ and so

$$\|\theta(t|x_0,t_0)-\hat{x}\|=\|\theta(t|x_2,t_2)-\hat{x}\|<\mu \quad \text{for all } t\geqslant t_2$$

and hence for all $t\geqslant t_0+T(\geqslant t_2)$.

(c) Proving global uniform asymptotic stability (Definition 5.7). Here $\|x_0 - \hat{x}\| < r$ where r is some finite (but arbitrarily large) constant. In view of [3], we can find a constant q such that $\alpha(q) > \beta(r)$ (refer to Figure 5.3; $\beta(r)$ is finite and $\alpha(q) \to \infty$ as $q \to \infty$). Asymptotic convergence follows by noting that $\|x_0 - \hat{x}\| < r \Rightarrow V(x_0, t_0) \leqslant \beta(r)$. The proof then follows (b).

Roughly speaking, the Liapunov function (which has a global minimum at $\hat{x}$) takes a value which depends upon the position $x(t)$ at a time t. It is then shown that V decreases over time. Thus V moves towards its global minimum value – and hence $x(t)$ must also converge on the equilibrium, $\hat{x}$.

Having established the basic Liapunov theorem, we now state the conditions on V required for various other forms of stability.

THEOREM 5.23 Sufficient conditions for stability. Given $\dot{x} = f(x, t)$ where $x \in \mathbb{R}^n$ and where $f(\hat{x}, t) = \theta$ for all t and $\hat{x}$ is a unique equilibrium, then:

Type of stability	*Conditions on V*
(a) Uniform global asymptotic	[1]–[4]
(b) Global asymptotic	[1], [2](a), [3], [4]
(c) Uniform asymptotic	[1], [2], [4]
(d) Asymptotic	[1], [2](a), [4]
(e) Uniform Liapunov	[1], [2], [4](a)
(f) Liapunov	[1], [2](a), [4](a)

Proof. With appropriate modifications, as for Theorem 5.22.

The relaxation of conditions is intuitive; clearly [2](b) is required for uniform types of stability, [3] for global and [4](b) for asymptotic convergence.

The autonomous system $\dot{x} = f(x)$, $x \in \mathbb{R}^n$, is an important special case (for which stability, if it exists, is always uniform).

THEOREM 5.22′ Consider the differential equations $\dot{x} = f(x)$, $x \in \mathbb{R}^n$ which has $f(\hat{x}) = \theta$ (where $\hat{x}$ denotes the equilibrium point).

If there exists a function $V(x) \in C_1$ such that:
[1] $V(\hat{x})=0$ and $V(x)>0$ if $x \neq \hat{x}$,
[2] $\dot{V}(\hat{x})=0$ and $\dot{V}(\hat{x})<0$ if $x \neq \hat{x}$,
[3] $V(x) \rightarrow \infty$ as $\|x-\hat{x}\| \rightarrow \infty$.
Then $\hat{x}$ is uniformly globally asymptotically stable and $V(x)$ is a **Liapunov function** for the differential equation system.

Actually, condition [2] can be relaxed by merely requiring $\dot{V}(x) \leqslant 0$ and $\dot{V}(\theta(t|x_0, t_0))$ not identically zero in t, $t \geqslant t_0$ for any t_0 and any $x_0 \neq \hat{x}$ (see Kalman and Bertram, 1960, corollary 1.3). The theorem may also be stated for a bounded subregion (see Takayama, 1985).

THEOREM 5.24 Given $\dot{x}=f(x)$, $x \in \mathbb{R}^n$, if there exists a function $V(x) \in C_1$ defined on a bounded region $U=\{x \in \mathbb{R}^n;\ V(x)<k\}$ such that:
[1] $\hat{x} \in U$, $f(\hat{x})=\theta$,
[2] $x_0 \in U$,
[3] $V(\hat{x})=0$,
[4] $V(x)>0$ if $x \neq \hat{x}$, $x \in U$,
[5] $\dot{V}(x)<0$ if $x \neq \hat{x}$, $x \in U$,
hold, then the solution converges asymptotically to $\hat{x}$ as $t \rightarrow \infty$ for all $x_0 \in U$.

Prior to this theorem, we have assumed the equilibrium was unique. However, it may be possible to analyze non-unique equilibria using Theorem 5.24 if the equilibria are isolated since this implies, for each equilibrium point, that there exists a neighbourhood containing no other equilibrium point (the case where equilibria are non-isolated is discussed in Section 5.7.1).

The technique may also be used to rule out stability.

THEOREM 5.25 If there exists a function V defined as in Theorem 5.22 (or 5.23, 5.24 as appropriate) such that:
[1] $V(\hat{x})=0$ and $V(x)>0$ if $x \neq \hat{x}$,
[2] $\dot{V}(x)>0$ for all $x \neq \hat{x}$,
then the equilibrium is globally unstable.

Construction of Liapunov functions

Norms satisfy most of the conditions required under the Liapunov theorems; only the properties of $\dot{V}$ need to be additionally investigated (this is often the most difficult part of the procedure). Thus, norms such as $(\sum_i x_i^2)^{1/2}$, $\max_i |x_i|$, $(\sum_i a_i x_i^2)^{1/2}$, $\sum_i a_i |x_i|$ (where a_i, $i = 1, \ldots, n$ are positive constants) may prove useful in applications. However, the set of possible Liapunov functions is much wider than this. For example, in engineering problems, energy often turns out to be a Liapunov function; it depends upon the state of the system and the physical reasoning is that, when energy attains its minimum, the system will be in a stable position. The problem of finding a Liapunov function is essentially one of trial and error, although experience of previous or parallel applications can simplify matters considerably. It has already been remarked that Liapunov functions are not always easy to find; the following theorem is useful since, if a globally stable equilibrium point is suspected to exist, it motivates the search for a Liapunov function (note however that it only holds for uniform global asymptotic stability).

THEOREM 5.26 If $\hat{x}$ is a unique uniformly globally asymptotically stable equilibrium point for the system $\dot{x} = f(x, t)$, where $x \in \mathbb{R}^n$, and f is Lipschitz continuous, then there exists a Liapunov function $V(x, t)$ (which is infinitely differentiable with respect to x, t) which satisfies the conditions of Theorem 5.22.

Proof. See Massera (1956) and Kalman and Bertram (1960).

Applications

EXAMPLE 5.10 Consider the non-linear differential equation $\dot{x} = f(x), x \in \mathbb{R}, f \in C_1$, with equilibrium $\hat{x}$ such that $f(\hat{x}) = 0$. This is globally stable if

$$f(x) \gtreqless 0 \quad \text{as} \quad x \lesseqgtr \hat{x}, \tag{i}$$

and globally unstable if

$$f(x) \gtreqless 0 \quad \text{as} \quad x \gtreqless \hat{x}. \tag{ii}$$

To prove this, consider the candidate Liapunov function

$$V(x)=(x-\hat{x})^2/2.$$

Suppose condition (i) holds. Theorem 5.22 conditions [1]–[3] are satisfied. [4] requires that $\dot{V}<0$ for all $x\neq\hat{x}$. Here $\dot{V}=V_x(x)\dot{x}=(x-\hat{x})f(x)$ so $\dot{V}=0$ if $x=\hat{x}$. If $x>\hat{x}$, then $f(x)<0$ and $\dot{V}<0$. If $x<\hat{x}$, then $f(x)>0$ and again $\dot{V}<0$. Hence [4] is satisfied and $\hat{x}$ is uniformly globally asymptotically stable. If condition (ii) holds, the conditions of Theorem 5.25 hold and the system is globally unstable.

EXAMPLE 5.11 Proof for Theorem 5.14.
The differential equation is $\dot{x}=Ax$, where $x\in\mathbb{R}^n$ and A is a real constant $n\times n$ matrix. Assume that $|A|\neq 0$ so the equilibrium $\hat{x}=\theta$ is unique. Consider the candidate Liapunov function (a quadratic form)

$$V=x'Bx,$$

where B is a positive definite matrix. Referring to Theorem 5.22, conditions [1]–[3] are immediately satisfied. [4] requires that $\dot{V}(x)<0$ for all $x\neq\hat{x}$. Here $\dot{V}=d/dt(x'Bx)=\dot{x}'Bx+x'B\dot{x}$. Now $\dot{x}'=x'A'$ so $\dot{V}=x'A'Bx+x'BAx=x'(A'B+BA)x$. Thus $\dot{V}<0$ for all $x\neq\hat{x}$ if $A'B+BA$ is negative definite.

Liapunov's direct method may also be used to prove the stability/instability Theorems 5.19, 5.20 for the local approximation system (see e.g. Lefschetz, 1957). A special case of Theorem 5.19 is treated below.

EXAMPLE 5.12 Consider the non-linear system

$$\dot{x}=f(x)=Ax+g(x) \qquad x\in\mathbb{R}^n, \tag{i}$$

where $|A|\neq 0, g=(g_1,\ldots,g_n)', g\in C_0, g(\theta)=\theta$ and $\|g(x)\|/\|x\|\to 0$ as $\|x\|\to 0$. Liapunov's second method may be used to show that the equilibrium point $\hat{x}=\theta$ is locally asymptotically stable if the real parts of the eigenvalues of A are negative (Theorem 5.19, see also

Kalman and Bertram, 1960). Consider the case where the eigenvalues of A are real and distinct. Let $y = W^{-1}x$ where W denotes the matrix of eigenvectors of A. From (i),

$$\dot{y} = W^{-1}AWy + W^{-1}g(x) = \Lambda y + W^{-1}g(x),$$

where Λ is the diagonal matrix of eigenvalues of A. The equilibrium point is $\hat{y} = \theta$. Consider the candidate Liapunov function $V = y'y$. This satisfies [1] and [2](a) in Theorem 5.22, so this leaves [4] to be checked.

$$\dot{V} = 2y'\dot{y} = 2\{y'\Lambda y + y'W^{-1}g(x)\}.$$

Let $a = \max_i \lambda_i$, where λ_i, $i = 1, \ldots, n$ denote the eigenvalues of A. Hence $a < 0$. Thus $y'\Lambda y < nay'y$ for all $y \neq \theta$. Hence

$$\dot{V} \leqslant 2[na\|y\|^2 + \|y\| . \|W^{-1}g(x)\|] \leqslant 2\|y\|^2 \left[na + \frac{\|W^{-1}g(x)\|}{\|y\|} \right].$$

Now $\|W^{-1}g\|^2 = \sum_i (w_i g)^2$, where w_i denotes the ith row of W^{-1} and $w_i g \leqslant \|w_i\| . \|g\|$, so $\|W^{-1}g\|^2 \leqslant \|g\|^2 \sum_i \|w_i\|^2$. Let $k_1 = (\sum_i \|w_i\|^2)^{1/2}$, hence $\|W^{-1}g\| \leqslant k_1 \|g\|$. Similarly, $x = Wy$, so $\|x\| \leqslant k_2 \|y\|$ for some $k_2 > 0$ (and $\|y\| \to 0 \Rightarrow \|x\| \to 0$). Hence

$$\dot{V} \leqslant 2\|y\|^2 [na + k_1 k_2 \|g(x)\| / \|x\|].$$

Since $\|g(x)\|/\|x\| \to 0$ as $\|y\| \to 0$ and $a < 0$, then, for sufficiently small $\|y\| \neq 0$, $\dot{V} < 0$. Hence [4] is satisfied. $\hat{y} = \theta$ is locally asymptotically stable and hence, so is $\hat{x}$.

EXAMPLE 5.13 The neoclassical economic growth model (see e.g. Burmeister, 1980) gives rise to the differential equation $\dot{k} = g(k) - c - \alpha k$, where c, k denote per capita consumption and capital stock and α is a positive constant (capital growth equals output, $g(k)$, minus consumption and capital depreciation). The production function $g(k)$ has $g \in C_2$, $g'(k) > 0$ and $g''(k) < 0$ for all k. Holding c constant, we may plot $\dot{k}$ against k as in Figure 5.4 (k_1, k_2 represent equilibria). Let $V = (k - \hat{k})^2/2$ and consider the equilibrium $\hat{k} = k_2$. Clearly, on the interval $U = (k_1, 2k_2 - k_1)$, $V(k) < (k_2 - k_1)^2/2$ and conditions [1]–[4] of Theorem 5.24 are

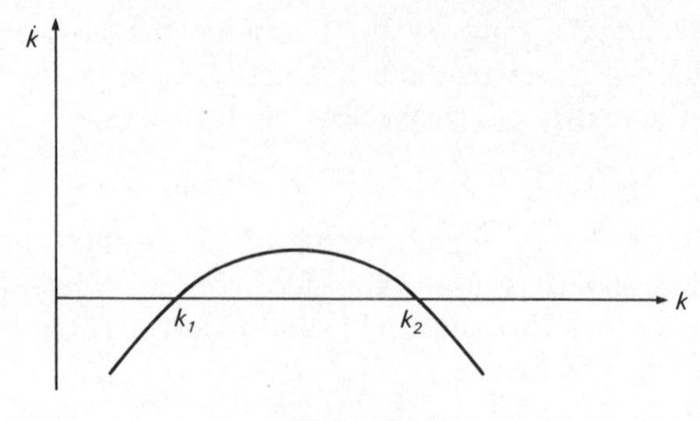

Figure 5.4

satisfied. $\dot{V}=V_k\dot{k}=(k-\hat{k})(g(k)-c-\alpha k)$ and, from Figure 5.4, $k\gtrless k_2\Rightarrow g(k)-c-\alpha k\lesseqgtr 0$, so $\dot{V}<0$ if $k\neq k_2, k\in U$. Hence [5] is also satisfied and the system is stable in this region. Similarly, Theorem 5.25 can be used to show that k_1 is an unstable equilibrium.

5.6.4 Non-unique equilibria

As remarked in Section 5.1, if there are multiple, but isolated, equilibria, then no single equilibrium can be globally asymptotically stable in the sense of Definition 5.6. However, it could be that, whatever the initial condition, trajectories converge on *some* equilibrium. Thus, although equilibrium points may not be, the **process** may be globally stable.

Consider the system $\dot{x}=f(x)$, $x\in X\subset\mathbb{R}^n$, where X is an open connected set and $f\in C_1$ on X. Assuming that a unique, continuous solution exists for all initial conditions $x_0\in X$, a solution $\theta(t|x_0)$ is said to **converge** on some point $\bar{x}$ if $\lim_{t\to\infty}\theta(t|x_0)=\bar{x}$.

THEOREM 5.27 Given the above system, and suppose that a solution converges on a point $\bar{x}$, then $\bar{x}$ must be an equilibrium such that $f(\bar{x})=\theta$.

Proof. Suppose that, at $\bar{x}$, $f^i(\bar{x})>0$ for some i (a similar proof can be constructed for $f^i(\bar{x})<0$). Now, f is continuous, hence

$$\lim_{t\to\infty} f^i(\theta(t|x_0)) = f^i(\lim_{t\to\infty} \theta(t|x_0)) = f^i(\bar{x})>0.$$

By continuity, there exists an $\varepsilon>0$ such that $f^i(\bar{x})>\varepsilon$. Then, for sufficiently large τ, by convergence, $\dot{\theta}_i = f^i(\theta(\tau|x_0)) \geqslant \varepsilon$. Hence, for $t>\tau$, θ_i grows without bound, contradicting convergence. Hence $f(\bar{x})=\theta$ (see Hahn, 1982).

DEFINITION 5.12 **The system or process $\dot{x}=f(x)$ is said to be globally stable** if, for all $x_0 \in X$, the solution $\theta(t|x_0)$ converges.

This natural definition of global stability for the process makes no reference to particular equilibrium points; if a solution converges, it necessarily converges to some equilibrium point.

Now, consider the trajectory in Figure 5.5. Suppose there is a set of non-isolated equilibria $\{\hat{x}_i\}$ (represented by the circle). The trajectory, starting from x_0 spirals ever closer. Such a trajectory does not converge on any particular equilibrium point yet meaningfully converges on the set of equilibria. If this happened for all $x_0 \in X \backslash \{\hat{x}_i\}$, we could meaningfully describe the process as 'stable'. The question

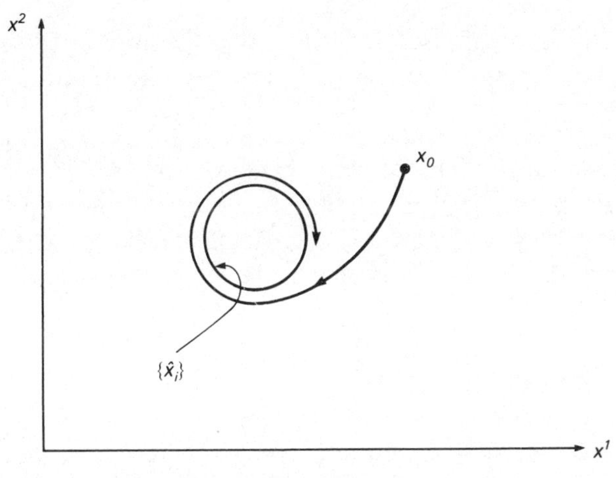

Figure 5.5

then, in this case, is whether a convergent subsequence exists (if $\theta(t|x_0)$ is bounded, it will have a converging subsequence – see Hahn, 1982 i.e. such that $\lim_{v\to\infty}\theta(t_v|x_0)=\hat{x}_i$).

DEFINITION 5.13 **The system or process** $\dot{x}=f(x)$ is said to be **quasiglobally stable** if:

(a) all limit points of the solution $\theta(t|x_0)$ are equilibrium points, i.e. $\lim\limits_{v\to\infty}\theta(t_v|x_0)=\hat{x}$, and

(b) the solution is uniformly bounded; for any $\mu>0$, there is some $\delta(\mu)>0$ such that $\|x_0-\hat{x}\|\leqslant\mu\Rightarrow\|\theta(t|x_0)-\hat{x}\|\leqslant\delta$, where $\hat{x}\in E$, the set of equilibria, defined by $E=\{\hat{x}:\hat{x}\in X,\ f(\hat{x})=\theta\}$.

Thus quasiglobal stability essentially requires that trajectories converge on some equilibrium point or other. We then have the following result:

THEOREM 5.28 If the system or process $\dot{x}=f(x)$ is quasiglobally stable, then equilibria are non-isolated.

Proof. See Hahn (1982) or Arrow and Hahn (1971).

Thus, if equilibria are isolated, the process cannot be quasiglobally stable. The principal result is as follows in Theorem 5.29.

THEOREM 5.29 The system $\dot{x}=f(x)$, $x\in X\subset\mathbb{R}^n$, where X is compact, is quasiglobally stable if there exists a function $V(x)$, $x\in X$ such that $V(\theta(t|x_0))$ is strictly decreasing in t, except where x_0 is an equilibrium. If equilibria are isolated, the system is globally stable.

Proof. See Uzawa (1961).

Uzawa gives several examples. The issue is raised again in an example in Section 5.8.2.

5.6.5 Phase diagrams

Phase diagrams may be used to study and illustrate the behaviour of first order differential equations in one or two variables. It is also possible to heuristically establish stability or instability.

The differential equation $\dot{x}=f(x)$, $x\in\mathbb{R}$

It may be possible to graph $\dot{x}$ against x (if only qualitatively); several examples are illustrated in Figure 5.6. Equilibrium points are defined by $\dot{x}=f(\hat{x})=0$. The arrows indicate the direction of motion over time. Suppose $f'(x)<0$ for all $x\in X$; this implies a phase diagram of type a; the arrows indicate the change in $\dot{x}$ and x as time passes; thus, if $\dot{x}>0$, x is increasing and as x increases, so $\dot{x}$ decreases. Thus the point $(x, \dot{x})$ follows the line toward the point a. Any displacement from a results in a convergence back to a, so a is a stable equilibrium. c is unstable. b is said to be left-stable and right-unstable (one sided stability), and d is left unstable and right stable.

EXAMPLE 5.14 Example 5.13 presented a neoclassical growth model. The accumulation equation is $\dot{k}=g(k)-c-\alpha k$. This may be

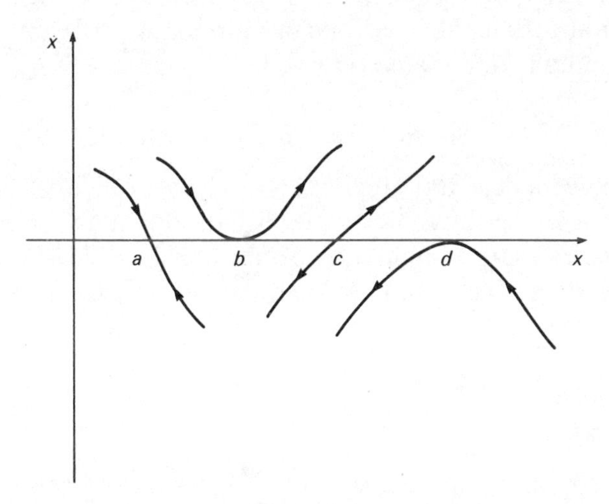

Figure 5.6

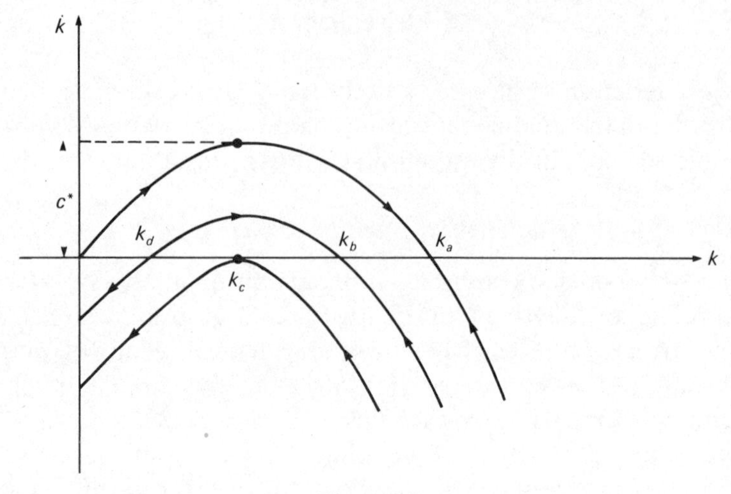

Figure 5.7

graphed assuming different but constant levels of consumption, c as in Figure 5.7. With $c=0$, k_a is a stable equilibrium point. With $c>0$, k_b is a stable equilibrium for all initial $k_0>k_d$. k_d is an unstable equilibrium, and, if initial $k_0<k_d$, the system cannot sustain the level of consumption; capital stock declines to zero. Increasing c, there is a maximum sustainable level of consumption, c^*; however, such a level can only be maintained if $k_0 \geqslant k_c$.

The differential equation $\dot{x}=f(x)$, $x \in \mathbb{R}^2$

In economic models, trajectories can rarely be explicitly calculated. However, it is often possible to establish the qualitative structure of the trajectory map around equilibrium points, at least for the 2-variable system. The method involves establishing the signs of $\dot{x}$ for all $x \in X$. Consider:

$$\dot{y}=f(y,z), \qquad (f \in C_1),$$

$$\dot{z}=g(y,z), \qquad (g \in C_1).$$

The equilibrium is given as the point(s) $(\hat{y},\hat{z})$ which satisfy $f(\hat{y},\hat{z})=0$ and $g(\hat{y},\hat{z})=0$. The phase diagram in yz-space is developed by first

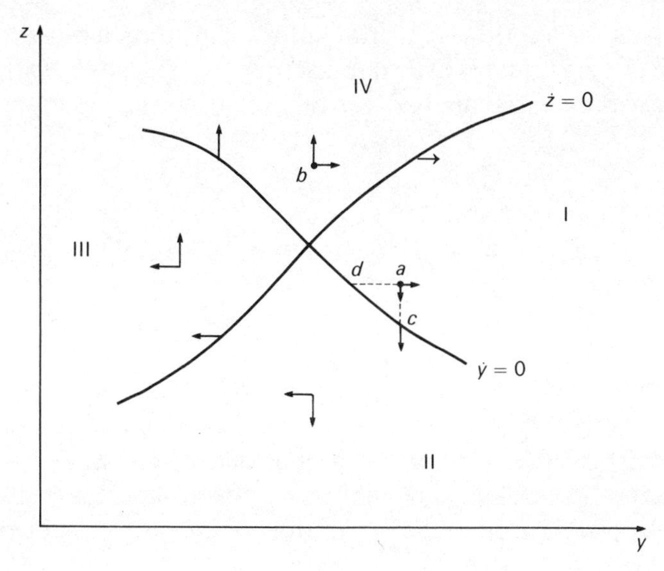

Figure 5.8

sketching $f(y,z)=0$ and $g(y,z)=0$. We shall refer to the graphs of these equations as **isokines**; their intersections constitute the equilibria for the system. Consider Figure 5.8 and, in particular, the point a. At such a point, the dynamic behaviour of the system is expressed by the rates of change $\dot{y}|_a = f(y_a, z_a)$, $\dot{z}|_a = g(y_a, z_a)$. Suppose, at $a, \dot{y}>0$ and $\dot{z}<0$. These directions of motion are indicated on the phase diagram by arrows in the appropriate direction. The isokines define regions I, II, III, IV, in Figure 5.8. In a region, the qualitative directions of motion remain the same (this follows from the assumed continuity of f, g); if, at a, $\dot{y}>0$ and $\dot{z}<0$, then this is true for all points in region I. Since $f, g \in C_1$, the sign of $\dot{y}$ and $\dot{z}$ changes on crossing the respective $\dot{y}, \dot{z}$ isokine. Thus, in moving from a to b, the $\dot{z}=0$ isokine is crossed and the sign of $\dot{z}$ changes. In this way, once the directions of motion for one region have been established, it is straightforward to fill in the arrows for the other regions. The directions of motion on the isokines have also been indicated in Figure 5.8; e.g. at c, the $\dot{y}$ component is zero.

Sometimes the functions are given analytically, so the isokines and directions of motion are straightforward to determine. More usually, only qualitative information is available (signs on derivatives etc.); the analysis for this situation is as follows:

(a) SLOPES OF ISOKINES

The $\dot{y}=0$ isokine is defined by $f(y,z)=0$. Totally differentiating and rearranging, we obtain

$$\left.\frac{\mathrm{d}z}{\mathrm{d}y}\right|_{\dot{y}=0} = \left.\frac{\mathrm{d}z}{\mathrm{d}y}\right|_{f(y,z)=0} = -f_y(y,z)/f_z(y,z) \tag{5.32}$$

(and a similar equation obtains for the $\dot{z}=0$ isokine). Thus, given knowledge of the signs of the partial derivatives f_y, f_z, g_y, g_z, the gradients of the isokines are qualitatively determined. Note that this may be possible around the equilibrium point even if not elsewhere.

(b) THE QUALITATIVE DIRECTIONS OF MOTION

Consider, for example, the isokine $f(y,z)=0$ which has negative slope in Figure 5.8; $\mathrm{d}z/\mathrm{d}y|_{\dot{y}=0}<0$. All points in a region have the same qualitative behaviour, so it suffices to consider a point just above the isokine such as a. Comparing a with points on the curve $\dot{y}=0$ yields the required information about the sign of $\dot{y}$ at a (and, similarly, comparisons with points on the $\dot{z}=0$ isokine affords information about $\dot{z}$). For $\dot{y}$ at a, the horizontal or vertical comparison points c or d may be used. Using d, we have

$$\dot{y}= f(y_d, z_d)=0 \text{ at } d.$$

The sign of $\dot{y}|_a$ can be obtained from the expansion around d:

$$\begin{aligned}\operatorname{sign}(\dot{y}|_a) &= \operatorname{sign}(f(y_d,z_d)+f_y(y_d,z_d).(y_a-y_d)\\ &\quad +f_z(y_d,z_d).(z_a-z_d)) = \operatorname{sign}(f_y(y_d,z_d)\,\Delta y),\end{aligned} \tag{5.33}$$

since $f(y_d,z_d)=0$, $\Delta z=z_a-z_d=0$ and $\Delta y=y_a-y_d>0$. Clearly then

$$\operatorname{sign}(\dot{y}|_a)=\operatorname{sign}(f_y(y_d,z_d)\Delta y). \tag{5.34}$$

Since $\Delta y>0$, if $f_y(y_d,z_d)\gtrless 0$, then $\dot{y}|_a \gtrless 0$. The comparison could equally have been made between a and c:

$$\text{sign}(\dot{y}|_a) = \text{sign}[f_z(y_c, z_c)\Delta z].$$

The conclusion on the sign of $\dot{y}|_a$ would, of course, be the same.

(c) SKETCHING TRAJECTORY PATHS

Having established the qualitative properties of trajectories in the various regions, the next step is to sketch illustrative trajectories. Suppose, for some point such as b in Figure 5.8, $\dot{y}$, $\dot{z} > 0$. Then the direction of the trajectory passing through b must lie in a north-easterly direction. The quantitative magnitudes of $\dot{y}, \dot{z}$ are unknown – hence the trajectory could lie close to either the northerly or easterly direction. However, trajectories must be sketched so that they nowhere violate the qualitative directions of motion. Note also that, as a trajectory crosses an isokine, the relevant component of $(\dot{y}, \dot{z})$ is zero. Across the $\dot{z}=0$ isokine, trajectories are horizontal, whilst across the $\dot{y}=0$ isokine they are vertical.

EXAMPLE 5.15 Consider the pollution/taxation problem of Example 5.8. Two differential equations were obtained:

$$\dot{s} = f(x) - \gamma s \qquad \equiv F(s, \tau),$$

$$\dot{\tau} = h'(s) - \tau \qquad \equiv G(s, \tau),$$

where $f' > 0$, $x = a(\tau)$, $a' < 0$, $h' > 0$, $h'' > 0$ and $\gamma > 0$. Assume an economically meaningful equilibrium point $(\hat{s}, \hat{\tau})$ such that $\hat{s}, \hat{\tau} > 0$. The isokines are given by $\dot{s} = 0$: $s = f(a(\tau))/\gamma$ and $\dot{\tau} = 0$: $\tau = h'(s)$. The slopes of these isokines are given as:

$$\dot{s} = 0: \qquad \left.\frac{d\tau}{ds}\right|_{\dot{s}=0} = \frac{\gamma}{f'a'} < 0 \text{ since } f'(a(\tau)),\ \gamma > 0 \text{ and } a'(\tau) < 0.$$

$$\dot{\tau} = 0: \qquad \left.\frac{d\tau}{ds}\right|_{\dot{\tau}=0} = h'' > 0.$$

The phase diagram (Figure 5.9) has four regions. For a point in region I, $\text{sign}(\dot{s}) = \text{sign}(F_s(s, \tau)|_{\dot{s}=0}\Delta s)$. Now $\Delta s > 0$ and $F_s = -\gamma < 0$, hence $\dot{s} < 0$. Hence $\dot{s} < 0$ in regions I, II and $\dot{s} > 0$ in

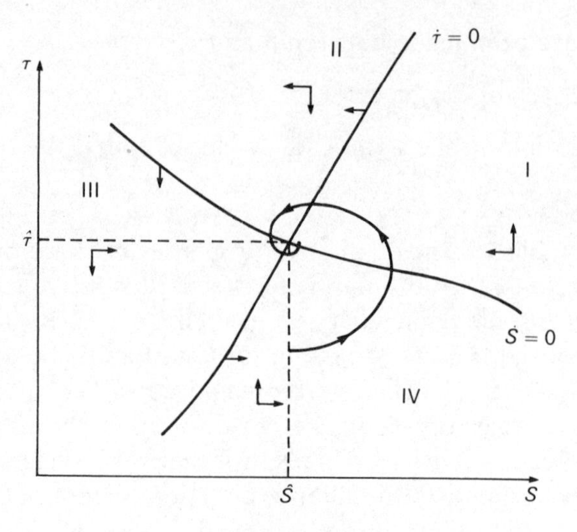

Figure 5.9

regions III, IV. For a point in region I, $\text{sign}(\dot{\tau})=\text{sign}(G_s(s,\tau)|_{\dot{\tau}=0}\Delta s)$. Here again, $\Delta s>0$ and $G_s(s,\tau)=h''>0$, hence $\dot{\tau}>0$ in I. Thus $\dot{\tau}>0$ in regions I, IV and $\dot{\tau}<0$ in II, III.

These directions of motion are depicted in Figure 5.9. The trajectories are clearly anticlockwise spirals, but it is not possible to tell whether the equilibrium is stable or unstable in this case. However, from the Olech conditions, the equilibrium is in fact stable, hence it is legitimate to sketch trajectories which converge on the equilibrium point. Even so, it is unclear whether the trajectory map looks like Figure 5.2 type 2(a) or 4(a). In the locality of the equilibrium, the structure of trajectories may be established by using the local approximation method; in Example 5.8, it was shown that it would be a stable spiral if $\gamma<1$. Hence we may sketch trajectories in accord with this information.

The above example illustrates the fact that phase diagram analysis is not always especially informative about the stability of equilibrium. However, we shall meet examples later where it is so.

5.7 ECONOMIC APPLICATIONS

In this section, the techniques discussed in the foregoing sections are applied to three economic examples.

5.7.1 A resource market model

The price of a natural resource adjusts in response to excess demand

$$\dot{p}(t) = f(D(t) - S(t)), \tag{5.35}$$

where $f \in C_2$, $f'(x) > 0$ for all x and $f(0) = 0$. $p(t)$, $D(t)$, $S(t)$ denote price, demand, supply at time t. There is a long-run expected price E, and supply adjusts according to whether price is above or below this level

$$\dot{S}(t) = g(p(t) - E), \tag{5.36}$$

where $g \in C_2$, $g'(.) > 0$ and $g(0) = 0$. The demand for the resource is given by

$$D(t) = h(p(t)), \tag{5.37}$$

where $h \in C_2$, and $h'(.) < 0$. Equations (5.35)–(5.37) give the following system in $p(t)$ and $S(t)$:

$$\dot{p} = f(h(p) - S), \tag{5.38}$$

$$\dot{S} = g(p - E) \tag{5.39}$$

which has an equilibrium at $(\hat{p}, \hat{S})' = (E, h(E))'$. Ignoring non-negativity restrictions ($D, S, p \geqslant 0$; discussed later), global uniform asymptotic stability may be established by appeal to Olech's Theorem (Theorem 5.21); this requires:

(a) $f_p + g_S < 0$,
(b) $f_p g_S - f_S g_p > 0$,
(c) $f_p g_S \neq 0$ or $f_S g_p \neq 0$

for all p, S. Here

$$f_p + g_S = f'(D-S)h'(p) < 0,$$

$$f_p g_S - f_S g_p = f'(D-S)g'(p-E) > 0,$$

$$f_p g_S = 0 \quad \text{but} \quad f_S g_p = -f'(.)g'(.) \neq 0.$$

Hence we have global stability.

We now illustrate the application of Liapunov's second method to a special case of the above problem. Let

$$f(D(t)-S(t)) = D(t)-S(t),$$

$$h(p(t)) = \beta/p(t),$$

$$g(p(t)-E) = \alpha(p(t)-E),$$

where α, $\beta > 0$ are given constants. Now let $V = \frac{1}{2}a(p-\hat{p})^2 + \frac{1}{2}b(S-\hat{S})^2$, where a, b are positive constants. This clearly satisfies [1] and [3] of Theorem 5.22′. Condition [2] requires that

$$\dot{V} = V_p\dot{p} + V_S\dot{S} = a(p-E)(\beta/p - s) + \alpha b(s - \beta/E)(p-E) < 0$$
$$\text{for } (p, S) \neq (\hat{p}, \hat{S}). \tag{5.40}$$

This does not hold since, whenever $p = E$, $\dot{V} = 0$ whatever the value of S. However, if the constants a, b are chosen such that $a = \alpha b$, then

$$\dot{V} = -a\beta(p-E)^2/pE.$$

So, assuming $p, E > 0$, $\dot{V} \leqslant 0$ for all p, S. Furthermore, except at the equilibrium point $(\hat{p}, \hat{S})$, $\dot{V}$ is not identically zero for all t; that is, if $p = E$ (so $\dot{V} = 0$) but $S \neq \hat{S} (= \beta/E)$, then $\dot{p} = \beta/p - S \neq 0$, so the trajectory moves away from the $\dot{V} = 0$ line. Hence the conditions of Theorem 5.22′ hold and the system is globally asymptotically stable.

The phase diagram is straightforward to construct. The isokines are $\dot{S} = 0 : p = E$ and $\dot{p} = 0 : s = h(p)$ (see Figure 5.10). The $\dot{p} = 0$ isokine has negative slope since $h'(.) < 0$. Consider a point such as a in region I. $\text{Sign}(\dot{p}|_a) = \text{sign}(f_s|_c \Delta S) < 0$ since $\Delta S > 0$ and $f_s = -f'(.) < 0$. $\text{Sign}(\dot{s}|_a) = \text{sign}(g_p|_b \Delta p) < 0$ since $\Delta p < 0$ and $g_p = g'(.) > 0$. The directions of motion for the other regions are then immediate. Given

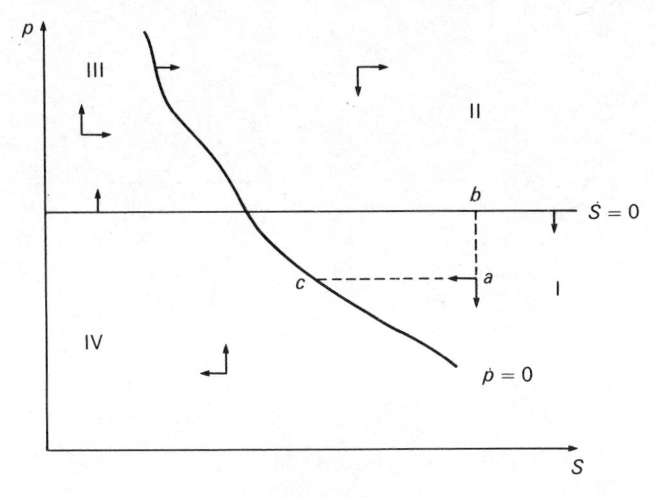

Figure 5.10

that the equilibrium is globally stable, the trajectories may converge or spirally converge to the equilibrium point. Note that the phase diagram does not, by itself, indicate whether the equilibrium is stable or not; trajectories consistent with the directions of motion could be sketched so as to converge toward or indeed to diverge away from the equilibrium.

The conditions under which a spiral or a node occurs (see Table 5.1) may be established in the neighbourhood of the equilibrium by examination of the local approximation system

$$\begin{pmatrix} \dot{p} \\ \dot{S} \end{pmatrix} = \begin{pmatrix} f'(0)h'(\hat{p}) & -f'(0) \\ g'(0) & 0 \end{pmatrix} \begin{pmatrix} p-\hat{p} \\ S-\hat{S} \end{pmatrix}.$$

Here trace $A = f'(0)h'(\hat{p}) < 0$ and $|A| = f'(0)g'(0) > 0$ indicating that the non-linear system is locally stable. Now

$$\Delta = (\text{trace } A)^2 - 4|A| = [f'(0)h'(\hat{p})]^2 - 4f'(0)g'(0).$$

If $\Delta < 0$, the system is a locally stable spiral; if $\Delta > 0$, a stable node, as illustrated in Figure 5.11. These diagrams indicate the nature of the price and quantity dynamics; thus, if $\Delta < 0$, there are damped oscillations around the equilibrium point.

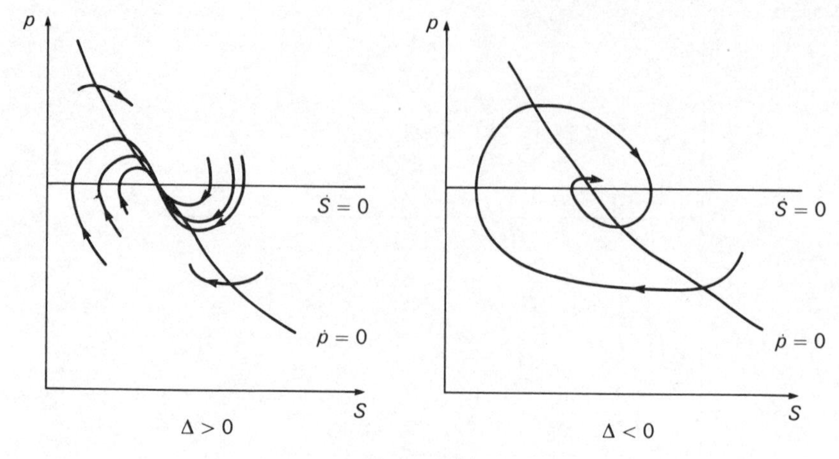

Figure 5.11

Finally, it is worth discussing the impact of non-negativity restrictions. Let us assume that $D>0$ for all $p \geqslant 0$. Note that, from (iv) and (v), price cannot go negative so long as $h(p)$ becomes sufficiently large as $p \to 0$, which can be assumed. However, S could go negative. Suppose we require that $S \geqslant 0$. Then one way to accommodate this is to modify (v) to

$$\dot{S}=G(p,S),$$

where

$$G(p,S)=g(p-E) \quad \text{for} \quad p \geqslant 0, S \geqslant 0,$$

and

$$G(p,S)=0 \quad \text{for} \quad p \geqslant 0, S \leqslant 0.$$

Here, $G(p,S)=0$ if $p=E$. Clearly S cannot go negative if the initial starting point is positive. Unfortunately, a consequence of this specification is that G is not Lipschitz continuous at points $(p,S) \in L=\{(p,S); p \neq E, S=0\}$. To see this, consider a point $(\bar{p},0) \in L$. Lipschitz continuity requires that $\|G(p,S)-G(\bar{p},0)\|/\|(p,S)-(\bar{p},0)\| \leqslant k$, where $k>0$ is some finite constant, for $(p,S) \neq (\bar{p},0)$ in the neighbourhood of $(\bar{p},0)$. Let $p=\bar{p}$ but $S>0$. Then $G(\bar{p},S)=g(\bar{p}-E)=c$, constant ($\neq 0$) and $G(\bar{p},0)=0$. So we require $|c|/S \leqslant k$. Clearly as $S \to 0$, no finite k exists.

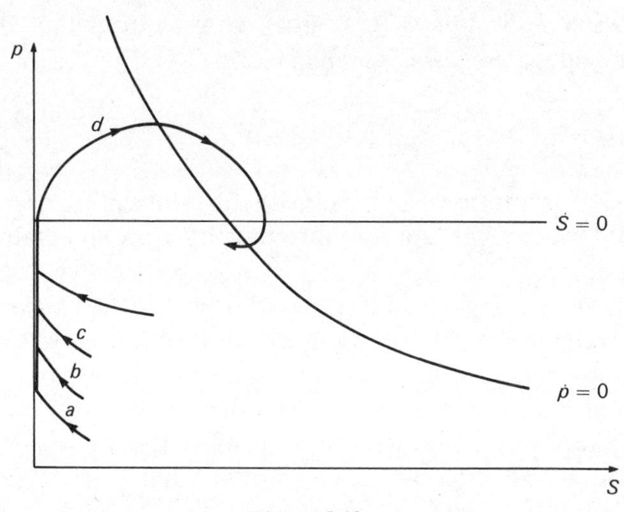

Figure 5.12

Where the function is not Lipschitz continuous, solutions may be non-unique – see Figure 5.12. A trajectory hitting the p-axis has $h(p) - S > 0$ and hence $\dot{p} > 0$. Thus p increases until $p = E$, when S starts to increase from zero. Thus any trajectory hitting the p-axis (e.g. a, b, c in Figure 5.12) ends up following trajectory d.

5.7.2 Stability of general equilibrium

Suppose there are $n+1$ commodities in an economy and for each commodity, i, there exists an associated excess demand function

$$E^i(P_1, \ldots, P_{n+1}) \qquad i = 1, \ldots, n+1.$$

A competitive equilibrium exists if there is a $\hat{P} > \theta$ such that

$$E^i(\hat{P}) = 0 \qquad i = 1, \ldots, n+1, \tag{5.41}$$

where $P = (P_1, \ldots, P_{n+1})'$. Such a $\hat{P}$ will exist if (a) the set of aggregate excess demands, at any set of prices, is a convex set, and the aggregate excess demand functions (which can be multi-valued) are continuous in prices, (b) the aggregate excess demand functions are homogeneous of degree zero in prices, and (c) Walras' Law holds (see Chapter 3, Section 3.3). Walras' Law requires that the total value of aggregate excess demands is zero at all prices:

$$\sum_{i=1}^{n+1} P_i E^i = 0. \tag{5.42}$$

Additional assumptions such as gross substitutability are required to guarantee uniqueness of equilibrium (see Arrow and Hahn, 1971, Chapter 9).

Assuming the above sufficient conditions for the existence of a general competitive equilibrium are met, a commodity (say $n+1$) may be chosen as numeraire and an equation dropped from (5.41) to give the normalized system of excess demands, $f^i(p)$, $i=1,\ldots,n$, where $p=(p_1,\ldots,p_n)'$ now denotes a vector of relative prices. To see this, note that homogeneity implies that we may write $E^i(P_1,\ldots,P_{n+1}) = E^i(P_1/P_{n+1},\ldots,P_n/P_{n+1},1) = f^i(p_1,\ldots,p_n)$ so long as $p_{n+1} \neq 0$. Walras' Law implies that, if $f^i=0, i=1,\ldots,n$, then $f^{n+1}=0$. Now, in this case, a competitive equilibrium exists if $\hat{p} > \theta$ and

$$f^i(\hat{p})=0, \qquad i=1,\ldots,n. \tag{5.43}$$

The stability of the 'tatonnement' price adjustment process is now considered. The 'auctioneer', or 'market manager', announces prices to the agents who then respond with their supply/demand plans, no trade actually taking place. Excess demands are computed and prices are assumed to be adjusted; upwards (downwards) for excess demand (supply). The question of stability is the question of whether, given an arbitrary (non-negative) starting price vector, such a process converges to the equilibrium price vector $\hat{p}$. The process of adjustment is assumed to be continuous and to satisfy $\dot{p}_i \gtreqless 0$ as $f^i(p) \gtreqless 0$ for the normalized system (or as $E^i(P) \gtreqless 0$ for the non-normalized system).

5.7.2(a) *Local stability*

Consider the normalized system, with an assumed unique equilibrium $\hat{p} > \theta$ and excess demand functions $f^i(p) \in C_0, i=1,\ldots,n$.

The adjustment process is taken to be given by

$$\dot{p}_i = g_i(f^i(p)) \qquad i=1,\ldots,n, \tag{5.44}$$

where $g_i \in C_1$ and $g_i(0)=0, g_i'(.)>0$. A Taylor series approximation of the right-hand side about $\hat{p}$ yields the local approximation system

$$\dot{p} = BA(p-\hat{p}), \tag{5.45}$$

where $B=[\delta_{ij}b_i]$ is an $n \times n$ diagonal matrix with $b_i = g_i'(f^i(\hat{p}))$ and A is an $n \times n$ matrix with elements $a_{ij} = \partial f^i(\hat{p})/\partial p_j$. Local asymptotic stability of (5.44) follows if BA in (5.45) is a stable matrix (by Theorem 5.19).

Consider the simple case where $n=2$, so that $BA = \begin{pmatrix} b_1 a_{11} & b_1 a_{12} \\ b_2 a_{21} & b_2 a_{22} \end{pmatrix}$; BA is stable if $\operatorname{tr} BA < 0$ (i.e. $b_1 a_{11} + b_2 a_{22} < 0$) and $|BA| > 0$ (i.e. $b_1 b_2 |A| > 0$). For example, suppose the own price effects are negative and the cross price effects positive but weaker; then a_{11}, $a_{22} < 0$, a_{12}, $a_{21} > 0$ and $|A| > 0$. Clearly $\operatorname{tr} BA < 0$ and $|BA| > 0$, so the system is locally stable.

In general, the coefficients $b_i (>0)$ may take any positive value, hence the concern is with so-called D-stability.

DEFINITION 5.14 A matrix A is said to be **D-stable** if BA is stable for *any* positive diagonal matrix B.

It can be shown that a necessary condition for D-stability is that the leading principal minors of A alternate in sign; $(-1)^i D_i \geq 0$, $i=1,\ldots,n$. This follows from a direct application of the Routh Hurwitz conditions. The following theorem gives some sufficient conditions for D-stability (see Hahn, 1982).

THEOREM 5.30 A matrix is D-stable if:

(1) there exists a positive diagonal matrix C such that $CA + A'C$ is negative definite,

(2) A is a Metzler matrix whose principal minors alternate in sign: $(-1)^n D_i \geq 0$, $i=1,\ldots,n$ – or equivalently, that A is Metzlerian and there exists an $h \in \mathbb{R}^n$ such that $Ah < 0$ (see Hahn, 1958),

(3) A has a negative dominant or negative quasidominant diagonal,
(4) A is quasinegative definite,
(5) A is symmetric and negative definite.

Naturally, if A is D-stable, it is stable (the special case where $B=\mathrm{I}$). (5) follows immediately from (4) which in turn follows from (1) by setting $C=\mathrm{I}$. (1) follows from Theorem 5.14 which states that BA is stable if there exists a symmetric positive definite matrix M such that $MBA+(BA)'M$ is negative definite. Let $M=CB^{-1}$. B is a positive diagonal matrix, hence $B=B'$, $B^{-1}=(B^{-1})'$. C is a positive diagonal matrix, hence so is M. Thus M is symmetric and positive definite. Also $MBA+(BA)'M=CA+A'C$ since $M=CM^{-1}$, and $CA+A'C$ is negative definite, by assumption.

Conditions (2), (3) are more interesting since they have greater economic content. A **Metzler matrix** is one having $a_{ii}<0$ all i, and $a_{ij}>0$, $i\neq j$. Since $a_{ij}=\partial f^i/\partial p_j$, A is the Jacobian of the set of excess demand functions, and the assumption of gross substitutability, $a_{ij}>0$, $i\neq j$, gives a Metzler matrix. The proof is omitted (see Hahn, 1958).

A matrix has a **negative dominant diagonal** if

$$a_{ii}<0 \qquad i=1,\ldots,n,$$

$$|a_{ii}|>\sum_{\substack{j\\ j\neq i}}|a_{ij}|, \qquad i=1,\ldots,n.$$

It is said to have a **negative quasi-dominant diagonal** if

$$a_{ii}<0 \qquad i=1,\ldots,n,$$

and there exist positive numbers w_i, $i=1,\ldots,n$ such that

$$w_i|a_{ii}|>\sum_{\substack{j\\ j\neq i}}w_j|a_{ij}| \qquad i=1,\ldots,n. \tag{5.46}$$

Diagonal dominance (dd) is a special case of quasidiagonal dominance (qdd) with $w_i=$constant, all i. Note that if A has a qdd, so does BA (suppose $w_1,\ldots,w_n$ have been found such that (5.46) holds for A. Then the numbers $w_i/b_{ii}, i=1,\ldots,n$ will satisfy (5.46) for the matrix BA). Hence it suffices to show that, if a matrix A has a qdd, it is stable. First we show that A must be non-singular. Suppose A is singular; then there exists a positive constant vector $c \neq \theta$ such that $c'A=\theta$(i.e $\sum_j c_j a_{ij}=0$, all i). Choose a k such that $c_k=\max_j c_j$, and write $c_k a_{kk}=-\sum_{j/j\neq k} c_j a_{kj}$. Taking absolute values, $c_k|a_{kk}|<\sum_{j/j\neq k} c_j|a_{kj}|$. This contradicts (5.46), hence A must be non-singular (hence there are no zero eigenvalues). Now consider the characteristic equation $|A-\lambda I|=0$. Note that

$$|a_{ii}-\lambda|=+\{(a_{ii}-\mathrm{Re}(\lambda))^2+\mathrm{Im}(\lambda)^2\}^{1/2}<|a_{ii}-\mathrm{Re}(\lambda)|.$$

Now, suppose $\mathrm{Re}(\lambda)>0$. Then, since A is qdd, $a_{ii}<0$, so $|a_{ii}|<|a_{ii}-\lambda|$ and $|a_{ii}|<|a_{ii}-\mathrm{Re}(\lambda)|$. Hence, if A is qdd, then so is $A-\lambda I$; this contradicts the fact that a matrix with a qdd is non-singular. Hence $\mathrm{Re}(\lambda)\not>0$ and, since $\mathrm{Re}(\lambda)\neq 0$, we must have $\mathrm{Re}(\lambda)<0$. Hence A is a stable matrix.

Diagonal dominance has a clear economic interpretation; own price effects are negative and dominate, in absolute value, the sum of the cross-price effects. Naturally, a matrix with a qdd need not be Metzlerian.

5.7.2(b) *Global stability*

(1) THE 3-COMMODITY CASE

This provides an excellent illustration of the phase diagram technique (see Arrow and Hurwitz, 1958, also Takayama, 1985). Assume gross substitutability, homogeneity (of degree zero for excess demand functions), and Walras' Law holds. The system is

$$\dot{P}_i=b_iE^i(P_0,P_1,P_2) \qquad i=0,1,2.$$

This is normalized by setting $P_0=1$, so leaving

$$\dot{P}_i=b_iE^i(1,P_1,P_2) \qquad i=1,2, \tag{5.47}$$

where $b_i>0$. Walras' Law implies that, if $E^1=E^2=0$, then $E^0=0$, hence the adjustment process need only be considered for markets 1 and 2. The slopes of the isokines in P_1P_2 space are given by:

(i) $\dot{P}_1=0 \Rightarrow E^1=0$. Total differentiation yields $\mathrm{d}P_2/\mathrm{d}P_1|_{\dot{P}_1=0}= -E_1^1/E_2^1>0$ (by gross substitutability; $E_i^i<0$, $E_j^i>0$, $i \neq j$). Here, $E_j^i \equiv \partial E^i/\partial P_j$.

(ii) $\dot{P}_2=0 \quad \mathrm{d}P_2/\mathrm{d}P_1|_{\dot{P}_2=0}= -E_1^2/E_2^2>0$.

Euler's Theorem implies that (since $P_0=1$)

$$E_0^1+P_1E_1^1+P_2E_2^1=0,$$

$$E_0^2+P_1E_1^2+P_2E_2^2=0.$$

By gross substitutability, E_0^1, $E_0^2>0$, hence

$$P_1E_1^1+P_2E_2^1<0 \qquad \text{and} \quad P_1E_1^2+P_2E_2^2<0.$$

These inequalities imply that

$$-E_1^2/E_2^2<P_2/P_1<-E_1^1/E_2^1,$$

hence the $\dot{P}_1=0$ isokine is the steeper of the two in Figure 5.13. Hence equilibrium, if it exists, is unique. The directions of motion in Figure 5.13 may be established as follows.

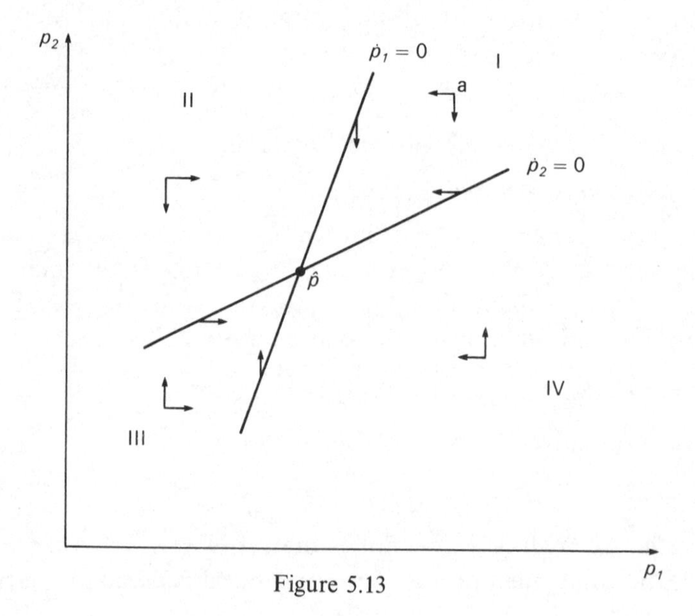

Figure 5.13

Consider a point such as a in Figure 5.13; $\text{sign}(\dot{P}_1|_a) = \text{sign}(E_1^1 \Delta P_1)$. Now $E_1^1 < 0$ and $\Delta P_1 > 0$, so $\dot{P}_1|_a < 0$. $\text{Sign}(\dot{P}_2|_a) = \text{sign}(E_2^2 \Delta P_2)$, where $E_2^2 < 0$ and $\Delta P_2 > 0$, so $\dot{P}_2|_a < 0$. The directions of motion in the other regions are then immediate. Notice that trajectories from regions II, IV either converge on the equilibrium or enter regions I or III, and that trajectories in regions I, III are clearly trapped within these regions and must converge on the equilibrium point $\hat{P}$. Hence the equilibrium must be globally stable.

(2) THE n-COMMODITY CASE

Again, we consider the non-normalized system $\dot{P} = BE(P)$ where $E(P)$ is the vector of excess demands and B, the diagonal matrix of speeds of adjustment. For simplicity, excess demand functions are assumed single valued and continuously differentiable. Again, gross substitutability, homogeneity and Walras' Law are assumed to hold. Consider the candidate Liapunov function

$$V(P) = \sum_i (P_i - \hat{P}_i)^2 / b_i. \tag{5.48}$$

This satisfies conditions [1]–[3] of Theorem 5.22, hence we need merely be concerned with [4]:

$$\begin{aligned} \dot{V}(P) &= 2 \sum_i (P_i - \hat{P}_i) \dot{P}_i / b_i \\ &= 2 \sum_i (P_i - \hat{P}_i) E^i(P) b_i / b_i = 2(P - \hat{P})' E(P). \end{aligned} \tag{5.49}$$

From Walras' Law, $P'E(P) = 0$ for all $P > \theta$ so

$$\dot{V}(P) = -2\hat{P}'E(P). \tag{5.50}$$

Notice that homogeneity implies that, if $\hat{P}$ is an equilibrium, any scalar multiple, $\alpha\hat{P}$ (with $\alpha > 0$), is too, i.e. $E(\hat{P}) = \theta \Rightarrow E(\alpha\hat{P}) = \theta$. V will be a Liapunov function if $\dot{V}(P) < 0$ for all $P \neq \alpha\hat{P}$ (for α some positive scalar). To show this, we first show that $\hat{P}'E(P) > 0$ for all $P \neq \alpha\hat{P}$ and so that $\hat{P}'E(P)$ attains a minimum at $P = \alpha\hat{P}$. The proof is by contradiction. Suppose $\hat{P}'E(P)$ attains a minimum at P^*, $P^* \neq \alpha\hat{P}$ (it is assumed that $P \gg 0$ and that $E(P)$ is bounded). A necessary condition for this is that

$$\frac{\partial}{\partial P_j}[\hat{P}'E(P)]=\sum_i \hat{P}_i \partial E^i(P^*)/\partial P_j=0 \text{ for all } j. \tag{5.51}$$

Now, find the kth price such that, letting z denote this relative price,

$$z=P_k^*/\hat{P}_k \geqslant P_i^*/\hat{P}_i \text{ all } i \tag{5.52}$$

(clearly at least one inequality must be strict). Then

$$0=E^k(\hat{P})=E^k(z\hat{P})>E^k(P^*) \tag{5.53}$$

since $z\hat{P}_k=P_k^*$ and $z\hat{P}_i \geqslant P_i^*$ all i with strict inequality for some i, $i \neq k$. Differentiating Walras' Law, $P^{*\prime}E(P^*)=0$, with respect to P_k yields

$$\sum_i P_i^* \, \partial E^i(P^*)/\partial P_k = -E^k(P^*)>0. \tag{5.54}$$

Divide (5.52) by z (where $z>0$) and note that $\hat{P}_i \geqslant P_i^*/z$, then

$$\sum_i \hat{P}_i \partial E^i(P^*)/\partial P_k \geqslant -E^k(P^*)/z>0. \tag{5.55}$$

(5.55) contradicts (5.51), hence $\hat{P}'E(P)$ attains its global minimum at $P=\alpha\hat{P}$, and so $\hat{P}'E(P)>0$ for all $P \neq \alpha\hat{P}$. Thus V is a Liapunov function satisfying the conditions of Theorem 5.22 and the system is uniformly globally asymptotically stable. The price vector converges on the equilibrium ray $\alpha\hat{P}$.

This completes the proof of global stability for this simple general equilibrium model. Naturally, a similar proof is possible for the normalized system. For a survey of these and more recent developments, see Hahn (1982).

(3) NON-UNIQUENESS

We have already considered the conceptual problem of stability when there are non-unique equilibria in Section 5.6.4. Here, we add only the following observations (which refer to the excess demand system $\dot{p}=E(p)$, where the speeds of adjustment have been normalized to unity for all goods).

DEFINITION 5.15 (Regularity) The economy is said to be **regular** if, for every equilibrium $\hat{P}$ such that $E(\hat{P})=\theta$, $\hat{P} \gg \theta$, there is an open neighbourhood, $N(\hat{P}) \in \mathbb{R}^n_{++}$ such that $|E_p(P)| \neq 0$ for all $P \in N(\hat{P})$

Debreu (1974) shows that almost all economies with continuously differentiable excess demand functions are regular (see also Hahn, 1982, Dierker, 1982). If $|E_p(P)| \neq 0$, the equilibria must be isolated. Expanding $E(P)$ around $\hat{P}$, we have $E(P) = E_p(\psi \hat{P} + (1-\psi)P)(P - \hat{P})$, where $0 < \psi < 1$. Hence, if $P \neq \hat{P}$, for all $P \gg \theta$, and $P \in N(\hat{P})$, then $E(P) \neq \theta$. Equilibrium $\hat{P}$ is isolated. With regular economies, the concern is with establishing stability rather than quasistability. For non-isolated equilibria, see Section 5.6.4. The assumption of regularity will play a role in Section 5.8.3.

Non-uniqueness of solutions to the differential equations may also be a problem. Thus, suppose the excess demand adjustment equation is

$$\dot{P}_i = G^i(E^i(P)) \qquad i = 1, \ldots, n. \tag{5.56}$$

So far, we have assumed $G^i \in C_1$, $G^i(0) = 0$ and $\partial G^i / \partial E^i > 0$. Now, suppose at a zero price, $p_i = 0$, we have $E_i < 0$ (excess supply). Then p_i will go negative. However, assuming **free disposability**, then prices should not go negative (excess supply can always be freely dumped). The G^i function must therefore be amended. The most natural formulation is to let

$$G^i(E^i) = 0 \qquad \text{if} \quad P_i = 0 \qquad \text{and} \quad E_i < 0$$

$$G^i(E^i) = g^i(E^i(P)) \quad \text{otherwise,}$$

where $g^i \in C_1$, $g^i(0) = 0$ and $\partial g^i / \partial E^i < 0$ as before.

This clearly prevents P_i going negative (see Arrow, Block and Hurwicz, 1959). The drawback is that G^i is no longer Lipschitz continuous. To see this, consider a $\bar{P} > 0$ such that $\bar{P}_i = 0$ and a $P_a \gg 0$, arbitrarily close to $\bar{P}$. Then $G^i(P_a) = g^i(E^i(P_a))$, whilst $G^i(\bar{P}) = 0$. Lipschitz continuity requires $|G^i(E^i(P_a)) - G^i(E^i(\bar{P}))| / \|P_a - \bar{P}\| \leqslant k$, where $k > 0$ is some finite constant. Now,

$$\lim_{\substack{\|P_a - \bar{P}\| \to 0 \\ P_a \gg 0}} |G^i(E^i(P_a)) - G^i(E^i(\bar{P}))| = \lim_{\substack{\|P_a - \bar{P}\| \to 0 \\ P_a \gg 0}} |G^i(E^i(P_a))|$$

$$= |g^i(E^i(\bar{P}))| > 0.$$

Hence no finite constant k exists.

In early work, this problem was either ignored or the adjustment process modified (in an economically rather arbitrary manner) so as to avoid the discontinuity (see Nikaido and Uzawa, 1960, Arrow and Hahn, 1971). More recently, it has been shown that systems with such discontinuities on the right-hand sides may still have unique solutions determined by the initial condition P_0 and continuous in P_0 (see Henry, 1972, 1973, 1974, Champsaur, Dreze and Henry, 1977).

5.7.3 A partly rational expectations macromodel

In this simplified version of Buiter and Miller (1981), the impact upon the macroeconomy of an unanticipated oil shock is examined. The model is a good example of a saddlepoint conditionally stable equilibrium.

The structural equations are as follows:

$$M = k(Y+R) - \lambda r + p \qquad (LM) \tag{5.57}$$

$$Y - \bar{Y} = \delta(e-p) + \psi R_\infty \qquad (IS), \tag{5.58}$$

$$\dot{p} = \theta(Y - \bar{Y}) + \pi, \tag{5.59}$$

$$\pi = \dot{M} \text{ (constant)}, \tag{5.60}$$

$$\dot{e} = r - r^*, \tag{5.61}$$

where k, λ, δ, ψ, $\theta > 0$ are given constants, $M = \ln$ (nominal money stock) (exogenous), $p = \ln$ (price level), $Y = \ln$ (real non-oil domestic income), $\bar{Y} = \ln$ (full employment non-oil income) (exogenous), R = oil production expressed as a fraction of real non-oil income (exogenous), R_∞ = permanent income equivalent of $R(R_\infty < R)$ (exogenous), r = domestic interest rate, r^* = foreign interest rate (exogenous), $e = \ln$ (exchange rate), π is the trend or 'core' inflation rate taken as equal to the rate of growth of money supply $\dot{M}$ in (5.60) (exogenous). Equation (5.59) is the Phillips curve.

Equation (5.57), the LM curve, describes the condition for equilibrium in the money market. The demand for nominal money balances depends on real income (non-oil and oil), the opportunity cost of holding money (r) and the price level, p. Equation (5.58), the

IS curve, gives the condition for equilibrium in the non-oil production market (i.e. its long-run value; with a finite stock, this would be zero). Buiter and Miller include the real interest rate as a determinant; for simplicity, this is ignored here, as is the dependence of domestic price level on exchange rate and the price of foreign goods. Money supply is assumed exogenous and (v) is the condition for equilibrium in the foreign exchange market (covered interest parity). In this model, prices are sluggish, but not so the exchange rate; the foreign exchange market is the 'rational' part of the model; 'speculators are smart in a world of high-speed capital movements'. This market is characterized by 'risk neutral speculators endowed with perfect information, and infinitely elastic covered interest arbitrageurs'. As a result, the interest differential must equal the forward discount on the currency, and the latter must accurately forecast the change in the spot rate.

It is convenient to work with the transformed variables $u = M - p$ and $c = e - p$. Thus u is a measure of the liquidity in the economy, whilst c is a measure of international competitiveness. Then, using (5.59)–(5.61),

$$\dot{u} = \dot{M} - \dot{p} = -\theta(Y - \bar{Y}), \tag{5.62}$$

$$\dot{c} = \dot{e} - \dot{p} = r - r^* - \theta(Y - \bar{Y}) - \pi, \tag{5.63}$$

r and Y may be obtained in terms of c, u and the other exogenous variables from (5.57) and (5.58):

$$Y = \delta c + \psi R_\infty + \bar{Y}, \tag{5.64}$$

$$r = \{kR - u + k(\delta c + \psi R_\infty + \bar{Y})\}/\lambda, \tag{5.65}$$

hence the system becomes

$$\begin{pmatrix} \dot{u} \\ \dot{c} \end{pmatrix} = \begin{pmatrix} 0 & -\theta\delta \\ -1/\lambda & (k/\lambda - \theta)\delta \end{pmatrix} \begin{pmatrix} u \\ c \end{pmatrix} + \begin{pmatrix} -\theta\psi R_\infty \\ \psi R_\infty (k/\lambda - \theta) - \pi - r^* + k(\bar{Y} + R)/\lambda \end{pmatrix}. \tag{5.66}$$

The equilibrium point is given by $\dot{u}=0$, $\dot{c}=0$

$$\begin{pmatrix}\hat{u}\\ \hat{c}\end{pmatrix}=\begin{pmatrix}k(\bar{Y}+R)-\pi-r^*\\ -\psi R_\infty/\delta\end{pmatrix}. \tag{5.67}$$

Since $|A|=-\theta\delta/\lambda<0$, the system exhibits saddlepoint stability. The isokines are

$$\dot{u}=0 \Rightarrow c=-\psi R_\infty/\delta \tag{5.68}$$

$$\dot{c}=0 \quad c=\frac{u}{(k-\lambda\theta)\delta}-\frac{\psi R_\infty}{\delta}+\frac{(\pi-r^*)\lambda-k(\bar{Y}+R)}{(k-\lambda\theta)\delta}. \tag{5.69}$$

The slope of the latter depends on the sign of $k-\lambda\theta$. If θ, the coefficient in the Phillips curve (5.59), is small, it will have positive slope; if large, it will have negative slope. The slope of the stable arm may be signed by examining the eigenvector associated with the negative root (see Theorem 5.15). Let $-\xi(\xi>0)$ denote this root. Then the associated eigenvector $v=(v_1,v_2)'$ is determined by

$$\begin{pmatrix}\xi & -\theta\delta\\ -1/\lambda & (k/\lambda-\theta)\delta+\xi\end{pmatrix}\begin{pmatrix}v_1\\ v_2\end{pmatrix}=\begin{pmatrix}0\\ 0\end{pmatrix}$$

hence $\xi v_1-\theta\delta v_2=0$, so v_1, v_2 have the same sign and the stable arm has positive slope. Figure 5.14 illustrates the case where $k-\theta\lambda<0$);

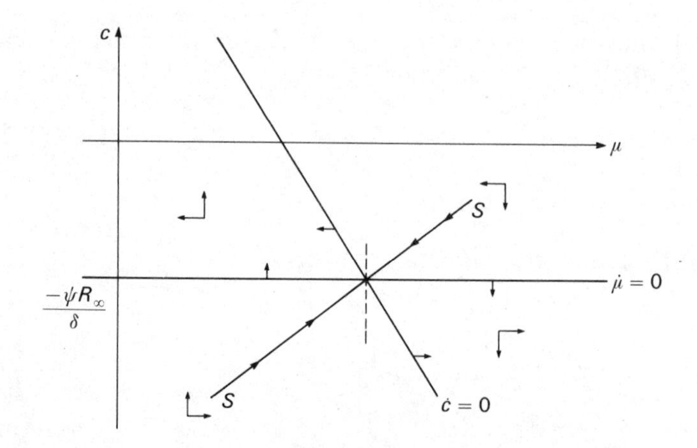

Figure 5.14

any trajectory left of the dotted line above the $\dot{u}=0$ line (or right of the dotted line below the $\dot{u}=0$ line) cannot converge to the equilibrium.

Now, the exchange rate, e, and hence competitiveness, c, is a forward looking variable. It is assumed free to make discrete jumps in response to 'news' (we can interpret e and c as right-hand time derivatives). The price level is sluggish in view of (5.59). If we make the assumption that the economy is stable, then it must be on the stable path SS (we could assume a 'terminal condition whereby, at some point in the future, the authorities step in to stabilize the economy – it then becomes rational for a point on the stable manifold to be chosen – rather than on one of the unstable, 'explosive' paths (or speculative bubbles)' – see Begg (1982), pp. 39–40 or Minford and Peel (1983), pp. 23–6.

Consider the case where, initially, there is no oil ($R_\infty = R = 0$). The system moves along SS in Figure 5.15. Now, suppose there is an unanticipated oil discovery which is undepletable and yields a constant output per period, $R = R_\infty$. Suppose the economy initially is at equilibrium A. The consequence of the discovery is an

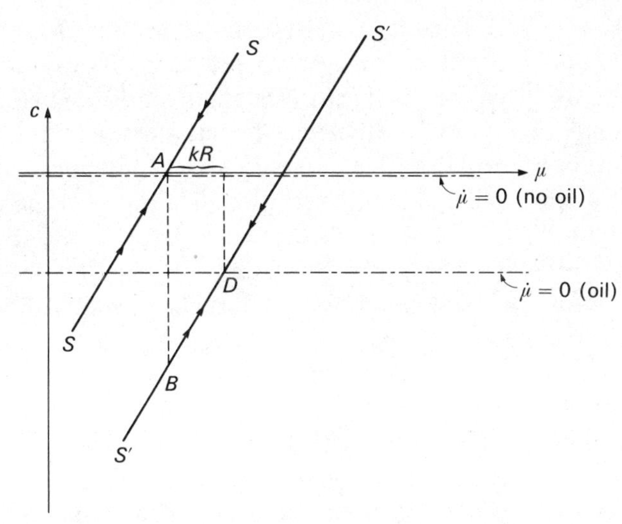

Figure 5.15

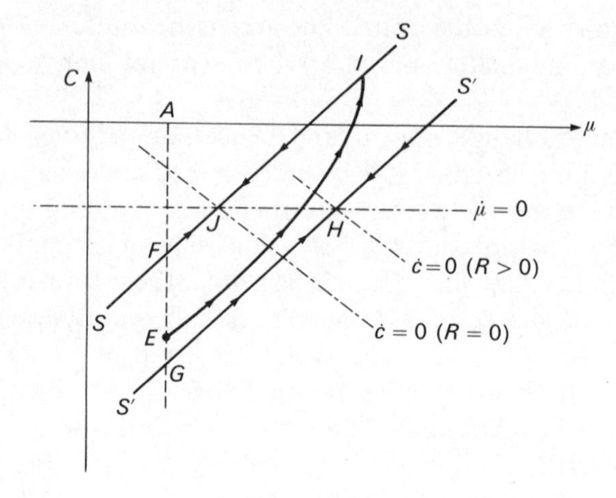

Figure 5.16

instantaneous drop in c from A to B, whereafter, adjustment proceeds along $S'S'$ to point D.

Now consider the case where the oil is depletable. The permanent income equivalent of the now finite stock of oil, R_∞, is that constant percentage of non-oil income which has the same present value as the finite flow of oil. Thus $R_\infty > 0$. But after some time, T, say, the oil runs out, and thereafter, $R=0$. Two saddle paths exist; one for $R>0(S'S')$ and one for $R=0(SS)$ (see Figure 5.16). The long-run equilibrium is the point J. However, for an interval of time T, the system follows the trajectory associated with saddlepoint $H(R>0)$. If, initially, the system is at A, then there is a jump to the point E intermediate between F, G such that, in a time T, the trajectory reaches the long-run stable arm associated with saddlepoint J (at I). At this point, the system changes to that in which $R=0$; the trajectory thereafter is thus IJ. The overall trajectory, EIF, is unique; that is, a continuum of solutions start with initial condition on the line FG and reach SS, but only one does this in precisely time T. Thus, there is an initial overshoot in the real exchange rate by the vertical distance EA, and then an undershoot by the vertical distance between I and A, before approach to the long-run equilibrium J.

The intuition is as follows: In the long run, the balance of payments constraint must be satisfied, hence production of oil must be at the expense of other, domestically produced goods (the 'Dutch disease', see Forsythe and Kay, 1980). This brings about a long-run decline in competitiveness (AJ). In the short run, oil output R raises the transactions component of money demand by kR. If oil was non-depletable, the liquidity would eventually have to accommodate this increase (and the path would be AGH in Figure 5.16); this is why $S'S'$ is kR units to the right of SS.

5.8 DYNAMICAL SYSTEMS: TOPOLOGICAL CONSIDERATIONS

The stability of given equilibria has so far been the primary focus. Such equilibria may be thought of as singularities of a vector field on a given manifold. In many cases, the nature of the manifold imposes restrictions upon the possible nature of the dynamical system, and a study of the topology of the system may help in establishing existence, uniqueness, non-uniqueness and possibly the number and type of equilibria. Closed orbits and limit cycles may also be analyzed, although the techniques here are essentially limited to planar ($x \in \mathbb{R}^2$) dynamical systems. The following treatment does not presume any great knowledge of differential topology, and proofs are generally omitted; for further details, see Milnor (1965), Hirsch and Smale (1974) and Guillemin and Pollak (1974).

5.8.1 Dynamical systems: introduction

Given the differential equation system

$$\dot{x} = f(x), \qquad x \in X \subset \mathbb{R}^n, \tag{5.70}$$

the way the system develops over time is given by the solution, denoted θ. The dynamical system consists of a specification of X and θ. θ is usually only known implicitly by the differential equations. However, assuming unique solutions exist, and, given an initial

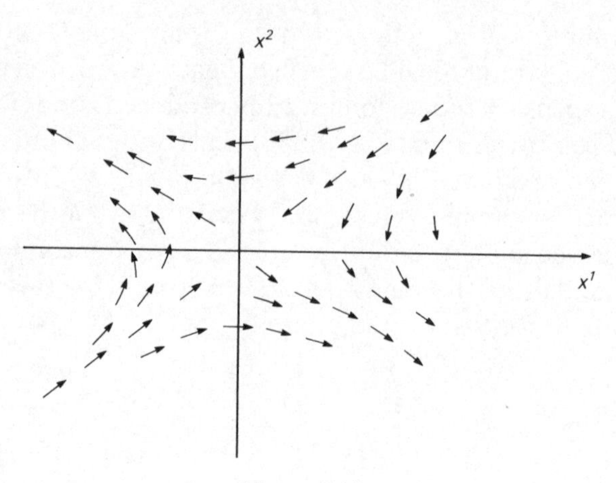

Figure 5.17

condition x_0, (5.70) defines a particular solution $\theta(t|x_0)$. Since $f(\theta(t))$ is a tangent vector to $\theta(t)$, a useful way to think of the trajectory 'flow' over time is by considering the vector field; $f(x)$ indicates the direction of motion at that point $x \in X$ (see Figure 5.17 where the tangent vectors are viewed as attached to each $x \in X$ for the case of a saddlepoint).

We now establish more precisely the concept of a limit cycle (see Figure 5.18). As usual, in what follows, X denotes an open connected set.

DEFINITION 5.16 Given a system $\dot{x} = f(x)$, $x \in X \subset \mathbb{R}^n$, for which a solution $\theta(t|x)$ exists for all $x \in X$. Then $x_0 \in X$ is an **ω-limit point** of $x \in X$ if there is a sequence $\{t_n\}$, $t_n \to \infty$ such that $\lim_{n \to \infty} \theta(t_n|x) = x_0$. x_0 is an **α-limit point** if $-\infty$ replaces $+\infty$ in the above.

DEFINITION 5.17 The set of all ω-limit points x_0 (resp. α-limit points) of a point $x \in X$ is the **ω-limit set** $L_\omega(x)$ (resp. **α-limit set** $L_\alpha(x)$).

From these definitions, it follows that, if $\hat{x}$ is asymptotically stable, it is the ω-limit set of all points in some neighbourhood of $\hat{x}$ – *if* $\hat{x}$ is

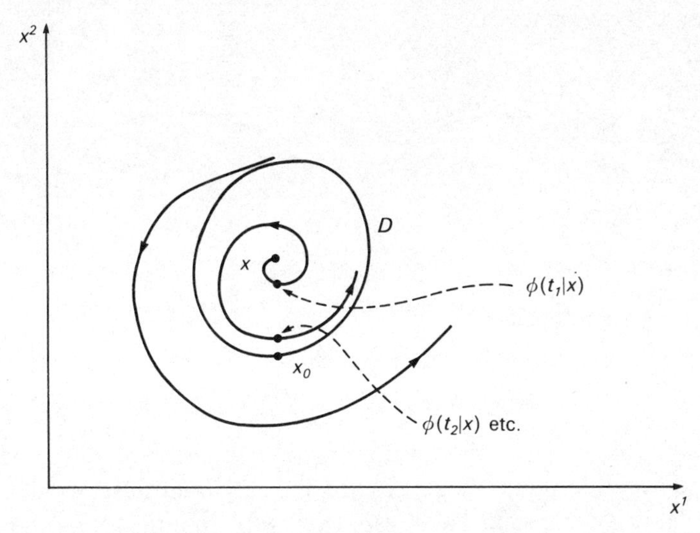

Figure 5.18

globally stable, then $\hat{x}$ is the ω-limit set of all points $x \in X$. If $\hat{x}$ is globally unstable, it is the α-limit set for all $x \in X$.

If D denotes a closed orbit (as in Figure 5.18), then clearly D is the α- and ω-limit set for all points $x \in D$. Clearly, a limit set must be connected.

DEFINITION 5.18 A trajectory D is a **closed orbit** if x is not an equilibrium for all $x \in D$ and for any given $x_0 \in D$ at time t_0, $\theta(t|x_0) = x_0$ for some $t > t_0$.

DEFINITION 5.19 A closed orbit D is a **limit cycle** if $D \subset L_\omega(x)$ or $D \subset L_\alpha(x)$ for some $x \notin D$. It is termed an **ω-limit** or **α-limit cycle** respectively.

Note that a limit set need not be an orbit or an equilibrium as the following example (due to Hirsch and Smale, 1974, p. 240) makes clear.

EXAMPLE 5.16 In Figure 5.19, there are two unstable spirals (x, z)

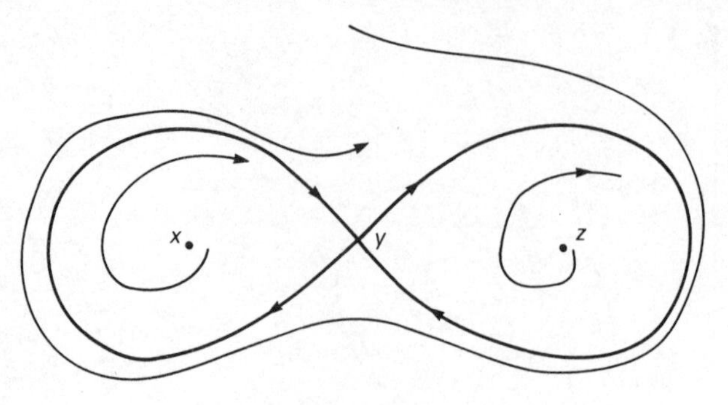

Figure 5.19

and a saddlepoint (y). The figure '8' is the ω-limit set of all points outside it. The right half (resp. left) is the ω-limit set of all points inside it, except the equilibrium z(resp. x)

Note the difference between a limit cycle and a closed orbit; a limit cycle is a closed orbit to which other trajectories converge as $t \to +\infty$ or $t \to -\infty$. Clearly, a closed orbit need not be a limit cycle (as with the centre for the linear system).

5.8.2 The system $\dot{x} = f(x)$, $x \in X \subset \mathbb{R}^2$

In this section, the concept of an index is developed, and the problem of detecting equilibria and limit cycles discussed. We assume familiarity with the concept of a line integral of a continuous real valued function f along a piecewise smooth curve, C, between two points A, B (in $\mathbb{R}^2$ in this section). The curve (or path of integration) is oriented by choosing a direction along C (see Figure 5.20: Note that C may be a closed curve). The opposite direction, or orientation, is denoted by $-C$. The curve may be given a parametric representation by defining arc length, τ, from a point such as A along the curve. The coordinates then depend upon τ. Thus, if we consider a real valued continuous function $f(y, z)$ where y, $z \in \mathbb{R}$, the line

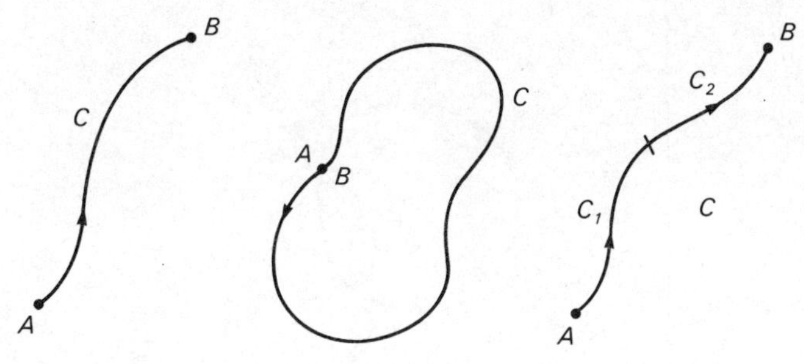

Figure 5.20

integral may be denoted as

$$\int_c f(y,z)d\tau = \int_{\tau_0}^{\tau_1} f[y(\tau), z(\tau)]d\tau.$$

When C is a closed curve, the notation $\oint_c$ is often adopted. Given real valued, continuous functions f, g, and C piecewise smooth, we have the following properties (see Figure 5.20):

$$\int_c f\,d\tau = -\int_{-c} f\,d\tau,$$

$$\int_c kf\,d\tau = k\int_c f\,d\tau,$$

$$\int_c (f+g)d\tau = \int_c f\,d\tau + \int_c g\,d\tau,$$

$$\int_c f\,d\tau = \int_{c_1} f\,d\tau + \int_{c_2} f\,d\tau.$$

The differential equation system to be examined is

$$\begin{aligned} \dot{y} &= f(y,z), \\ \dot{z} &= g(y,z), \end{aligned} \tag{5.71}$$

where $f, g \in C_1$. For any point (y,z) for which $f(y,z) \neq 0$, we may write

$$\frac{dz}{dy} = \frac{\dot{z}}{\dot{y}} = h(y,z) \quad \text{where } h(y,z) = g(y,z)/f(y,z).$$

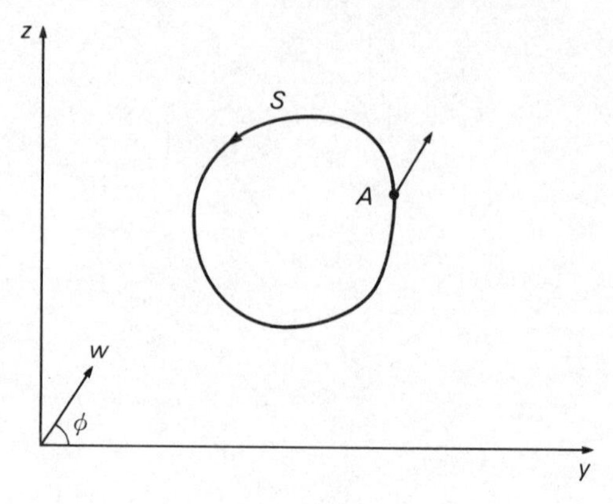

Figure 5.21

Let S denote a closed curve of non-equilibrium points as in Figure 5.21. The direction of the trajectory passing through any point (y, z) may be represented by an arrow at (y, z) in the direction $w = (\dot{y}, \dot{z})$. The angle θ is given by

$$\tan\theta = h(y, z). \tag{5.72}$$

Thus, in principle, the value of θ can be computed at all points on S. If we arbitrarily choose a direction, or orientation, for the path around S (conventionally anticlockwise) and take an arbitrary starting point A, the value of θ on completing the circuit will differ from its initial value by $2\pi n$ (n integer). n is termed the **index of the circuit *S*** and is denoted I_S. Let τ measure arc length from A in the given direction, and let τ_S denote the length of the complete circuit. Then, on S, (y, z) are given parametrically by $(y(\tau),\ z(\tau))$. Differentiating (5.72) with respect to τ gives

$$(1+h^2)\frac{\mathrm{d}\theta}{\mathrm{d}\tau} = h_1\frac{\mathrm{d}y}{\mathrm{d}\tau} + h_2\frac{\mathrm{d}z}{\mathrm{d}\tau} \tag{5.73}$$

since

$$\frac{\mathrm{d}}{\mathrm{d}\tau}(\tan\theta) = \frac{\mathrm{d}}{\mathrm{d}\tau}(\sin\theta/\cos\theta) = (1+\tan^2\theta)\mathrm{d}\theta/\mathrm{d}\tau.$$

The change in θ can thus be calculated as $\int_0^{\tau_s}(\mathrm{d}\phi/\mathrm{d}\tau)\mathrm{d}\tau$ and the index as

$$I_S = \frac{1}{2\pi}\int_0^{\tau_S} \frac{\mathrm{d}\theta}{\mathrm{d}\tau}\mathrm{d}\tau = \frac{1}{2\pi}\int_0^{\tau_S}\left\{\frac{h_1\dfrac{\mathrm{d}y}{\mathrm{d}\tau}+h_2\dfrac{dz}{\mathrm{d}\tau}}{(1+h^2)}\right\}\mathrm{d}\tau$$

$$= \frac{1}{2\pi}\oint_S \frac{h_1\mathrm{d}y + h_2\mathrm{d}z}{(1+h^2)}. \tag{5.74}$$

The notation $\oint$ denotes a line integral evaluated anticlockwise around a closed curve.

The line integral in (5.74) will be more compactly expressed as $\oint_S \mathrm{d}\theta$ when all that is required is an expression for the change in θ over the circuit S. It is often easier to compute the index indirectly rather than by using (5.74).

DEFINITION 5.20 I_S is called the index of the circuit S(S being described anti-clockwise).

EXAMPLE 5.17 Consider the system $\dot{x} = Ax$ where $\hat{x} = \theta$ and $A = \begin{pmatrix} 0 & -1 \\ -1 & 0 \end{pmatrix}$ (a saddlepoint equilibrium). Let S be the unit circle (Figure 5.22). Using the parameterization $x_1 = \cos\gamma$,

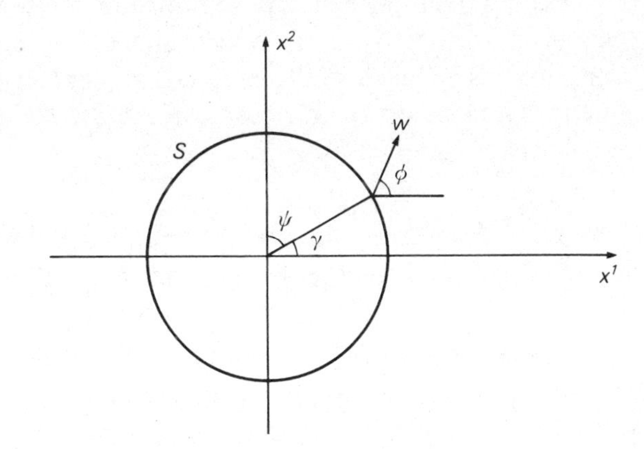

Figure 5.22

$x_2 = \sin\gamma$, then $\tan\theta = \dot{x}_2/\dot{x}_1 = x_1/x_2 = \tan\psi \Rightarrow \theta = \psi = \pi/2 - \gamma$. As γ goes from $0 \to 2\pi$, θ goes from $\pi/2 \to 0 \to -3\pi/2$. Hence $I_S = -1$.

It turns out that the index has several interesting properties; these are stated without proof in the form of a theorem (see Hirsch and Smale, 1974). It is assumed that both f and f_x are continuous, so the existence and uniqueness theorems (for solutions) hold.

THEOREM 5.31

(i) If, on and inside a closed circuit S, there is no equilibrium, then $I_S = 0$.
(ii) Consider two closed circuits S and S' with S' everywhere within S. If, between S and S', there is no equilibrium point, then $I_S = I_{S'}$.
(iii) In a region containing a single equilibrium point, any closed circuit S surrounding the point generates the same index.
(iv) If S is a closed circuit with respect to which the field vectors are directed either inwards (resp. outwards), then $I_S = 1$.

From Theorem 5.3(iii), the concept of the index of a point derives, since the value of the index no longer depends upon the particular path S (variation in S has no effect so long as the variation does not include further singular points).

DEFINITION 5.21 Given an isolated equilibrium point $\hat{x}$, there exists a neighbourhood $N(\hat{x})$ in which there are no other equilibrium points. The **index of the equilibrium point** $\hat{x}$, denoted $I_{\hat{x}}$, is defined as the index of a closed circuit $S \subset N(\hat{x})$ which encloses $\hat{x}$ (S being described anti-clockwise).

THEOREM 5.32 If the closed circuit S surrounds n equilibria, $E_1, \ldots, E_n$, then $I_S = \sum_{i=1}^{n} I_i$, where I_i is the index of equilibrium E_i.

The index of a closed path is I_S, where $I_S = 1/2\pi \oint_S d\theta$. Consider Figure 5.23; clearly $\oint_S d\theta = \oint_{S_1} d\theta + \oint_{S_2} d\theta$ since $\int_{AA'} d\theta = -\int_{A'A} d\theta$, and hence the additive property of Theorem 5.35; see Figure 5.24 where the integrals over the 'bridges' clearly cancel, hence $I_S = I_{S_1} + I_{S_2}$. The

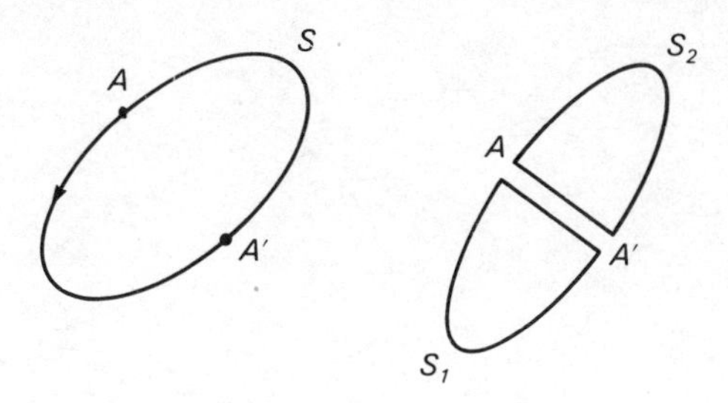

Figure 5.23

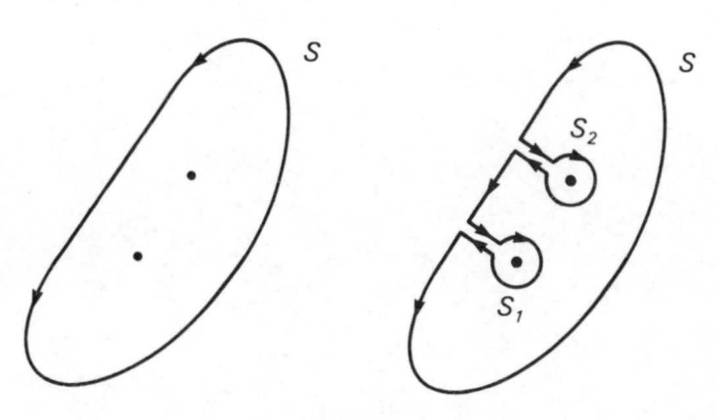

Figure 5.24

index may be computed by examining the vector field as follows:

(a) Let p be the number of times g/f changes sign from $+\infty$ to $-\infty$, and q, the number of times from $-\infty$ to $+\infty$. Then $I_S=(p-q)/2$. Alternatively,

(b) $\tan\theta=0$ when $g=0$ – let p, q be the number of changes in sign$(\tan\theta)$ from $-/+$, $+/-$ respectively; $I_S=(p-q)/2$.

Type of equilibrium	I_S
None	0
Saddle	-1
Centre	$+1$
Spiral (stable or unstable)	$+1$
Node (stable or unstable)	$+1$

Note that, for all types of equilibria, $I_S \neq 0$. Hence, if, for some closed circuit S, the vector field points inward (or outward) at every point on S (so $I_S=1$), then S contains at least one equilibrium point.

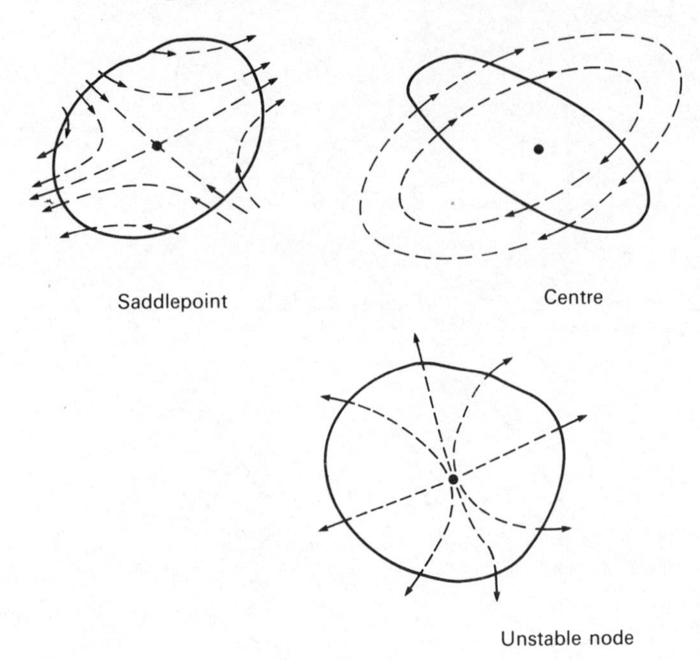

Figure 5.25

This observation extends to $\mathbb{R}^n$ and may be used to establish the existence of general equilibrium (Example 5.21).

Finally, notice how $|f_x(\hat{x})|$ reports on the index of an equilibrium point; if $|f_x(\hat{x})| \neq 0$, the trajectory map of the local approximation system is similar to that of the non-linear system in the neighbourhood of the equilibrium point. Thus, if there is a saddlepoint, $|f_x(\hat{x})| < 0$, whilst, if there is a spiral (stable or unstable) or node (stable or unstable), $|f_x(\hat{x})| > 0$. Thus $I_S = -1, +1$ as $|f_x(\hat{x})| <, > 0$.

Detection of limit cycles

Linear systems with constant coefficients do not exhibit limit cycles. With the non-linear system, in the absence of an explicit solution, an indirect approach to the identification of limit cycles and closed orbits is necessary. Various theorems exist which prove helpful in this respect. Again, it is assumed that $f \in C_1$.

THEOREM 5.33 (Bendixson's negative criterion) The system $\dot{x}=f(x)$, $x \in X \subset \mathbb{R}^2$, where $f(x)=(f^1(x), f^2(x))'$, has no closed paths in its domain X if $f_1^1(x)+f_2^2(x)$ has the same sign for all $x \in X$.

This is a sufficient condition for the non-existence of a closed orbit. It can be applied to many of our earlier examples to show the non-existence of such closed paths.

THEOREM 5.34

(i) If S is a closed orbit, then the index of S is $I_S=1$.
(ii) If S is a closed orbit, the sum of indices of equilibria enclosed by S is $+1$.
(iii) If S is a closed orbit enclosing an open set U, then U contains at least one equilibrium point.

Part (i) of this theorem follows from the definition of the index and that of a closed orbit (the vector field is tangential to S at all points, hence the index is $+1$) – see Figure 5.26. Thus, if a system has a unique equilibrium whose index $=1$, then periodic motions are impossible.

A system of three equilibria comprising one saddlepoint and two stable nodes would satisfy the necessary condition of part (ii) of the theorem. A unique saddlepoint would rule out the possibility of a surrounding limit cycle or closed orbit.

Theorem 5.34(i) and (ii) imply (iii); that is, a closed orbit, if it exists, must contain at least one equilibrium point.

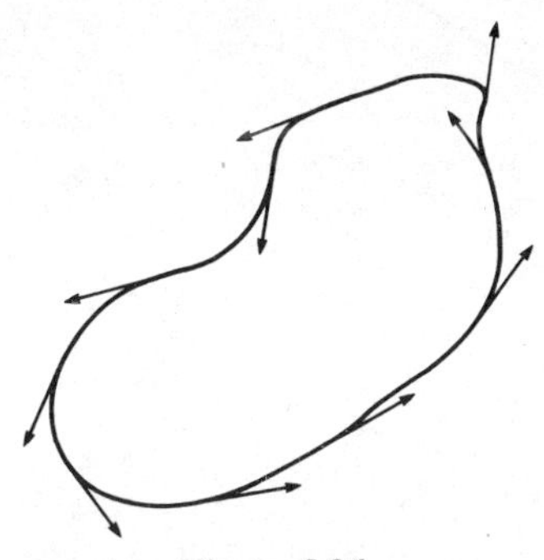

Figure 5.26

Although useful, the above theorems are not particularly powerful tools for the positive identification of closed orbits. The following is perhaps the most empirically useful theorem (see Hirsch and Smale, 1974).

THEOREM 5.35 (Poincaré-Bendixson) A non-empty compact limit set of a planar dynamical system which contains no equilibrium point is a closed orbit.

Suppose there are a finite number of equilibria. Then clearly, every non-empty limit set of such a system must be either a closed orbit – or the union of equilibria and trajectories $\theta(t|x)$ such that $\lim_{t\to\infty}\theta(t|x)$ and $\lim_{t\to-\infty}\theta(t|x)$ are equilibria (Hirsch and Smale, 1974, p. 249) – see, for example, Figure 5.19.

These observations are useful in the following way. If a region W can be found such that some trajectory P lies entirely within W, then either P will be a closed orbit, or limit cycle, or will approach a limit cycle or an equilibrium point (the existence properties essentially imply that a trajectory cannot 'wander about forever' in a closed bounded region). Now, if W contains no equilibria, and P remains in W, then W must contain a closed orbit. One way to identify such a closed orbit is thus to identify such a region. Figure 5.27 illustrates this; if the vector field traps trajectories in W, then there must be a

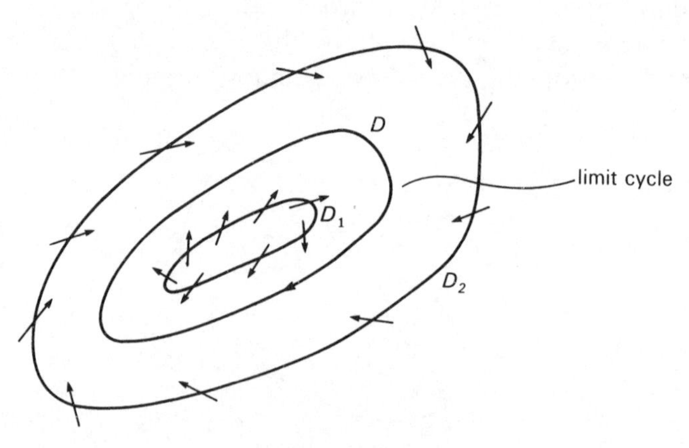

Figure 5.27

closed orbit in W. The problem in practice is that of identifying the closed curves D_1, D_2.

A potentially useful technique for identifying whether a vector field points inwards on curves like D_2, or outwards on D_1, is to use a Liapunov function. Consider the system (5.71) for which $\dot{y}=f(y,z)$ and $\dot{z}=g(y,z)$. Suppose the curve to be considered is denoted D. Define the function $V(x,y)=k$, constant with $k>0$, to describe closed curves such that the outer curves correspond to larger k values. Let $V(z,y)=k_0$ be the curve D in Figure 5.27. Now

$$\frac{\mathrm{d}V}{\mathrm{d}t}=\frac{\partial V}{\partial y}\dot{y}+\frac{\partial V}{\partial z}\dot{z}=V_y f(y,z)+V_z g(y,z). \tag{5.75}$$

If $\dot{V}>0$ on D, the trajectories pass outward (as for D_1 in Figure 5.27) and if $\dot{V}<0$ on D, trajectories pass inward (as in D_2). If $\dot{V}=0$ for a curve, then this is a closed trajectory – it may be a closed orbit or limit cycle.

In practice it is usually difficult to make use of the above technique, although there are several scientific and engineering applications (e.g. Van der Pol and Lienards equations). The following example of a growth cycle turns out to be of the Lotka-Volterra type; a Liapunov approach establishes the existence of a cycle.

EXAMPLE 5.18 (Goodwin's growth cycle model) The original paper uses phase diagram techniques – for this and a more detailed discussion of the economic assumptions and results, see Goodwin (1967), Gandolfo (1980). The assumptions are:
Technical progress; output/man grows exponentially

$$Y/L=y=a\,\mathrm{e}^{bt} \qquad a,b>0. \tag{i}$$

Growth rate of the available work force;

$$N=N_0\mathrm{e}^{\gamma t} \qquad N_0,\gamma>0. \tag{ii}$$

Employed labour is L and there are two factors of production, K, L. The real wage is w. All wages are consumed, all profits reinvested, there is no depreciation and the capital-output ratio $k=K/Y$ is assumed contant. The workers' share of the product is defined as

$$v = w/y. \tag{iii}$$

The capitalists' share, for reinvestment, is thus $1 - v$, hence

$$\dot{K} = (1 - v)Y. \tag{iv}$$

The rate of profit is given by

$$r = \dot{K}/K = \dot{Y}/Y. \tag{v}$$

The employment rate is

$$m = L/N. \tag{vi}$$

The rate of change of wage depends upon the employment rate and may be approximated by a linear function

$$\dot{w}/w = -\alpha + \beta m \qquad \alpha, \beta > 0. \tag{vii}$$

First, a pair of differential equations in v, m are obtained. Note that $\dot{y}/y = b$, $\dot{N}/N = \gamma$. Equations (iii), (vii) give

$$\dot{v}/v = \dot{w}/w - \dot{y}/y = \dot{w}/w - b = -(\alpha + b) + \beta m, \tag{viii}$$

while (vi) yields

$$\dot{m}/m = \dot{L}/L - \dot{N}/N = \dot{L}/L - \gamma. \tag{ix}$$

From (i) and (iv)

$$\begin{aligned} \dot{y}/y = b = \dot{Y}/Y - \dot{L}/L = \dot{K}/K - \dot{L}/L = (1 - v)Y/K - \dot{L}/L \\ = (1 - v)/k - \dot{L}/L. \end{aligned} \tag{x}$$

Substituting (x) into (ix) gives

$$\dot{m}/m = (1 - v)/k - (b + \gamma). \tag{xi}$$

The system is then

$$\dot{v} = v\{-(\alpha + b) + \beta m\} = v(A + \beta m), \tag{xii}$$

$$\dot{m} = m\{(1/k - b - \gamma) - v/k\} = m(B - Cv), \tag{xiii}$$

where $A = -(\alpha + b)$, $B = (1/k) - b - \gamma$, $C = 1/k$.
These are Lotka-Volterra equations; there are equilibria at $(0, 0)$

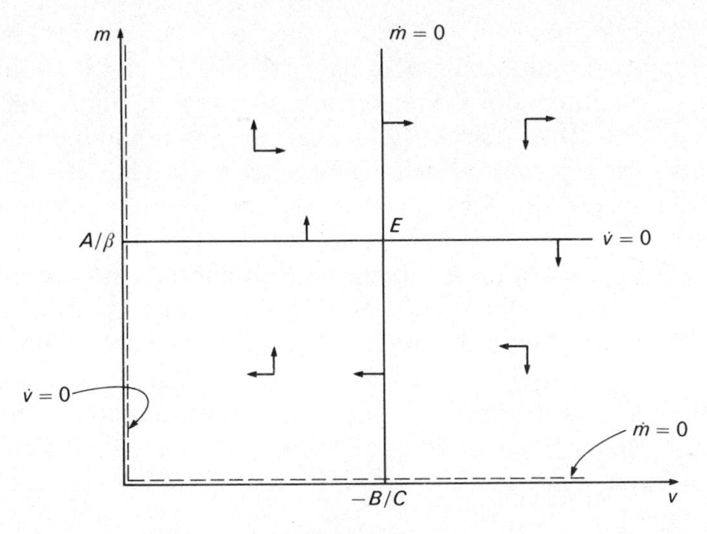

Figure 5.28

and $(-B/C, +A/\beta)$. Phase plane analysis tells us that the trajectories tend to move clockwise – which gives little away about stability (see Figure 5.28).

Consider the function $V(m, v) = k$. Then

$$\dot{V} = V_m \dot{m} + V_v \dot{v} = V_m m(B + Cv) + V_v v(A + \beta m). \tag{xiv}$$

If there is a closed trajectory, then, on such a trajectory, $\dot{V} = 0$. From (xiv), this would imply

$$\frac{vV_v}{-B + Cv} = \frac{mV_m}{(A + \beta m)}. \tag{xv}$$

Noticing the form of (xv), suppose we set this equal to a constant which we can choose to be unity. Thus, integrating (f, g denote arbitrary functions),

$$V_v = -B/v - C \Rightarrow V = -B \ln(v) - Cv + f(m),$$

$$V_m = A/m + \beta \Rightarrow V = A \ln(m) + \beta m + g(v),$$

so

$$V = A \ln(m) - B \ln(v) + \beta m - Cv, \tag{xvi}$$

V here satisfies (xv) and $\dot{V}=0$. Hence V defined by (xvi) is constant on solution curves of the system (xii), (xiii). V attains its global minimum (for the region $m, v>0$) at the equilibrium point $E=(-B/C, A/\beta)$ – since V_v, $V_m=0$ at this point and the Hessian matrix for V is positive definite for all m, $v>0$. Thus $V-V|_E$ is a Liapunov function for the system; the equilibrium E is Liapunov stable.

The next question is whether the equilibrium is a centre or whether limit cycles exist. It may be shown that trajectories are confined to a compact region (see Hirsch and Smale, 1974, Chapter 12), so Theorem 5.35 may be applied. A trajectory is either a closed orbit or approaches such a path – or the equilibrium point. Consider the point $w=(v, m)$, $v, m>0$ such that $w \neq E$ and is not in a closed orbit – then there is a sequence

$$-\infty < \ldots < t_{-1} < t_0 < t_1 < \ldots < +\infty$$

such that a solution $\theta(t_s|w)$ lies on the line $v=-B/C$ and $t_s \to \pm\infty$ as $s \to \pm\infty$. The points $\theta(t_s)$ lie monotonically on the line. Thus, necessarily, $\theta(t_s|w) \to E$ as $s \to +\infty$ or as $s \to -\infty$. This implies a contradiction – since $V(w)$ is constant on the trajectory of w whilst the limit process implies $V(w) \to V(E)$. But $V(E)$ at E is the global unique minimum of V for the region, so $V(w) \neq V(E)$. Hence every trajectory is a closed orbit (see Figure 5.29).

In economic terms, there is a cycle in the employment ratio and in the growth rate of income, as depicted in Figure 5.30. High

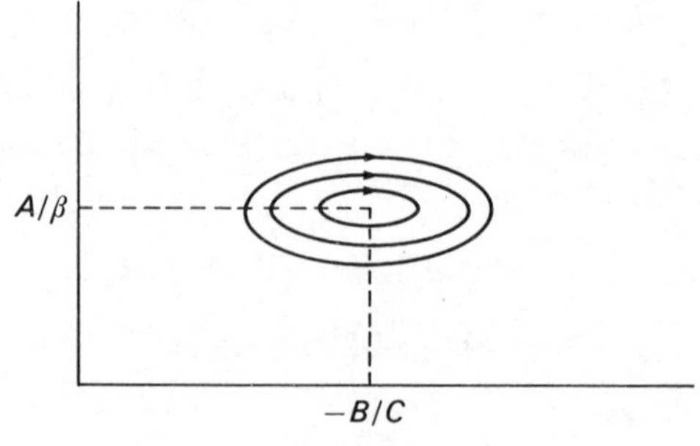

Figure 5.29

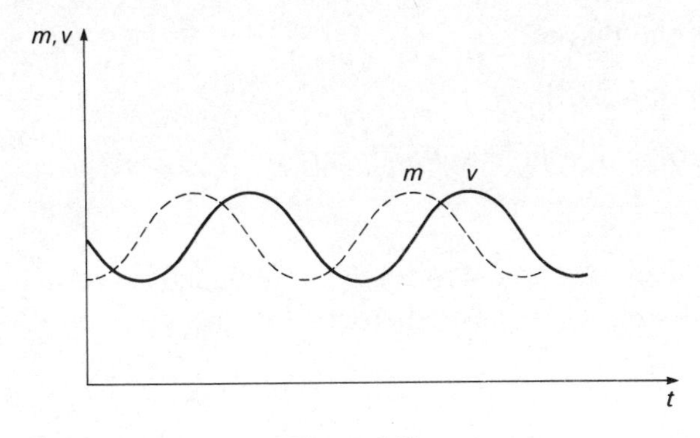

Figure 5.30

profits lead to high growth rates, increasing wages, and so profit rates are squeezed. This then lowers employment, lowers growth rates and output, so restoring profits because productivity exceeds wage rate growth, and so on.

Most economic examples feature constant amplitude cycles only as special cases. That is, the cycles are structurally unstable. The following macroeconomic example (a slightly simplified version of Eckalbar (1985), is interesting in that a limit cycle exists for a much wider range of parameter values than is usual. It also features switching between a pair of linear dynamical systems; it is this switching which enables the existence of a limit cycle, since linear systems do not normally allow such phenomena.

EXAMPLE 5.19 (Inventory fluctuations in disequilibrium macroeconomics, Eckalbar, 1985) There are two sectors; a household and a firm, and three goods; money (M), labour, (L), and output (Q). Output can be stored by the firm (stock level v) but not by the household. w denotes the nominal wage; p,output price; s, the household's commodity (output) demand, and s^e, the firm's expected sales. L is the (fixed) amount of labour the household wishes to supply, and L^d, the amount the firm tries to buy. The amount actually traded, L, is determined by the short side of the market:

$$L = \min[\bar{L}, L^d]. \tag{i}$$

The production function is linear

$$Q = \bar{d}L \tag{ii}$$

and the firm is assumed to try to maintain a fixed ratio of expected sales to stocks; i.e. desired stocks (v^*) are given by

$$v^* = ks^e. \tag{iii}$$

Given initial stock v, and expected demand s^e, the firm wishes to produce

$$Q^* = s^e + v^* - v = s^e(1+k) - v \tag{iv}$$

and hence wishes to employ

$$L^d = (1/d)Q^*.$$

Actual output may be constrained by labour supply, and is given by

$$Q = \min[s^e(1+k) - v, d\bar{L}]. \tag{v}$$

The labour market transaction is assumed to occur before the household decides upon its demand for output, s. The latter is determined by utility maximization subject to a budget constraint. Start of period money balances are $\bar{M}$, labour income is wL so, assuming a Cobb-Douglas utility function, the problem is

$$\underset{M,\,s}{\text{Maximize}}\ As^{\alpha}(M/p)^{1-\alpha} \qquad (0<\alpha<1) \tag{vi}$$

subject to $\bar{M} + wL \geqslant M + ps$.

Beginning of period balances are always $\bar{M}$ by assuming that all the firm's profits are passed to the household as dividends. It is routine to derive the following demand equation from (vi):

$$\text{s} = \beta + \text{cQ}, \tag{vii}$$

where $\beta = \alpha\bar{M}/p$ and $c = w\alpha/pd$. For the firm to be profitable, $0<c<1$.

For simplicity, it is assumed that stocks are sufficient for the firm to supply the household's demand, s. Hence the change in stocks, $\dot{v}$, is given by $Q-s$.

$$\dot{v}=Q-s. \tag{viii}$$

Sales expectations are adaptive (e.g. due to rational least squares learning; Friedman (1979)):

$$\dot{s}^e=s-s^e.$$

Thus we have the dynamical system

$$\begin{aligned}\dot{v}&=-\beta+(1-c)\min[(1+k)s^e-v,d\bar{L}],\\ \dot{s}^e&=\beta+c\min[(1+k)s^e-v,d\bar{L}]-s^e.\end{aligned} \tag{ix}$$

This system is subject to regime (denoted I_i) switching as follows:

$$(I_i)\qquad \begin{bmatrix}\dot{v}\\ \dot{s}^e\end{bmatrix}=A_i\begin{pmatrix}v\\ s^e\end{pmatrix}+B_i\qquad i=1,2, \tag{x}$$

where

$$A_1\equiv\begin{pmatrix}-(1-c) & (1+k)(1-c)\\ -c & c(1+k)-1\end{pmatrix},\qquad B_1\equiv\begin{pmatrix}-\beta\\ \beta\end{pmatrix},$$

$$A_2\equiv\begin{pmatrix}0 & 0\\ 0 & -1\end{pmatrix},\qquad B_2\equiv\begin{pmatrix}(1-c)d\bar{L}-\beta\\ cd\bar{L}+\beta\end{pmatrix}$$

and $i=1,2$ as $(1+k)s^e-v-d\bar{L}>(<)0$. The reader may find it instructive to examine the system when $\operatorname{tr} A_1<0, |A_1|>0$ and the equilibrium E is locally stable. This, and several other cases are examined in detail by Eckalbar. We shall focus on the case where I_1 is unstable, with $\operatorname{tr} A_1>0$. The switching line $W=\{(v,s^e)\colon (1+k)s^e-v=d\bar{L}\}$ divides the phase diagram (Figure 5.31) into two regions $W_1=\{(v,s^e)\colon (1+k)s^e-v<d\bar{L}\}$ (below the W-line) and $W_2=\{(v,s^e)\colon (1+k)s^e-v>d\bar{L}\}$ (above the W-line). In W_1, I_1 is operative and in W_2, I_2. The equilibrium in W_1 is $(\hat{v},\hat{s}^e)=(k\beta/1-c',\beta/1-c)$ (denoted E in Figure 5.31). The associated output level is $\hat{Q}=\hat{s}^e=\beta/1-c$. We have assumed in Figure 5.31 that there is Keynesian underemployment at the

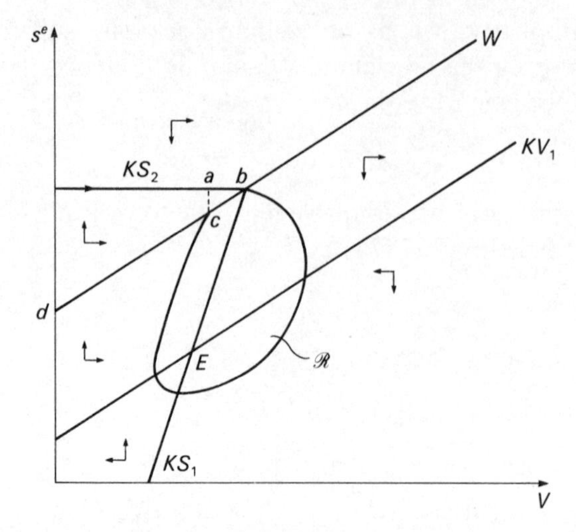

Figure 5.31

equilibrium, i.e. $\hat{Q} < d\bar{L}$. The trajectories are everywhere unique and continuous (why?). The isokines are defined as follows:

$$KS_i = \{(v, s^e) \colon \dot{s}^e = 0, (v, s^e) \in W_i\} \quad i = 1, 2, \tag{xi}$$

$$KV_1 = \{(v, s^e) \colon \dot{v} = 0, (v, s^e) \in W_1\}. \tag{xii}$$

There is no KV_2 isokine. The equations of the isokines are given by setting each equation in (x) equal to 0 respectively, e.g. KV_1 has positive slope of less than unity and is given by

$$KV_1 : s^e = \frac{\beta}{(1+k)(1-c)} + \frac{v}{(1+k)}. \tag{xiii}$$

With tr $A_1 > 0$, KS_1 has positive slope greater than that of KV_1

$$KS_1 : s^e = -\frac{\beta}{c(1+k)-1} + \frac{v}{1+k-1/c}. \tag{xiv}$$

The directions of motion are left as an exercise. Consider a trajectory starting from b, spiralling away from E; it arrives at some point c on W. Clearly the trajectory is thereafter trapped within the region $\mathscr{R} - E$ (since E is unstable). The Poincaré-

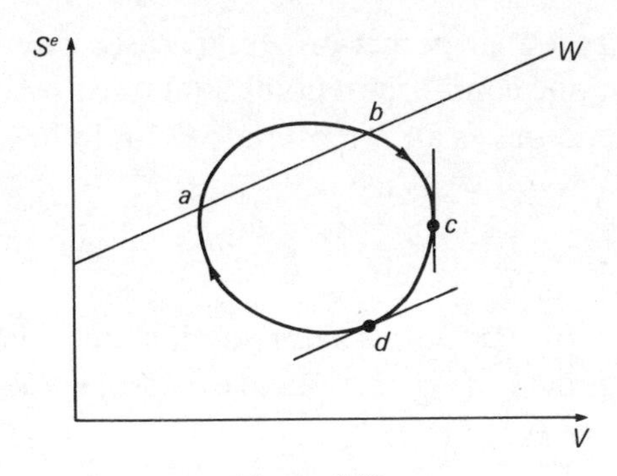

Figure 5.32

Bendixson Theorem (Theorem 5.35) may now be applied: There must be a closed orbit in $\mathscr{R}-E$ which is also a ω-limit cycle. Considering the global behaviour of the system, we require some modification to ensure $\dot{v} \geqslant 0$ as $v \to 0$ and $\dot{s}^e \geqslant 0$ as $s^e \to 0$ (see Section 5.7) – however, ignoring non-negativity issues, it is clear that trajectories eventually cross db and enter the region $\mathscr{R}$. The limit cycle is thus globally stable.

The economic interpretation is that, from (v), peak output Q comes at the point a in Figure 5.32 as the cycle enters full employment that lasts until b; v reaches its maximum at c before the minimum output point d (lines tangent to W are iso-quantity lines, from (v)).

5.8.3 The system $\dot{x} = f(x)$, $x \in X \subset \mathbb{R}^n$

Differential topology has proved quite fruitful in the analysis of existence and uniqueness in general equilibrium models (see Mas Colell, 1985). A full understanding of this recent work requires a knowledge of differential topology we do not wish to presume (for an introduction, see Guillemin and Pollak, 1974). The essential idea is that the nature of the set X involved imposes global restrictions on the possible behaviour of the vector field around the set of equilibrium points.

5.8.3(*a*) *Existence of equilibria*

In two dimensions, Theorem 5.34 indicated that, if the vector field pointed inwards for all points on some closed circuit, then there must be at least one equilibrium point contained within that closed circuit. This theorem has an extension to $\mathbb{R}^n$ as follows (see Spanier, 1966, Varian, 1982). Let

$$D^n = \{x \in \mathbb{R}^n, \|x\| \leqslant 1\}.$$

THEOREM 5.36 If $f: D^n \to \mathbb{R}^n$ is a continuous vector field that points in on the boundary of D^n, then there exists an equilibrium point $\hat{x}$ in D^n such that $f(\hat{x}) = \theta$.

The vector field points in on the boundary of D^n if $x'f(x) < 0$ for all $x \in \mathbb{R}^n$, $\|x\| = 1$. Intuitively, the idea is that, if a region is found from which the trajectory cannot escape, the solution trajectory must converge on a closed orbit or an equilibrium (in $\mathbb{R}^2$, closed orbits always contain equilibria). Varian (1982) gives the following example.

EXAMPLE 5.20 Consider the general equilibrium model where $f(p)$ is the excess demand vector. Assume prices are normalized by setting $\sum_{i=1}^{n} p_i^2 = 1$, so the state space becomes the non-negative orthant of the unit sphere

$$S_+^{n-1} = \{p \in D^n; \|p\| = 1, p_i \geqslant 0\},$$

where p denotes the normalized price vector. Prices adjust according to $\dot{p} = f(p)$. Assuming continuity of f on S_+^{n-1}, Walras' Law ($p'f(p) = 0$ for all $p \in S_+^{n-1}$) and desirability ($f^i(p) > 0$ if $p_i = 0$, all i), then there exists an equilibrium $\hat{p} \in S_+^{n-1}$ such that $f(\hat{p}) = \theta$. To prove this, note that Walras' Law implies $f(p)$ normal to p, and desirability implies that $f(p)$ points in on the boundary of S_+^{n-1} (on the boundary, some $p_i = 0$, so $f^i(p) > 0$ and $\dot{p}_i > 0$ at this p; p_i tends to increase with t). Hence, Theorem 5.36 may be applied to obtain the result. This is illustrated in Figure 5.33 for $p \in S_+^3$.

5.8.3(*b*) *Uniqueness and related questions*

The concept of an index may be generalized to higher dimensions and, as in $\mathbb{R}^2$, the Jacobian of an equilibrium point $\hat{x}$ may often be

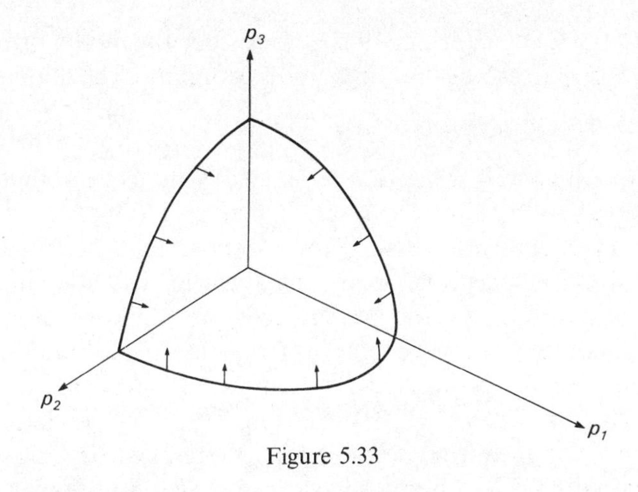

Figure 5.33

used to report on the index of such a point. In $\mathbb{R}^2$, Theorem 5.31 gave $I_S = 0$ if S contained no critical points, whilst Theorem 5.32 gave that a circuit S enclosing n equilibrium points has index $I_S = \sum_{i=1}^{n} I_i$, where I_i is the index of the ith equilibrium point. Furthermore, if the vector field points in, $I_S = 1$ and so $\sum_{i=1}^{n} I_i = 1$ for this case. These ideas generalize to $\mathbb{R}^n$; in particular, it can be shown that, under reasonable technical assumptions about f and X, that, if the vector field points in (or out) everywhere on the boundary of X, then $\sum_{i=1}^{n} I_i = +1$. This is, in fact, a special case of the Poincaré-Hopf Theorem (see Mas Colell, 1985).

The Jacobian of a critical point finds use as follows.

THEOREM 5.37 Let $\hat{x}$ be an equilibrium point of a vector field f on a manifold $X \subset \mathbb{R}^n$. If $|f_x(\hat{x})| \neq 0$, then $\hat{x}$ is an isolated equilibrium point, with index $I_{\hat{x}} = +1$ or -1 as $|f_x(\hat{x})| > 0$ or < 0.

If $|f_x(\hat{x})| = 0$, the problem of computing the index of $\hat{x}$ depends on further topological considerations beyond the scope of this text.

Suppose, in some application, we show that the vector field points in on the boundary, and we discover a saddlepoint ($I = -1$); then there must be at least two other equilibrium points. Thus equilibrium is non-unique.

EXAMPLE 5.21 (Varian, 1982) Consider the model of Example 5.20 for the case of the three good economy. The state space is

$$S^2=\{p\in D^3;\ \|p\|=1, p_i>0\}.$$

Prices adjust according to $\dot{p}=f(p)$ as before. Desirability implies that $f(p)$ points in on the boundary of S^2. Example 5.20 showed that an equilibrium exists. Now suppose that $\hat{p}$ is an unstable equilibrium – unconditionally unstable (not a saddlepoint). Then the local approximation system is also unstable. Since $\hat{p}$ is not a saddlepoint, $|f_p(\hat{p})|>0$. Hence $I_{\hat{p}}=+1$. Hence equilibrium is unique. However, this gives a system as in Figure 5.33. In this case, referring to Figure 5.27, D_2 is S^2 and D_1 is $\hat{p}$. A trajectory $Q\in S^2$, $Q\neq\hat{p}$, is trapped in $S^2\backslash\hat{p}$; therefore, there must be a closed orbit surrounding $\hat{p}$. Furthermore, because the equilibrium is unstable, this cycle of prices over time will be the main feature of the system.

5.9 DIFFERENCE EQUATION SYSTEMS

5.9.1 Introduction

The use of stochastic difference equation systems in economic analysis is now widespread but is beyond the scope of the present text. In this section, consideration of basic existence, solution and stability results for deterministic difference equation systems are considered, exploiting, as far as possible, the similarities between difference and differential equation systems.

Time is divided into periods and variables may vary in value between periods but not within. The period chosen is mathematically arbitrary and is typically chosen based upon economic considerations. The choice of period *vis-à-vis* continuous time analysis may be motivated by economic, analytic or computational reasons. Conclusions can be sensitive to the form of analysis adopted, so the choice is by no means purely cosmetic.

Consider the vector $x(t)$, $t=1,2,\ldots$, where $x(t)=(x_1(t),\ldots,x_n(t))'$ and the integer t indexes time periods. *A* **first difference** is defined as

$$\Delta x(t) = x(t+1) - x(t), \tag{5.76}$$

where Δ is termed the difference operator, sometimes forward difference (to distinguish it from the backward difference $x(t)-x(t-1)$). The second difference is $\Delta^2 x(t) = \Delta[\Delta x(t)] = \Delta[x(t+1)-x(t)] = x(t+2)-2x(t+1)+x(t)$ and in general

$$\Delta^n x(t) = \sum_{r=0}^{n} (-1)^n C_r^n x(t+n-r), \tag{5.77}$$

where $C_r^n = n!/r!(n-r)!$, $r=0,\ldots,n$ are binomial coefficients. The description and terminology for difference equations parallels that for differential equations. Thus, for example, with $x(t) \in \mathbb{R}$,

$$\Delta^n x(t) = f[t, x(t), \Delta x(t), \ldots, \Delta^{n-1} x(t)] \tag{5.78}$$

is a non-linear nth order difference equation (and so on). A solution to (5.78) clearly exists if f is single valued since (5.77) may be used to write (5.78) in the form

$$x(t+n) = g[t, x(t), x(t+1), \ldots, x(t+n-1)]. \tag{5.79}$$

Given initial conditions $x(0), \ldots, x(n-1)$, the solution can be found by repeated direct ennumeration (a process easily handled by computer for particular solutions).

The general first order non-linear difference equation system is

$$\Delta x(t) = f[x(t), t] \tag{5.80}$$

and the linear first order system is

$$\Delta x(t) = A(t)x(t) + u(t), \tag{5.81}$$

where $x(t) \in \mathbb{R}^n$, $u(t) = (u_1(t), \ldots, u_n(t))' \in \mathbb{R}^n$, $f = (f', \ldots, f^n)'$, and $A(t) = \{a_{ij}(t)\}$, $(i,j = 1, \ldots, n)$.

Higher order difference equations can be reduced to first order by suitable choice of variables. Thus, (5.78) may be written as (5.79) and then, setting

$$z_1(t) = x(t),$$

$$z_2(t) = z_1(t+1) = x(t+1),$$

$$\ldots$$

$$z_n(t) = z_{n-1}(t+1) = x(t+n-1),$$

where $x(t)$, $z_i(t) \in \mathbb{R}$. The equivalent first order system to (5.79) is

$$\Delta z_i(t) = z_{i+1}(t) - z_i(t), \qquad i = 1, \ldots, n-1,$$

$$\Delta z_n(t) = g[t, z_1(t), z_2(t), \ldots, z_n(t)] - z_n(t).$$

5.9.2 The linear difference equation system

No general analytic solution exists for non-linear difference equation systems, but the solution to linear autonomous systems parallels that for $\dot{x} = Ax$ – as might be expected. The system

$$\Delta x(t) = Ax(t) \tag{5.82}$$

is now considered. A is an $n \times n$ real non-singular matrix which is assumed to possess distinct eigenvalues (the somewhat more complicated case where there are repeated roots is not discussed in view of the approximation Theorem 5.9). In explicit form, (5.82) may be written as

$$x(t+1) = (A+I)x(t), \tag{5.83}$$

hence, given an initial condition $x(0) = x_0$, the solution is

$$x(t) = (A+I)^t x_0. \tag{5.84}$$

This solution is not particularly revealing of the qualitative structure of the solution, so the diagonalization procedure is adopted; $A = P\Lambda P^{-1}$, where Λ is the diagonal matrix $\{\delta_{ij}\lambda_i\}$ and P is the matrix of column eigenvectors. Since $A = P\Lambda P^{-1}$, then $A + I = P(\Lambda + I)P^{-1}$ and $(A+I)^t = P(\Lambda + I)^t P^{-1}$ so the solution may be written in the form

$$x(t) = P(\Lambda + I)^t P^{-1} x_0.$$

Writing $k = P^{-1} x_0$, this becomes

$$x(t) = P(\Lambda + I)^t k = \sum_{i=1}^{n} k_i (1 + \lambda_i)^t. \tag{5.85}$$

This is in fact the general solution and clearly parallels that for the linear differential equation system. In exactly the same way, if the original system had been non-homogeneous, so that

$$\Delta x(t)=Ax(t)+D, \tag{5.86}$$

where $D \in \mathbb{R}^n$ is a column vector of constants, then, assuming $|A| \neq 0$, a particular (equilibrium) solution is

$$x(t)=-A^{-1}D, \tag{5.87}$$

so a general solution to (5.86) is given by the sum of general solution to the homogeneous system plus the particular solution (5.87) i.e.

$$x(t)=P(\Lambda+I)^t k-A^{-1}D. \tag{5.88}$$

5.9.3 Equilibrium and stability for linear systems

The concepts of equilibrium and stability are identically those given for differential equations (except that Δx replaces $\dot{x}$). Thus $\hat{x}=\theta$ is the equilibrium point for (5.82) since it entails $x(t)=\theta$ for all t.

It is convenient to define the matrix B by $B \equiv A+I$ and to work with (5.82) in the form

$$x(t+1)=Bx(t). \tag{5.89}$$

The eigenvalues of B, denoted μ_i, $i=1,\ldots,n$ are related to those of A by

$$\mu_i=1+\lambda_i \qquad i=1,\ldots,n.$$

THEOREM 5.38 The equilibrium point $\hat{x}=\theta$ for the system $x(t+1)=Bx(t)$, where B is an $n \times n$ real non-singular matrix, is uniformly asymptotically globally stable if, and only if,

$$|\mu_i|<1 \qquad i=1,\ldots,n,$$

where $|\mu_i|$ denotes the modulus of μ_i.

The system is stable, therefore, if the roots μ_i lie within the unit circle in the complex domain. Necessary and sufficient conditions for Theorem 5.38 to hold have been developed (see Samuelson, 1947, Chipman, 1950). Space precludes a detailed consideration of these conditions. We simply state the implied conditions for all cases up to where B is 4×4. Note that, in what follows, $|z|$ denotes the modulus of z if z is a scalar, and the determinant of z if z is a matrix; the context makes this clear.

THEOREM 5.39 (Necessary and sufficient conditions) Necessary and sufficient conditions for stability of the difference equation $x(t+1)=Bx(t)$, where B is an $n \times n$ real constant matrix with characteristic equation $\mu^n + d_{n-1}\mu^{n-1} + \ldots + d_0 = 0$:

[1] $n=1$

$x(t+1)=bx(t)$ is stable if, and only if, $|d_0|<1$

Equivalently, if $|b|<1$.

[2] $n=2$

$x(t+1)=Bx(t)$ is stable if, and only if,

$$1+d_1+d_0>0, \tag{i}$$

$$1-d_1+d_0>0, \tag{ii}$$

$$d_0<1. \tag{iii}$$

(i), (ii) are equivalent to

$$1+d_0>|d_1|$$

Equivalently, $|B|<1$ and $1+|B|>$ 1 trace $B|$.

[3] $n=3$

$x(t+1)=Bx(t)$ is stable if, and only if,

$$1+d_1-|d_2+d_0|>0,$$

$$1-d_1+d_2d_0-d_0^2>0,$$

$$d_1<3.$$

[4] $n=4$

$x(t+1)=Bx(t)$ is stable if, and only if,

$$d_0<1,$$

$$3+3d_0-d_2>0,$$

$$1+d_2+d_0-|d_3+d_1|>0,$$

$$(1-d_0)(1-d_0^2)-d_2(1-d_0)^2+(d_3-d_1)(d_1-d_3d_0)>0.$$

Proof. See Farebrother (1973).

There is little symmetry in these conditions, and it is difficult to extract much intuition from problems where $n>2$. To illustrate the use of the above conditions, consider the following simple cobweb model where B is 2×2.

EXAMPLE 5.22 (A cobweb model) For agricultural produce, the delay between the production decision and bringing the produce to market may be up to a year or more. For products which are perishable or expensive to store, the farmer must accept whatever price he can get in the market. In making his production decision, the farmer must therefore form an expectation of what the future price will be. Assume then that

$$D(t)=a_1+b_1p(t), \tag{i}$$

$$S(t)=a_2+b_2p^e(t), \tag{ii}$$

where $S(t)$, $D(t)$, $p(t)$, $p^e(t)$ denote, respectively, demand, supply, price and expected price of the good at time t, and a_1, a_2, b_1, b_2 are constants. Assume market equilibrium

$$D(t)=S(t) \tag{iii}$$

and that price expectations follow the adaptive scheme

$$p^e(t)=p(t-1)+\alpha\{p(t-1)-p(t-2)\}. \tag{iv}$$

Thus prices are expected to follow a trend if $\alpha>0$, or to reverse it if $\alpha<0$. Substituting (i), (ii), (iv) into (iii) and rearranging yields the second order difference equation

$$p(t)+\gamma_1p(t-1)+\gamma_2p(t-2)+\gamma_3=0, \tag{v}$$

where $\gamma_1=-b_2(1+\alpha)/b_1$, $\gamma_2=\alpha b_2/b_1$, $\gamma_3=(a_1-a_2)/b_1$.

This may be reduced to first order by introducing $z(t)$ such that

$$z(t)=p(t-1), \tag{vi}$$

so (v) may be written as

$$p(t)=-\gamma_1 p(t-1)-\gamma_2 z(t-1)-\gamma_3. \tag{vii}$$

Equations (vi), (vii) form a pair of first order difference equations

$$\begin{pmatrix} p(t) \\ z(t) \end{pmatrix} = \begin{pmatrix} -\gamma_1 & -\gamma_2 \\ 1 & 0 \end{pmatrix} \begin{pmatrix} p(t-1) \\ z(t-1) \end{pmatrix} + \begin{pmatrix} -\gamma_3 \\ 0 \end{pmatrix} \tag{viii}$$

The equilibrium point is $\hat{p}=\hat{z}=(a_2-a_1)/(b_2-b_1)$. Thus, the matrix B, in terms of the original parameters, is

$$B=\begin{pmatrix} \dfrac{b_2(1+\alpha)}{b_1} & \dfrac{-\alpha b_2}{b_1} \\ 1 & 0 \end{pmatrix}.$$

Suppose demand and supply have their usual slopes; $b_1<0$, $b_2>0$, and write $D=b_2/b_1<0$. From Theorem 5.39, the necessary and sufficient conditions for global asymptotic stability are that

$$|B|=\alpha D<1 \tag{ix}$$

and

$$1+\alpha D-|D(1+\alpha)|>0. \tag{x}$$

(a) *Suppose* $\alpha>0$ (Extrapolative expectations).
(ix) is satisfied. (x) gives $D>-1/(1+2\alpha)$. This is a restriction on the ratio of the gradients, and is more restrictive than under myopia ($\alpha=0$); that is, stability is less likely if expectations are extrapolative.

(b) *Suppose* $\alpha<0$ (Regressive expectations).
Manipulation of (ix), (x), yields the results:

for $0>\alpha\geqslant -1/3$: $\quad D>-1/(1+2\alpha)$,

for $-1/3\geqslant\alpha$: $\quad D>1/\alpha$.

Thus, if $\alpha<-1$, the condition is more restrictive than under myopia, if $\alpha=1$, they are the same, and, if $0>\alpha>-1$, then the condition is less restrictive. For further discussion of various types of cobweb model, see Gandolfo (1980).

CHAPTER 6

INTRODUCTION TO DYNAMIC OPTIMIZATION AND THE CALCULUS OF VARIATIONS

6.1 INTRODUCTION

Sections 6.1–6.3 provide an introduction to Control theory and the Calculus of Variations. The theory of the latter is developed in Sections 6.4–6.12. Chapter 7 then deals with Control theory. As we shall see, Calculus of Variations problems can usually be transposed into the Optimal Control format and solved with little (if any) additional difficulty and the latter is better able to deal with the inequality constraints so often manifest in economic applications. Indeed, the trend in economic analysis is undoubtedly towards the Control theory format although the Calculus of Variations continues to find application. However, the variational approach is, perhaps, the more accessible and it also provides a useful way of introducing Optimal Control results.

Intertemporal constrained optimization problems can in principle be handled by the techniques of non-linear programming developed in Chapter 2 – if change can be modelled as occurring at discrete time intervals. The control vector in each period may be given a time subscript, and, with a finite horizon, T, the problem may be cast in the form

$$\underset{x_1,\ldots,x_T}{\text{Maximize}}\ f(x_1,\ldots,x_T) \tag{6.1a}$$

$$\text{subject to} \qquad g(x_1,\ldots,x_T) \geqslant \theta. \tag{6.1b}$$

Most intertemporal problems feature special kinds of relationships, particularly between variables in consecutive time periods (stock and flow relationships), and this usually allows more to be said about the nature of solutions than would otherwise be the case. Notice that, for a given time horizon, the smaller the time period, the more subscripted variables must be defined. In the limit, 'continuous time', x_t has to be specified for all t, $0 \leqslant t \leqslant T$. The problem thus becomes one of choosing a control function defined over this time interval, typically so as to maximize the value of an integral objective function. In the limit then, the choice set is no longer finite, but, rather infinite dimensional (see Takayama, 1985, pp. 419–31 for an introductory discussion).

Consider the following Control problem.

PROBLEM 6.1

$$\underset{u}{\text{Maximize}}\ J(u) = \int_{t_0}^{t_1} f^0(x(t), u(t), t)\mathrm{d}t + B[x(t_1), t_1] \tag{6.2a}$$

subject to

$$x(t) \in X, \qquad X \subset \mathbb{R}^n, \qquad t_0 \leqslant t \leqslant t_1, \tag{6.2b}$$

$$x(t_0) = x_0, \qquad x_0 \in X, \tag{6.2c}$$

$$x(t_1) = x_1, \qquad x_1 \in X, \tag{6.2d}$$

$$u(t) \in U \subset \mathbb{R}^m, \qquad t_0 \leqslant t \leqslant t_1, \tag{6.2e}$$

$$\dot{x} = f(x(t), u(t), t) \qquad t_0 \leqslant t \leqslant t_1. \tag{6.2f}$$

In this Control problem, the concern is with some time interval $t_0 \leqslant t \leqslant t_1$, where **initial** ($t_0$) and **final** ($t_1$) time may be exogeneously fixed, or may be themselves choice variables (e.g. when to begin and when to terminate mining of a resource deposit). Note that t need not be 'time' as such, but commonly is so in economic applications (occasional examples will occur in which t bears some other interpretation; see Example 6.4). The **State vector** for the system is $x(t)=(x_1(t),\ldots,x_n(t))'$. It fully describes the 'status' of the system at time t. The range of the function $x(t)$ is denoted X and is assumed to be an open, connected, subset of $\mathbb{R}^n$. x_0, x_1 are initial and final (or terminal) states, and again may be exogeneously fixed or may have to be chosen as part of the optimization process. For notational simplicity, we shall denote the state trajectory as $x\equiv\{x(t), t_0\leqslant t\leqslant t_1\}$. Thus x denotes a function. Often, in the interests of notational compactness, we shall also write x in place of $x(t)$ – the context makes clear which interpretation is appropriate. The **control vector** is $u(t)=(u_1(t),\ldots,u_m(t))'$ and the associated **control trajectory** is $u\equiv\{u(t),\ t_0\leqslant t\leqslant t_1\}$ and the same notational comments apply. Controls, both here and in Chapter 7, are assumed to be piecewise continuous functions of time (so, there are a possible finite number of points in time at which $u(t^-)\neq u(t^+)$). The control vector may be restricted, as above, to some subset U *of* $\mathbb{R}^m$ which may be required to have properties such as compactness, convexity, time invariance etc. (see later). Clearly the definition of U may have important consequences for the existence of an optimal solution (and its characteristics). The classic example of a control problem is that of a rocket; this is a guidance problem in which there are various controls and state variables might be position, fuel stocks etc.

We shall also assume throughout this and Chapter 7 that our piecewise continuous functions satisfy

$$\theta(t^-)=\theta(\mathrm{t}) \qquad \text{for} \quad t\in[t_0,t_1].$$

This is so that we may avoid adding 'except possibly on a countable subset' to many of our statements about $u(t)$ and $x(t)$ (notice that a perturbation in $u(t)$ at any single point t has no impact upon the state trajectory or the value of the integral; any concept of a piecewise continuous optimal solution being unique can only be 'almost everywhere' in the absence of the above assumption).

DEFINITION 6.1 A continuous state trajectory x which satisfies the state variable constraints on $[t_0,t_1]$ *is* called **admissible**. A piecewise continuous control trajectory u satisfying the control variable constraints on $[t_0, t_1]$ is called an **admissible control**.

The problem is one of selecting an admissible control so as to maximize $J(u)$.

The specification of Problem 6.1 is highly appropriate for many economic applications; for example, firms are often concerned with maximizing discounted profits; $J=\int_0^T \pi(t)\mathrm{e}^{-rt}\mathrm{d}t$ (where $\pi(t)$ denotes instantaneous profits, r is the discount rate and T, the time horizon), individuals (or society) may be concerned with maximizing discounted utility; $J=\int_0^\infty U[C(t)]\mathrm{e}^{-rt}\mathrm{d}t$ where $U[.]$ is the instantaneous utility function, $C(t)$, instantaneous consumption and r, the time preference rate. Note that e^{-rt} is the continuous time discount factor:

$$\lim_{n\to\infty}(1+r/n)^{-nt}=\mathrm{e}^{-rt}$$

(\$1 next year has present value $\$1/(1+r)$ if compounding is annual; it is $\$1/(1+r/2)^2$ if six monthly, and so on).

Equation (6.2f) specifies how the choice of u affects and so determines x. Given a choice of control trajectory, (6.2f) is a system of first order differential equations subject to initial conditions (6.2c). This system will usually have a unique solution (refer to Chapter 5, Section 5.2). We shall call a trajectory x **feasible** if it may be attained through a choice of some admissible u. Clearly, some trajectories x may not be feasible and, indeed, there may be no admissible control trajectory which 'carries' the system between the given endpoints. This is the problem of 'Controllability' and is clearly important to the question of whether an optimal control trajectory exists. Space precludes a detailed consideration of controllability (see Athans and Falb, 1966) and in this and Chapter 7, the principal concern will be with deriving necessary and sufficient conditions for the dynamic optimization problem.

$f^0(x(t), u(t), t)$ is commonly termed the **intermediate function** and $B[x(t_1), t_1]$, the **final** or **bequest function** (the latter for obvious reasons in economic problems of lifetime consumption planning). The bequest function plays an active role only if t_1 and/or $x(t_1)$ are variables to be chosen. When the objective function has the form (6.2a), the problem is referred to as one of **Bolza**. If $J(u)=\int_{t_0}^{t_1} f^0(x,u,t)\mathrm{d}t$ or $J(u)=B[x(t_1),t_1]$, these are termed the problems of **Lagrange** and **Mayer**, respectively. These may appear to be special cases of the former problem but, by suitable definition of variables, the three forms may be shown to be equivalent. Such interrelationships will be outlined *inter alia* as we develop necessary conditions for the above problem(s) – (see Chapter 7, Sections 7.2, 7.3).

Now consider the following Calculus of Variations problem.

PROBLEM 6.2

$$\text{Maximize } J(x)=\int_{t_0}^{t_1} f^0(x(t),\dot{x}(t),t)\mathrm{d}t+B[x(t_1),t_1] \tag{6.3a}$$

subject to

$$x(t)\in\mathbb{R}^n, \qquad t_0\leqslant t\leqslant t_1, \tag{6.3b}$$

$$x(t_0)=x_0, \tag{6.3c}$$

$$x(t_1)=x_1, \tag{6.3d}$$

$$x \text{ piecewise smooth on } [t_0,t_1]. \tag{6.3e}$$

(We shall refer to a function x which is continuous on $[t_0,t_1]$ with piecewise continuous derivative as **piecewise smooth**.)

Comparing (6.3) with (6.2), notice that (6.3) is a special case of (6.2) in which the control trajectory is given by $u=\dot{x}$. Here, as in (6.2), $\dot{x}$ is required to be real and piecewise continuous; however, no other restriction is placed on it, so $U=\mathbb{R}^n$. From the control viewpoint, we 'choose' u. Here, this is $\dot{x}$; however, clearly, this is equivalent to choosing x. In the Calculus of Variations, the problem is viewed as one of choosing an admissible x so as to maximize J.

Both (6.2) and (6.3) involve first order derivatives. However, the formulations are more general than might appear at first sight since suitable transformations allow problems involving higher order derivatives to be reduced to equivalent problems involving only first order derivatives; the method is exactly that used for reducing the order of differential equation systems (see Example 5.4) and is illustrated here in Example 6.9 (page 271).

The bulk of the analysis in this chapter and Chapter 7 is concerned with establishing necessary conditions and sufficient conditions for optimality. As in static optimization, if an optimal solution exists, it must satisfy the necessary conditions; necessary conditions are useful in so far as they characterize the properties of, and reduce the number of, candidate solutions for optimality. However, it should be noted that, unless an optimal solution has been shown to exist, none of the candidates need in fact be optimal. Existence may sometimes be established by appeal to existence theorems (see for example, Seierstad and Sydsaeter, 1987), or by analysis of the specific problem at hand (it is also not uncommon to simply assume that an optimal solution exists). Alternatively, if the problem is such that a solution can be found which satisfies the sufficient conditions, then this will be an optimal solution.

6.2 MATHEMATICAL PRELIMINARIES

Let z denote a trajectory in the calculus of variations problem. The concept of a local maximum requires that we define what it means to say that one trajectory is 'close' to another.

DEFINITION 6.2 If x is a bounded function from $[t_0, t_1]$ into $\mathbb{R}^n$ and is of class C_r on $[t_0, t_1]$, then a norm of x may be defined as, for $K \leqslant r$

$$\|x\|_K = \max_{0 \leqslant i \leqslant K} \{ \sup_{t \in [t_0, t_1]} \|x^{(i)}(t)\| \},$$

where $\|x(t)\|$ denotes the Euclidean norm and $x^{(i)}(t) \equiv d^i x(t)/dt^i, x^{(0)}(t) \equiv x(t)$

We may thus identify $\|x-y\|_K$ as the 'distance' (of order K) between two functions x and y. Distance measures of orders $K=0$ and, in particular, $K=1$, are most important for our purposes. $\|x-y\|_0$

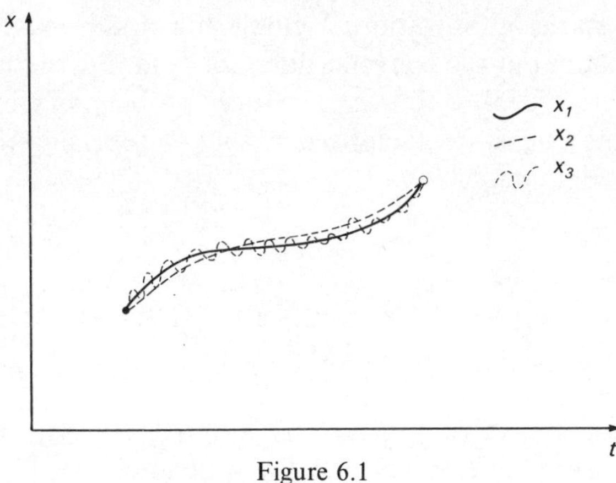

Figure 6.1

focusses upon the point of maximum distance between the functions x, y, whilst $\|x-y\|_1$ considers also the maximum distance between $\dot{x}$ and $\dot{y}$. If functions are close in the sense of $\| \ \|_1$, they are close in the sense of $\| \ \|_0$ but the converse need not hold (in Figure 6.1 above, functions x_1, x_2, x_3 are 'close' in the sense of $\| \ \|_0$ but only the first two are 'close' in the sense of $\| \ \|_1$).

We are now in a position to define the concepts of a maximum; these parallel those defined in Chapter 2.

DEFINITION 6.3 The objective function J has a **global maximum** at x if x is admissible and if $J(x) \geqslant J(z)$ for all admissible z. If the strict inequality holds for all $z \neq x$, then J has a **strict global maximum**.

DEFINITION 6.4 The objective function J has a **local** (or **relative**) **maximum** at x if x is admissible and if there exists a $\mu > 0$ (however small) such that $J(x) \geqslant J(z)$ for all admissible z such that $\|z-x\|_k < \mu$. If the strict inequality holds for $z \neq x$ then J has a **strict local** (or **strict relative**) **maximum**.

If $K=1$, we speak of a **weak** maximum, whilst if $K=0$, it is termed **strong**. Thus, in Definition 6.4, if $K=1$, we are considering a weak local or relative maximum.

As with static optimization, a global maximum must also be a local maximum but the converse does not hold. Furthermore, as in Chapter 2, we deal with maximization problems; minimization problems may be so characterized simply by reversing the sign on the objective function.

6.3 SEPARABILITY AND THE PRINCIPLE OF OPTIMALITY

The integral objective function is additively separable in the following sense: Let $[a,b] \subset [t_0, t_1]$ and denote

$$J(x,a,b) \equiv \int_a^b f^0(x(t), u(t), t)\mathrm{d}t, \tag{6.4a}$$

$$\text{where } x(a) = x_a, \tag{6.4b}$$

$$x(b) = x_b. \tag{6.4c}$$

Then, for given x and u,

$$J(x, t_0, t_1) = J(x, t_0, a) + J(x, a, b) + J(x, b, t_1). \tag{6.5}$$

Let x^* denote the optimal trajectory for Problem (6.4) on the interval $[t_0, t_1]$ between given endpoints x_0, x_1. Then we have

THEOREM 6.1 (The principle of optimality) The optimal solution to (6.4) between endpoints $x_a = x^*(a)$, $x_b = x^*(b)$ on the interval $[a,b] \subset [t_0, t_1]$ is given by x^* on $[a,b]$. That is, any portion of an optimal trajectory is also an optimal trajectory to an appropriate subproblem.

Proof. Suppose that x^* associated with u^* gives a global maximum for J over $[t_0, t_1]$. Hence from Definition 6.3 we have $J(x^*, t_0, t_1) \geqslant J(x, t_0, t_1)$ for all admissible x. The principle of optimality states that, if x^* is optimal for the overall problem on $[t_0, t_1]$, then x^* must also maximize J for all subproblems of type (6.4) on $[a,b] \subset [t_0, t_1]$, where $x(a) = x^*(a)$ and $x(b) = x^*(b)$.

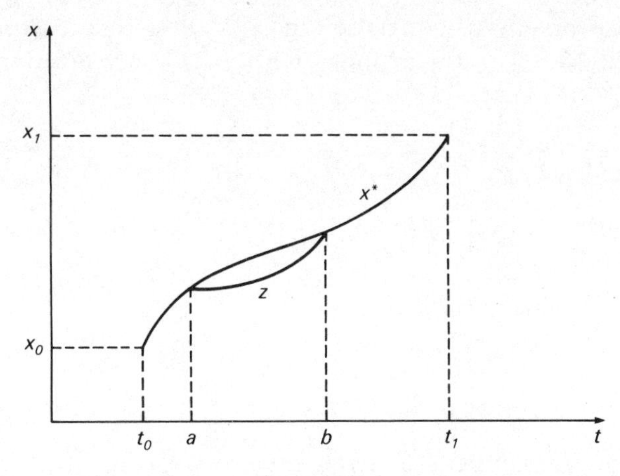

Figure 6.2

This may be proved by contradiction (see Figure 6.2). Let $V(a,b)$ denote the set of admissible variations for the problem (6.4) between the above endpoints and let $x^*(a,b)$, $u^*(a,b)$ denote the trajectories x^*, u^* on (a,b). Clearly $V(a,b)$ is non-empty since $x^*(a,b) \in V(a,b)$. Now suppose that $x^*(a,b)$ is not optimal for the subproblem. Then there exists a $z(a,b) \in V(a,b)$ such that $J(z,a,b) > J(x^*,a,b)$. Given the additive separability of the integral, this implies that

$$J(x^*,t_0,a)+J(z,a,b)+J(x^*,b,t_1)>J(x^*,t_0,t_1). \tag{6.6}$$

Since the trajectory $x^*(t_0,a) \cup z(a,b) \cup x^*(b,t_1)$ is an admissible trajectory on $[t_0,t_1]$, this contradicts the premis that x^* is optimal on $[t_0,t_1]$.

This relationship between subproblem and overall problem solutions finds application at various points in subsequent analysis both in this chapter and Chapter 7.

6.4 CALCULUS OF VARIATIONS: NECESSARY CONDITIONS

We proceed constructively by starting with the simplest, fixed endpoint problem, where $x(t)\in\mathbb{R}$ and t_0, t_1, x_0, x_1 are exogenously given. Extensions to the problem with $x(t)\in\mathbb{R}^n$ are given in Section 6.8.

PROBLEM 6.3

$$\text{Maximize } J(x)=\int_{t_0}^{t_1} f^0(x(t),\dot{x}(t),t)\mathrm{d}t \tag{6.7a}$$

$$\text{subject to } x(t_0)=x_0, \tag{6.7b}$$

$$x(t_1)=x_1, \tag{6.7c}$$

$$x \text{ piecewise smooth on } [t_0,t_1]. \tag{6.7d}$$

Note that $f^0(a,b,c)$ is a given function of three variables. We consider the choice of a function x which satisfies (6.7b)–(6.7d) such that (6.7a) is maximized. It is assumed that $f^0\in C_2$ for all $(x,\dot{x},t)\in\mathbb{R}^3$. Partial derivatives with respect to first, second, third arguments are denoted f_x^0, $f_{\dot{x}}^0$, f_t^0 etc. For example, if $f^0=x^2\dot{x}+t\dot{x}^2+t$, then $f_x^0=2x\dot{x}$, $f_{\dot{x}}^0=x^2+2t\dot{x}$, $f_{\dot{x}\dot{x}}^0=2t$ (etc.).

6.4.1 First order necessary conditions

The approach parallels that of Chapter 2. Let x denote the optimal solution, which is assumed to exist, and z, some other admissible trajectory. Then, for a local maximum, by Definition 6.4, $J(x)\geqslant J(z)$ for all z 'close' to x.

DEFINITION 6.5 The **increment** in $J(x)$, denoted ΔJ, is defined by

$$\Delta J(x,\delta x)=J(x+\delta x)-J(x),$$

where δx is called the **variation** in the function x.

Thus $\delta x(t)\equiv z(t)-x(t)$. If x constitutes a local maximum for J, then $\Delta J\leqslant 0$ for all admissible δx satisfying $\|\delta x\|_k<\mu$ for some $\mu>0$. Variations for which $\|\delta x\|_0<\mu$ are termed strong, whilst variations such that $\|\delta x\|_1<\mu$ are termed weak. For the derivation of the principal necessary conditions, it suffices to consider weak variations. Let

$$z(t)=x(t)+\varepsilon\eta(t), \tag{6.8}$$

so that $\delta x(t)=\varepsilon\eta(t)$. Here, ε is a constant and the displacement function $\eta(t)$ is an arbitrary piecewise smooth function on $[t_0, t_1]$ (hence $z(t)$ is piecewise smooth). Differentiating (6.8) gives $\dot{z}(t)=\dot{x}(t)+\varepsilon\dot{\eta}(t)$ (where, at corners, this equation is understood in terms of the appropriate one-sided derivatives). Clearly, we are dealing with weak variations since, as $\varepsilon\to 0$, $z(t)\to x(t)$ and $\dot{z}(t)\to\dot{x}(t)$. The only restriction otherwise placed on $\eta(t)$ is that $\eta(t_0)=\eta(t_1)=0$ since, to be admissible, $z(t_0)=x_0$ and $z(t_1)=x_1$. For any given arbitrary displacement function $\eta(t)$, variation in ε generates a one parameter family of variations. With η fixed, J may be viewed simply as a function of the parameter ε:

$$J(\varepsilon)=\int_{t_0}^{t_1} f^0(x(t)+\varepsilon\eta(t), \dot{x}+\varepsilon\dot{\eta}(t), t)\mathrm{d}t, \tag{6.9}$$

where the integrand is continuous except at corners of $x(t)$ and $\eta(t)$. Since $z(t)\to x(t)$ as $\varepsilon\to 0$, clearly we must have

$$\Delta J=J(\varepsilon)-J(0)\leqslant 0 \tag{6.10}$$

for $|\varepsilon|$ sufficiently small. Now, $J(\varepsilon)\in C_2$ since $f^0\in C_2$, so necessary conditions are that

$$J'(0)=0, \tag{6.11}$$

$$\text{and } J''(0)\leqslant 0. \tag{6.12}$$

Let the range of integration be divided into a finite number of segments by the points in time at which $\dot{x}(t)$, $\dot{\eta}(t)$ are discontinuous – the theorem (appendix) on differentiation (with respect to ε) under the integral sign may be applied to each of these and the results summed to obtain

$$J'(0)=\int_{t_0}^{t_1} [f_x^0\eta(t)+f_{\dot{x}}^0\dot{\eta}(t)]\mathrm{d}t=0, \tag{6.13}$$

$$\text{and } J''(0)=\int_{t_0}^{t_1} [f_{xx}^0\eta(t)^2+2f_{x\dot{x}}^0\eta(t)\dot{\eta}(t)+f_{\dot{x}\dot{x}}^0\dot{\eta}(t)^2]\mathrm{d}t\leqslant 0, \tag{6.14}$$

where the arguments of f_x^0 etc. are $(x(t), \dot{x}(t), t)$, the elements of the optimal solution.

[1] *Suppose* $x(t) \in C_2$ *on* $[t_0, t_1]$

Before proceeding with the above analysis for the general case, consider the special case where the optimal trajectory $x(t) \in C_2$ (i.e. has no corners). This implies that $\mathrm{d}/\mathrm{d}t\, f_{\dot{x}}^0$ exists on $[t_0, t_1]$ and hence that the second term in (6.13) may be integrated by parts;

$$\int_{t_0}^{t_1} f_{\dot{x}}^0 \dot{\eta}\, \mathrm{d}t = [f_{\dot{x}}^0 \eta(t)]_{t_0}^{t_1} - \int_{t_0}^{t_1} \left(\frac{\mathrm{d}}{\mathrm{d}t} f_{\dot{x}}^0\right) \eta(t) \mathrm{d}t. \tag{6.15}$$

The first term on the right-hand side is zero since $\eta(t_0) = \eta(t_1) = 0$. Hence using (6.15) in (6.13) gives

$$\int_{t_0}^{t_1} \left(f_x^0 - \frac{\mathrm{d}}{\mathrm{d}t} f_{\dot{x}}^0\right) \eta(t) \mathrm{d}t = 0. \tag{6.16}$$

With $x(t) \in C_2$, both the term in brackets and $\eta(t)$ are continuous on $[t_0, t_1]$. Now (6.16) must hold for arbitrary admissible functions η; intuition suggests that this must imply that $f_x^0 - \mathrm{d}/\mathrm{d}t\, f_{\dot{x}}^0 = 0$. The following theorem (a fundamental lemma in the calculus of variations) establishes this intuition.

THEOREM 6.2 If $G(t)$ is a given continuous function on $[t_0, t_1]$ and $\int_{t_0}^{t_1} G(t)\alpha(t)\mathrm{d}t = 0$ holds for all continuous functions $\alpha(t)$ on $[t_0, t_1]$ satisfying $\alpha(t_0) = \alpha(t_1) = 0$, then $G(t) = 0$ on $[t_0, t_1]$.

Proof. Suppose, at $\hat{t} \in [t_0, t_1]$, $G(t) \neq 0$. If $G(\hat{t}) > 0$ (resp. < 0), then, by continuity, $G(t) > 0$ (resp. < 0) on an open subinterval T including $\hat{t}$ (half open if $\hat{t} = t_0$ or t_1). Select an $\alpha(t)$ such that $\alpha(t) > 0$ (resp. < 0) for $t \in T \cap (t_0, t_1)$ and $\alpha(t) = 0$ for $t \in [t_0, t_1] \backslash T$ (note $\alpha(t_0) = \alpha(t_1) = 0$). Then

$$\int_{t_0}^{t_1} G(t)\alpha(t)\mathrm{d}t = \int_T G(t)\alpha(t)\mathrm{d}t > 0$$

which contradicts the premis that the integral $= 0$.

Hence, if $x(t)\in C_2$, we have the **EULER** necessary condition that

$$f_x^0(x(t),\dot{x}(t),t)-\frac{\mathrm{d}}{\mathrm{d}t}f_{\dot{x}}^0(x(t),\dot{x}(t),t)=0, \qquad t\in[t_0,t_1]. \tag{6.17}$$

This is a non-linear second order differential equation; expanding the right-hand term, (6.17) becomes

$$f_{\dot{x}x}^0\dot{x}(t)+f_{\dot{x}\dot{x}}^0\ddot{x}(t)+f_{\dot{x}t}^0=f_x^0 \tag{6.17'}$$

where $f_{\dot{x}x}^0=f_{\dot{x}x}^0(x(t),\dot{x}(t),t)$ etc.

[2] *Suppose* $x(t)$ *piecewise smooth*
Now, at a corner of x, there are two elements; $(x(t),\dot{x}(t^-),t)$ and $(x(t),\dot{x}(t^+),t)$; $\dot{x}$ is piecewise continuous.

Returning to the original problem, integration by parts in (6.13) requires that $f_{\dot{x}}^0$ be differentiable – and this need no longer be the case. We therefore operate on the first term in (6.13), integrating by parts between the points at which $\dot{x}(t)$, $\dot{\eta}(t)$ are discontinuous (and then summing the results); in detail, let $D=\{t^i\}$ denote the set of points of discontinuity and let

$$W(t)\equiv\int_{t_0}^{t}f_x^0(x(\tau),\dot{x}(\tau),\tau)\mathrm{d}\tau, \tag{6.18}$$

so that $W(t_0)=0$,

and $W'(t)=f_x^0(x(t),\dot{x}(t),t)$ for $t\in[t_0,t_1]\backslash D$. (6.19)

Hence $\mathrm{d}/\mathrm{d}t[W(t)\eta(t)]=f_x^0(x(t),\dot{x}(t),t)\eta(t)+W(t)\dot{\eta}(t)$ for $t\in[t_0,t_1]\backslash D$, or $f_x^0\eta=\mathrm{d}/\mathrm{d}t[W(t)\eta(t)]-W(t)\dot{\eta}(t))$ for $t\in[t_0,t_1]\backslash D$.
Using this in (6.13) gives

$$\int_{t_0}^{t_1}(f_x^0\eta+f_{\dot{x}}^0\dot{\eta})\mathrm{d}t=[W(t)\eta(t)]_{t_0}^{t_1}+\int_{t_0}^{t_1}[f_{\dot{x}}^0-W(t)]\dot{\eta}(t)\mathrm{d}t=0. \tag{6.20}$$

The first term is zero since $\eta(t_0)=\eta(t_1)=0$. Now, $f_{\dot{x}}^0-W(t)$ and $\dot{\eta}(t)$ are both piecewise continuous functions – for which we have the following result.

THEOREM 6.3 If $G(t)$ is a piecewise continuous function on

$[t_0, t_1]$ and if $\int_{t_0}^{t_1} G(t)\dot{\eta}(t)\mathrm{d}t = 0$ for any $\eta(t)$ which is piecewise smooth and for which $\eta(t_0) = \eta(t_1) = 0$, then there is a constant k such that $G(t) = k$ for every t at which $G(t)$ is continuous.

Proof. Since $G(t)$ is piecewise continuous on $[t_0, t_1]$ (hence bounded), there is a constant $k \in \mathbb{R}$ such that $\int_{t_0}^{t_1} G(t)\mathrm{d}t = k(t_1 - t_0)$ (so that $\int_{t_0}^{t_1}[G(t) - k]\mathrm{d}t = 0$). Select an $\eta(t)$ such that $\dot{\eta}(t) = G(t) - k$ (this clearly satisfies the assumptions about $\eta(t)$). So $\int_{t_0}^{t_1} G(t)\dot{\eta}(t)dt = \int_{t_0}^{t_1} G(t))[G(t) - k]\mathrm{d}t$ and, of course, $\int_{t_0}^{t_1} -k[G(t) - k]\mathrm{d}t = 0$. Putting these together, clearly $\int_{t_0}^{t_1} G(t)\dot{\eta}(t)\mathrm{d}t = 0$ if, and only if, $\int_{t_0}^{t_1}[G(t) - k]^2\mathrm{d}t = 0$. The latter equals zero if and only if $G(t) = k$ almost everywhere on $[t_0, t_1]$

Applying Theorem 6.3 to (6.20) gives $f_{\dot{x}}^0 - W(t) = \text{constant}$, or

$$f_{\dot{x}}^0(x(t), \dot{x}(t), t) = \int_{t_0}^{t} f_x^0(x(\tau), \dot{x}(\tau), \tau)\mathrm{d}\tau + k \qquad t_0 \leqslant t \leqslant t_1. \tag{6.21}$$

Equation (6.21) is known as the **Du Bois-Reymond** necessary condition. The constant k is defined by

$$k = f_{\dot{x}}^0(x(t_0), \dot{x}(t_0), t_0). \tag{6.22}$$

At corners in x, (6.21) holds for both $(x(t), \dot{x}(t^-), t)$ and $(x(t), \dot{x}(t^+), t)$. Between corners, the integrand in (6.21) is continuous, so may be differentiated to obtain the Euler equation

$$f_x^0(x(t), \dot{x}(t), t) - \frac{\mathrm{d}}{\mathrm{d}t} f_{\dot{x}}^0(x(t), \dot{x}(t), t) = 0. \tag{6.23}$$

This holds for all t so long as, at a corner point, the derivative $\mathrm{d}/\mathrm{d}t$ is interpreted as a left- or right-hand derivative. Thus (6.23) holds even if $x(t)$ fails to have a second derivative. However, if, in addition, $x \in C_2$, then (6.23) can be put into the form (6.17′) as before. Between corners on x, we know that $\dot{x}(t)$ is continuous, but what of $\ddot{x}(t)$? (under [1], $x(t) \in C_2$ was simply assumed). Theorem 6.4 is informative in this respect.

DEFINITION 6.6 If a trajectory x satisfies the above necessary conditions and has $f^0_{\dot{x}\dot{x}} \neq 0$ between corners, then x is termed **non-singular** or **regular** on that interval.

THEOREM 6.4 If a trajectory x satisfying the above necessary conditions is non-singular between corners, then, between corners, it is of class C_m if f^0 is of class $C_m (m \geqslant 2)$. Furthermore, if it is non-singular or regular on $[t_0, t_1]$, then, on $[t_0, t_1]$, it is of class $C_m (m \geqslant 2)$.

Proof. See Hestenes (1966), pp. 60–1.

Hence if $f^0 \in C_2$ and $f^0_{\dot{x}\dot{x}} \neq 0$ between corners, then $x(t) \in C_2$ on this interval. Notice that the extremal in this case has a higher level of differentiability than was assumed for the variations.

The following first **Weierstrass-Erdmann corner condition** also follows from (6.21). Since the right-hand side of (6.21) is continuous, so is the left-hand side; hence, at a corner

$$f^0_{\dot{x}}(x(t), \dot{x}(t^-), t) = f^0_{\dot{x}}(x(t), \dot{x}(t^+), t). \tag{6.24}$$

Corners are discussed in more detail in Section 6.7. We now summarize the above results in the following theorem.

THEOREM 6.5 If x yields a local maximum for the Problem 6.1 and if f^0 is of class C_2, then (du Bois-Reymond)

$$f^0_{\dot{x}}(x(t), \dot{x}(t), t) = \int_{t_0}^{t} f^0_x(x(\tau), \dot{x}(\tau), \tau) \mathrm{d}\tau + k \qquad t_0 \leqslant t \leqslant t_1,$$

and (Euler)

$$\frac{\mathrm{d}}{\mathrm{d}t}\{f^0_{\dot{x}}(x(t), \dot{x}(t), t)\} = f^0_x(x(t), \dot{x}(t), t) \qquad t_0 \leqslant t \leqslant t_1,$$

where, at points of discontinuity, the appropriate left- or right-hand limit is taken. Furthermore (Weierstrass-Erdmann corner condition) $f^0_{\dot{x}}[x(t), \dot{x}(t), t]$ is continuous on $[t_0, t_1]$ (i.e. including across corners).

We shall often refer to trajectories which satisfy the above necessary conditions as **Extremals**. As already noted, in the absence of corners, the necessary condition yields a second order non-linear differential equation with split boundary conditions $x(t_0)=x_0$, $x(t_1)=x_1$. So, even in the absence of corners, analytic solutions will be straightforward only in certain special cases. Furthermore, the split boundary conditions complicate the problem of numerical solution. In economic analysis, however, the concern is less with determining analytic or explicit solutions and more with characterizing the qualitative properties of extremals. The qualitative analysis of differential equations of the above type has been studied in some detail in Chapter 5 and the tools developed there prove of considerable importance for the study of necessary conditions of the above type.

Clearly if f^0 does not depend on x, the Euler equation becomes $\mathrm{d}/\mathrm{d}t\, f^0_{\dot{x}}=0$ so $f^0_{\dot{x}}=\text{constant}$. If f^0 depends only on $x, \dot{x}$ but not on t, it becomes

$$f^0_x - f^0_{\dot{x}x}\dot{x} - f^0_{\dot{x}\dot{x}}\ddot{x}=0.$$

Multiplying through by $\dot{x}$ yields an exact differential on the left side so giving

$$\frac{\mathrm{d}}{\mathrm{d}t}(f^0 - \dot{x}f^0_{\dot{x}})=0,$$

or $f^0 - \dot{x}f^0_{\dot{x}}=\text{constant}$.

The following problem illustrates the qualitative information which may be derived from this necessary condition for the classic Hotelling monopolistic mineowner's problem.

EXAMPLE 6.1 A mineowner possesses resource stock, S, extractable at zero cost. He wishes to maximize discounted profits over a finite time horizon $[0, T]$. The rate of production is $x(t)$ and price is given by the inverse demand function $p(t)=g(x(t))$, where

$g \in C_2$ and $g'(x) < 0$ for $x \geqslant 0$. The problem is

$$\text{Maximize} \int_0^T R(t)\mathrm{e}^{-rt}\mathrm{d}t = \int_0^T x(t)g[x(t)]\mathrm{e}^{-rt}\mathrm{d}t$$

$$\text{subject to} \int_0^T x(t)\mathrm{d}t \leqslant S.$$

$R(t)$ denotes revenue at time t. A useful trick worth noting converts this constrained problem into one of the unconstrained type analyzed above (for an explicit treatment of integral constraints, see Section 6.10). Define $y(t) = \int_0^t x(\tau)\mathrm{d}\tau$. Then $\dot{y}(t) = x(t)$, and the problem becomes

$$\text{Maximize} \int_0^T \dot{y}g(\dot{y})\mathrm{e}^{-rt}\mathrm{d}t$$

$$\text{subject to } y(0) = 0$$
$$y(T) \leqslant S.$$

At this stage, we simply assume there is sufficient demand to guarantee that the stock will be exhausted, so that $y(T) = S$. The problem is then of the standard form discussed above (the inequality constraint is discussed in Example 7.1). The intermediate function depends only on $\dot{y}$, t, so from the du Bois-Reymond equation, $f_{\dot{y}}^0 = c$, a constant. Hence

$$\{g(\dot{y}) + \dot{y}g(\dot{y})\}\mathrm{e}^{-rt} = c$$

or

$$MR = \mathrm{d}R/\mathrm{d}x = \mathrm{d}R/\mathrm{d}\dot{y} = \{g(\dot{y}) + \dot{y}g(\dot{y})\} = ce^{rt}. \tag{i}$$

Thus, a necessary condition for discounted profit maximization is that discounted marginal revenue ($MR\,\mathrm{e}^{-rt}$) must remain constant over time. If this were not so, discounted profits could be increased by shifting sales from one point in time to another. from (6.22), $c = MR(0)$. If we assume sufficient demand, a solution involving $c = 0$ will not satisfy the constraints – and so we may assume $c > 0$. Hence undiscounted MR rises exponentially. Furthermore, if we assume $\mathrm{d}MR/\mathrm{d}x < 0$, then (i) implies that the optimal extraction rate declines over time, since

$$\mathrm{d}/\mathrm{d}t\{MR\}=(\mathrm{d}MR/\mathrm{d}x)\dot{x}=cr\,\mathrm{e}^{rt}>0 \Rightarrow \dot{x}(t)<0.$$

An explicit solution may be obtained if demand is assumed to be linear; letting $p(t)=a-bx(t)$, then the Euler equation may be integrated to yield the solution $x(t)=\dot{y}(t)=(a-c\,\mathrm{e}^{rt})/2b$ and

$$y(t)=\{a/2b\}t-c\,\mathrm{e}^{rt}/(2br)+d,$$

where c, d, are arbitrary constants determined by the boundary conditions $y(0)=0$ and $y(T)=S$; thus

$$c=r(2bS-aT)/(1-\mathrm{e}^{rT}), \qquad d=(2bS-aT)/\{2b(1-\mathrm{e}^{rT})\}.$$

So, given values for a, b, r, T and S, an explicit solution can be calculated. (Many variants of this problem are possible. For further examples, see Kamien and Schwartz, 1981). The reader may care to examine the case of the price-taking firm, or one in which extraction costs are present. Some variants are also considered later in illustrating further points.)

6.4.2 Second order necessary conditions

Returning to the second order necessary condition of equation (6.14), notice that the integrand is a quadratic form. It is thus non-positive if f^0 is concave in $(x, \dot{x})$. Such concavity will play a role in sufficient conditions (Section 6.9). The following necessary condition may also be derived from (6.14).

THEOREM 6.6 (Legendre necessary condition) If x yields a local maximum for Problem 6.1, then $f^0_{\dot{x}\dot{x}}(x(t), \dot{x}(t), t)\leqslant 0 (t_0\leqslant t\leqslant t_1)$ (where at corners, the equation holds for both left- and right-hand limits).

Proof. See Hestenes (1966).

EXAMPLE 6.2 Consider the resource depletion problem of Example 6.1. The Legendre condition requires that

$$f^0_{\dot{y}\dot{y}}=\{2g'(\dot{y})+\dot{y}g''(\dot{y})\}\mathrm{e}^{-rt}\leqslant 0 \qquad \text{for} \quad 0\leqslant t\leqslant T.$$

Since $d^2R/dx^2 = f^0_{\dot{y}\dot{y}} e^{rt}$, this implies the firm must operate at a point of non-increasing marginal revenue. Note that this does not suffice to establish that the extraction rate cannot rise over time since it is only a weak inequality for d^2R/dx^2 (see Example 6.1).

6.4.3 The Weierstrass necessary condition

The following Weierstrass condition must be satisfied (note that strong variations are admitted here). The **Weierstrass excess function** is defined by

$$E(x, \dot{x}, \dot{z}, t) = f^0(x, \dot{z}, t) - f^0(x, \dot{x}, t) - f^0_{\dot{x}}(x, \dot{x}, t)(\dot{z} - \dot{x}),$$

where x denotes the extremal and z, some other admissible trajectory.

THEOREM 6.7 If x yields a strong local or global maximum for J in Problem 6.1 and if z denotes another admissible trajectory, then

$$E(x, \dot{x}, \dot{z}, t) \leqslant 0 \qquad t_0 \leqslant t \leqslant t_1$$

must hold.

Proof. See Hestenes (1966), pp. 66–8.

The Weierstrass excess function E measures the difference between $f^0(x, \dot{z}, t)$ and its first order expansion in $\dot{z}$ around $(x, \dot{x}, t)$. A sufficient (but not necessary) condition for $E \leqslant 0$ is that f^0 is concave in $\dot{z}$. To see this, note that $f^0 \in C_2$ and expand $f^0(x, \dot{z}, t)$ in $\dot{z}$ to second order around $(x, \dot{x}, t)$; then

$$E = \tfrac{1}{2} f^0_{\dot{x}\dot{x}}(x, \psi, t)(\dot{z} - \dot{x})^2, \tag{6.25}$$

where $\psi = \alpha\dot{x} + (1 - \alpha)\dot{z}$, for some α, $0 < \alpha < 1$. If $f^0(x, \psi, t)$ is concave in ψ, then $f^0_{\dot{x}\dot{x}}(x, \psi, t) \leqslant 0$ for all ψ and hence $E \leqslant 0$ for all $\dot{z} \in \mathbb{R}$. Notice also that, if the Weirstrass condition holds, the Legendre condition holds a fortiori (but not the reverse). Thus, let $\dot{z} \to \dot{x}$ in (6.25); then $\psi \to \dot{x}$ and so $E \leqslant 0 \Rightarrow f^0_{\dot{x}\dot{x}}(x, \dot{x}, t) \leqslant 0$.

EXAMPLE 6.3 Consider the problem Max $J = -\int_0^1 \dot{x}^3 dt$ subject to $x(0)=0$, $x(1)=1$. The Euler necessary condition can be solved to yield the extremal $x=t$. The Legendre condition is $f^0_{\dot{x}\dot{x}} = -6\dot{x} = -6$ on $[0,1]$($\dot{x}=1$ at all points in the interval). Hence the Legendre necessary condition (for a maximum) is satisfied. The Weierstrass condition is

$$E = -\dot{z}^3 + \dot{x}^3 + 3\dot{x}^2(\dot{z}-\dot{x}) = -\dot{z}^3 + 3\dot{x}^2\dot{z} - 2\dot{x}^3 = -\dot{z}^3 + 3\dot{z} - 2.$$

For a maximum with respect to strong variations, we require $E \leqslant 0$. Clearly this does not hold for all $\dot{z}$ (since $E > 0$ for $\dot{z} < -2$). It follows that x cannot yield a strong maximum for the problem. It is worth noting that $E \leqslant 0$ for $\dot{z} \geqslant -2$ whilst f^0 is concave in $\dot{z}$ only for $z \geqslant 0 [f^0_{\dot{x}\dot{x}}(x, \dot{z}, t) = -6\dot{z}]$.

For further details, and a consideration of the Jacobi necessary condition (omitted here), see Hestenes (1966). From an economist's viewpoint, the Euler and Legendre conditions prove to be the most useful for characterizing extremals.

6.5 VARIABLE ENDPOINTS AND TRANSVERSALITY CONDITIONS

In this section, the endpoints are no longer exogenously fixed. Given the symmetry of the problem, it suffices to treat only the terminal condition, leaving the initial condition as before. The following terminal conditions will be considered:

(i) x_1 fixed, t_1 free,
(ii) x_1 free, t_1 fixed,
(iii) x_1, t_1 free,
(iv) Terminal surface constraints,
(v) Inequality constraints.

We consider first how perturbations in the endpoint affect J, and then apply this analysis to the above conditions.

PROBLEM 6.4

Maximize $\int_{t_0}^{t_1} f^0(x(t), \dot{x}(t), t)\mathrm{d}t$

subject to $x(t_0) = x_0$,
x_0, t_0 fixed,
x_1, t_1 (to be specified),
x piecewise smooth.

As usual, x denotes the optimal solution (i.e. provides a local maximum for J) which is assumed to exist and terminate at a given point x_1, t_1 which satisfies the terminal conditions. z denotes another admissible trajectory (see Figure 6.3), which terminates at $z_1 = x_1 + \delta x_1$ at time $t_1 + \delta t_1$.

Notice that the domain of x and z may not coincide, given that δt_1 may be >0 or <0. However, we can extend the domain of definition (using a Taylor series expansion of appropriate order) as follows:

$$x(t) = x_1 + \dot{x}(t_1)(t - t_1) + \ldots + x^{(p)}(t_1)(t - t_1)^p/p! \quad \text{for} \quad t > t_1 \qquad (6.26)$$

if $x \in C_p$ at t_1. The extension then has the same differentiability

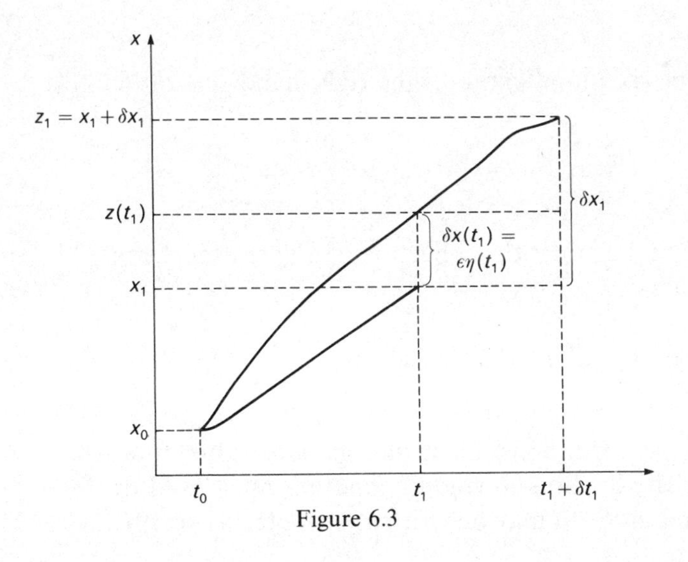

Figure 6.3

properties as assumed for x (we shall be principally concerned with the first order extension in what follows).

Since we are considering perturbations both to the trajectory and its endpoint, let us define the following norm

$$\|z-x\|'=\operatorname*{Sup}_t|\varepsilon\eta|+\operatorname*{Sup}_t|\varepsilon\dot{\eta}|+|\delta t_1|+|\delta x_1|,$$

so that both δt_1 and δx_1 are brought explicitly into the measure of closeness.

$$\Delta J=J(z)-J(x)=\int_{t_0}^{t_1}[f^0(z,\dot{z},t)-f^0(x,\dot{x},t)]\mathrm{d}t$$
$$+\int_{t_1}^{t_1+\delta t_1} f(z,\dot{z},t)\mathrm{d}t\leqslant 0 \qquad (6.27)$$

is a necessary condition if x is optimal (for $\|z-x\|'$ sufficiently small).

$$\int_{t_0}^{t_1} f^0(x+\varepsilon\eta,\dot{x}+\varepsilon\dot{\eta},t)-f^0(x,\dot{x},t)\mathrm{d}t=\varepsilon\int_{t_0}^{t_1}[f_x^0\eta+f_{\dot{x}}^0\dot{\eta}]\mathrm{d}t+0(.) \qquad (6.28)$$

and the second integral on the right-hand side of (6.27) is

$$\int_{t_1}^{t_1+\delta t_1} f^0(z,\dot{z},t)\mathrm{d}t=f^0(z(t_1),\dot{z}(t_1),t_1)\delta t_1+0(.),$$

where $0(.)$ denotes the collection of higher order terms. Thus (6.27) becomes

$$\Delta J=\varepsilon\int_{t_0}^{t_1}[f_x^0\eta+f_{\dot{x}}^0\dot{\eta}]\mathrm{d}t+f^0(z(t_1),\dot{z}(t_1),t_1)+0(.). \qquad (6.29)$$

Now, there are admissible variations which also terminate at (x_1,t_1), hence the du Bois-Reymond equation must hold on $[t_0,t_1]$. The integral in (6.29) may be written, using (6.18), (6.19), and (6.21), as

$$\varepsilon \int_{t_0}^{t_1} \{W'(t)\eta + [W(t)+k]\dot{\eta}\}\mathrm{d}t$$

$$= \varepsilon \int_{t_0}^{t_1} \frac{\mathrm{d}}{\mathrm{d}t}\{[W(t)+k]\eta\}\mathrm{d}t = [W(t)+k]\varepsilon\eta\big|_{t_0}^{t_1}$$

$$= f_{\dot{x}}^0(x(t_1), \dot{x}(t_1), t_1)\delta x(t_1),$$

since at t_0, $\eta(t_0)=0$, $\delta x(t_1)=\varepsilon\eta(t_1)$ and, from the du Bois-Reymond equation, $W(t_1)+k = f_{\dot{x}}^0|_{t_1}$. So (6.29) becomes

$$\Delta J = f_{\dot{x}}^0(x(t_1), \dot{x}(t_1), t_1)\delta x(t_1) + f^0(z(t_1), \dot{z}(t_1), t)\delta t_1 + 0(.).$$

ΔJ may now be obtained in terms of δx_1 and δt_1 by using (see Figure 6.3)

$$\delta x(t_1) = z(t_1) - x_1,$$

$$\text{and } x_1 + \delta x_1 = z(t_1) + \dot{x}(t_1)\delta t_1 + 0(.),$$

so

$$\delta x(t_1) = \delta x_1 - \dot{x}\,\delta t_1 + 0(.). \tag{6.30}$$

Finally, therefore,

$$\Delta J = f_{\dot{x}}^0(x(t_1), \dot{x}(t_1), t)(\delta x_1 - \dot{x}\,\delta t_1) + f^0(x(t_1), \dot{x}(t_1), t_1)\delta t_1 + 0(.). \tag{6.31}$$

If x is optimal, then, since $0(.)/\|z-x\|' \to 0$ as $\|z-x\|' \to 0$, we must require that

$$\{f^0(x(t_1), \dot{x}(t_1), t_1) - \dot{x}(t_1) f_{\dot{x}}^0(x(t_1), \dot{x}(t_1), t_1)\}\delta t_1 + f_{\dot{x}}^0(x(t_1), \dot{x}(t_1), t_1)\delta x_1 \leqslant 0. \tag{6.32}$$

This is the basic condition from which the following results follow:

[1] *Free horizon*
x_1 fixed, hence $\delta x_1 = 0$. t_1 is free, so δt_1 is arbitrary and unrestricted in sign. From (6.32), a necessary condition is that

$$[f^0(x(t_1), \dot{x}(t_1), t_1) - \dot{x}(t_1) f_{\dot{x}}^0(x(t_1), \dot{x}(t_1), t_1)] = 0. \tag{6.33}$$

[2] *Free terminal state*
t_1 fixed, hence $\delta t_1=0$. x_1 is free, hence δx_1 is unrestricted in sign From (6.32), a necessary condition is that

$$f_{\dot{x}}^0(x(t_1),\dot{x}(t_1),t_1)=0. \tag{6.34}$$

[3] *Completely free endpoint*
x_1, t_1 free, hence δx_1, δt_1 are unrestricted in sign. Both (6.33) and (6.34) must hold, and so the necessary conditions are that

$$f_{\dot{x}}^0(x(t_1),\dot{x}(t_1),t_1)=0$$
$$f^0(x(t_1),\dot{x}(t_1),t_1)=0.$$

[4] *Endpoint constrained to lie on a terminal surface*
Suppose that this constraint may be represented by a differentiable function $G(x_1,t_1)=0$ (see Figure 6.4). In this case, variations in x_1,t_1 cannot be independent. Expanding G around (x_1,t_1) gives

$$\begin{aligned}G(x_1+\delta x_1,t_1+\delta t_1)&=G(x_1,t_1)+G_1(x_1,t_1)\delta x_1+G_2(x_1,t_1)\delta t_1+0(.)\\&=G_1(x_1,t_1)\delta x_1+G_2(x_1,t_1)\delta t_1+0(.)=0,\end{aligned}$$

where G_1,G_2 denote the partial derivatives of G. Assume $G_1(x_1,t_1)$, $G_2(x_1,t_1)\neq 0$ (otherwise the situation is one of those described above). Substituting for δx_1 in (6.31) gives

$$(f^0-\dot{x}f_{\dot{x}}^0)\delta t_1-f_{\dot{x}}^0G_2\,\delta t_1/G_1+0(.)\leqslant 0. \tag{6.35}$$

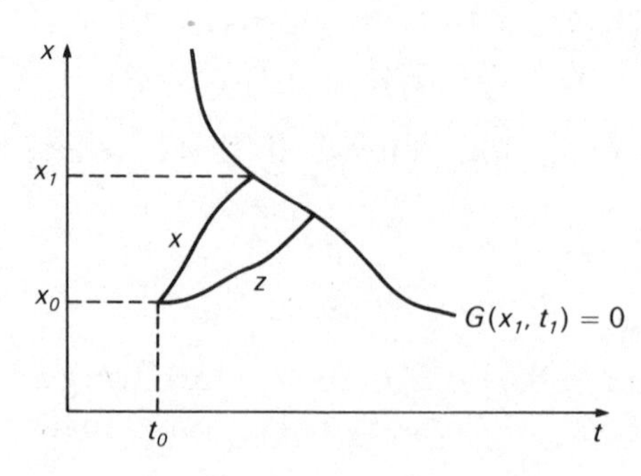

Figure 6.4

Here, δt_1 is arbitrary, hence a necessary condition is that

$$G_1(f^0 - \dot{x}f_{\dot{x}}^0) - G_2 f_{\dot{x}}^0 = 0, \tag{6.36}$$

where, in (6.36), all functions are evaluated at x_1, t_1.

[5] *Inequality endpoint constraints*

Inequalities give rise to conditions analogous to those developed for static optimization. The basic condition (6.32) may be derived as before (with some appropriate restrictions now on the set of admissible functions). Just one case is considered here; others may be developed by analogous reasoning; t_0, x_0, x_1 are fixed and t_1 must satisfy $t_1 \leqslant T$, where T is a given constant. So, $\delta t_0, \delta x_0, \delta x_1 = 0$ and $t_1 + \delta t_1$ must satisfy $t_1 + \delta t_1 \leqslant T$.

On the extremal x, either $t_1 < T$ or the constraint binds. If $t_1 < T$, variation about x may terminate before or after t_1 so δt_1 is unrestricted, and it follows, as before, that (6.33) must hold. However, if the constraint binds, $t_1 = T$ and admissible variations must terminate at, or to the left of, t_1, so $\delta t_1 \leqslant 0$; from (6.32), this implies

$$f^0(x(t_1), \dot{x}(t_1), t_1) - \dot{x}(t_1) f_{\dot{x}}^0(x(t_1), \dot{x}_1(t_1), t) \geqslant 0. \tag{6.37}$$

In the standard format, we may express these conditions as

$$t_1 \leqslant T, f^0 - \dot{x}f_{\dot{x}}^0|_{t_1} \geqslant 0, (t_1 - T)(f^0 - \dot{x}f_{\dot{x}}^0)|_{t_1} = 0. \tag{6.38}$$

If the constraint were $t_1 \geqslant T$, the inequalities are simply reversed.

Other inequalities give rise to similar conditions; for example, if $x_1 \leqslant \bar{x}$, with x_0, t_0, t_1 fixed, the transversality condition is

$$x_1 \leqslant \bar{x}, f_{\dot{x}}^0|_{t_1} \geqslant 0, (x_1 - \bar{x}) f_{\dot{x}}^0|_{t_1} = 0. \tag{6.39}$$

Second order transversality conditions also exist but rarely find use in economic analysis (see Dreyfus, 1965, pp. 116–19).

We have discussed only the final endpoint; naturally, if the initial endpoint is subject to similar conditions, analogous transversality conditions arise (with the functions evaluated at the initial point). Example 6.4 illustrates how the analysis of the transversality condition, terminal surface and extremal equations may be used to determine a particular solution.

EXAMPLE 6.4

Maximize $\int_0^T \dot{x}^3 \, dt$ subject to $x(0)=1$ and $x(T)$, T satisfying the terminal surface condition

$$G(x(T), T) = x(T).T - 2 = 0. \tag{i}$$

The Euler equation for this problem gives

$$x = at + b, \tag{ii}$$

where a, b are arbitrary constants. From the initial condition, $b=1$. The transversality condition (6.36) gives

$$T(\dot{x}^3 - 3\dot{x}^2.\dot{x}) - x.3\dot{x}^2 = 0, \tag{iii}$$

where $x, \dot{x}$ are evaluated at T. Since $x = at+1$, $\dot{x}=a$, and so (iii) becomes

$$-2a^3T - 3a^2x(T) = 0.$$

However, $x(T) = aT+1$, hence (i)–(iii) may be solved to yield $x(T)=2/5$, $T=5$ and $a=-3/25$. The extremal is thus $x(t) = -3t/25+1$. The Legendre condition is satisfied; $f^0_{\dot{x}\dot{x}} = 6\dot{x} = 6a = -18/25 < 0$. As in Example 6.3, the Weierstrass condition is not satisfied; the solution thus cannot yield a strong maximum.

EXAMPLE 6.5 (The Hotelling problem, Example 6.1, revisited) Suppose terminal time is free. Then, from (6.33), the terminal transversality condition is $\dot{y}g(\dot{y})e^{-rT} - \dot{y}\{g(\dot{y}) + \dot{y}g'(\dot{y})\}e^{-rT} = 0$, where $\dot{y}$ denotes $\dot{y}(T)$. Hence $\dot{y}^2 g'(\dot{y})e^{-rT} = 0$. Since by assumption $g'(.) < 0$ then, if T is finite, $\dot{y}(T) = 0$ (i.e. $x(T)=0$). Thus in this case, production must just cease as terminal time arrives. For example, in the case of linear demand analyzed in Example 6.1, the solution was

$$y(t) = (a/2b)t + (2bS - aT)(1 - e^{rt})/\{2b(1-e^{rT})\},$$

and $\dot{y}(t) = (a - c\,e^{rt})/2b = (a/2b) - (2bS - aT)r\,e^{rt}/\{2b(1-e^{rT})\}$.

If T is finite, $y(T)=0$ so T may be calculated from

$$0=(a/2b)-(2bS-aT)r\,\mathrm{e}^{rT}/\{2b(1-\mathrm{e}^{rT})\}. \tag{i}$$

Notice that, at T, $p(T)=a-by(T)=a$ and $MR(T)=a-2by(T)=a$. Thus price rises so as to just choke off demand at time T. Further study of transversality conditions may be of interest; for example, routine comparative statics analysis of (i) reveals that $\mathrm{d}T/\mathrm{d}S>0$, $\mathrm{d}T/\mathrm{d}a>0$. Thus increases in reserves or demand increases the optimal economic life of the mine.

6.6 THE BEQUEST OR FINAL VALUE FUNCTION

In this section, we briefly consider the problem

$$\text{Maximize}\, J=\int_{t_0}^{t_1} f^0(x,\dot{x},t)\mathrm{d}t+B(x_1,t_1) \tag{6.40}$$

subject to $x(t_0)=x_0$ and various endpoint conditions. $B\in C_2$ is assumed. The kind of economic applications which involve such a function include the individual's lifetime consumption and bequest problem, terminal capital stock mazimizations and so on. The approach is via the analysis of Section 6.5. The variational analysis yields an obvious modification in equation (6.32) as follows:

$$(f^0-\dot{x}f^0_{\dot{x}})|_{t_1}\delta t_1+f^0_{\dot{x}}|_{t_1}\delta x_1+B_1\delta x_1+B_2\delta t_1\leqslant 0, \tag{6.41}$$

so the only modification lies in the transversality condition, which becomes

$$(f^0-\dot{x}f^0_{\dot{x}}+B_2)|_{t_1}\delta t_1+(f^0_{\dot{x}}+B_1)|_{t_1}\delta x_1\leqslant 0, \tag{6.42}$$

where B_1, B_2 denote the partial derivatives of B evaluated at x_1, t_1. The analysis then proceeds as before. If there is an initial value function, this would entail analogous modifications to the initial transversality conditions. Section 6.11 provides an extended application involving a bequest function.

6.7 CORNERS

In this section, the possibility that a trajectory may have corners is considered in more detail. As usual, let x denote the optimal trajectory and z, some other admissible trajectory.

At a corner, $x(t^+)=x(t^-)$ but $\dot{x}(t^+)\neq\dot{x}(t^-)$. Let $\{t^i\}$, $i=1,\ldots,m$, denote the set of times at which the extremal x has a corner. We can divide the problem into subproblems on which the extremal is differentiable. Consider the problems

$$\text{Maximize} \int_{t^j}^{t^{j+1}} f^0(z,\dot{z},t)$$

$$\text{subject to } z(t^j)=x(t^j),$$

$$z(t^{j+1})=x(t^{j+1}).$$

The solution to each of these subproblems must be given by x on $[t^j,t^{j+1}]$ if x is optimal on $[t_0,t_1]$ by the principle of optimality (Section 6.3).

The properties which must hold at a corner are given in the following theorem.

THEOREM 6.8 (Weierstrass-Erdmann corner conditions) If x is an optimal solution to Problem 6.1, then $f_{\dot{x}}^0$ and $f^0-\dot{x}f_{\dot{x}}^0$ must be continuous on $[t_0,t_1]$.

Clearly these functions are continuous if x is smooth. Consider then the Problem 6.1 where there is just one corner at some point $(\hat{x},\hat{t})$, $t_0<\hat{t}<t_1$. An admissible variation may shift the corner, as in Figure 6.5. Thus we have

$$\begin{aligned} J(z)-J(x)= & \int_{t_0}^{\hat{t}+\delta\hat{t}} f^0(z,\dot{z},t)\mathrm{d}t+\int_{\hat{t}+\delta\hat{t}}^{t_1} f^0(z,\dot{z},t)\mathrm{d}t \\ & -\int_{t_0}^{\hat{t}} f^0(x,\dot{x},t)\mathrm{d}t-\int_{\hat{t}}^{t_1} f^0(x,\dot{x},t)\mathrm{d}t. \end{aligned} \tag{6.43}$$

Referring to the analysis yielding equation (6.32), we can immedi-

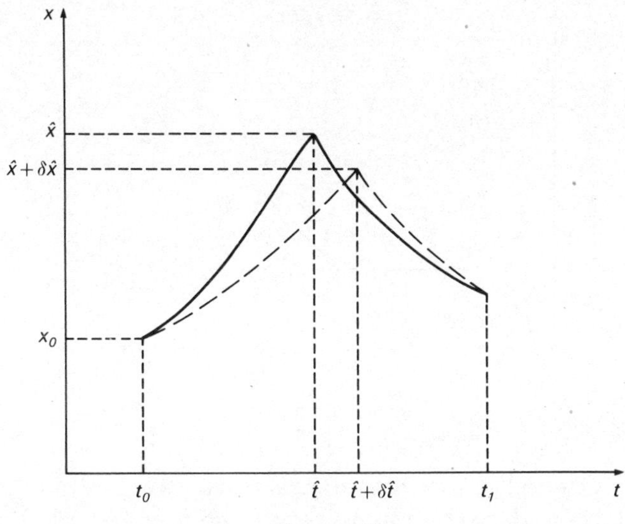

Figure 6.5

ately write the necessary condition as

$$\varepsilon\int_{t_0}^{\hat{t}}\{f_x^0\eta+f_{\dot{x}}^0\dot{\eta}\}\mathrm{d}t+(f^0-\dot{x}f_{\dot{x}}^0)_{\hat{t}^-}\delta\hat{t}+f_{\dot{x}}^0|_{\hat{t}^-}\delta\hat{x}$$

$$+\varepsilon\int_{\hat{t}}^{t_1}\{f_x^0\eta+f_{\dot{x}}^0\dot{\eta}\}\mathrm{d}t-(f^0-\dot{x}f_{\dot{x}}^0)_{\hat{t}^+}\delta\hat{t}-f_{\dot{x}}^0|_{\hat{t}^+}\delta\hat{x}\leqslant 0. \tag{6.44}$$

The admissible variation could have a corner at $(\hat{x},\hat{t})$ so, as usual, the integrals must vanish. Furthermore, the shifts in the corner by $\delta\hat{t}$, $\delta\hat{x}$ are arbitrary, hence it follows that

$$(f^0-xf_{\dot{x}})_{\hat{t}^-}=(f^0-\dot{x}f_{\dot{x}})_{\hat{t}^+}, \tag{6.45}$$

$$f_{\dot{x}}^0|_{\hat{t}^-}=f_{\dot{x}}^0|_{\hat{t}^+} \tag{6.46}$$

Equations (6.45) and (6.46) imply that $f^0-\dot{x}f_{\dot{x}}$, $f_{\dot{x}}^0$ must be continuous across corners – and so must be continuous at all points.

Corners will typically be induced where there are constraints involving $x,\dot{x}$ which bind at some point – although corners can sometimes occur at interior points of the domain (as in the following

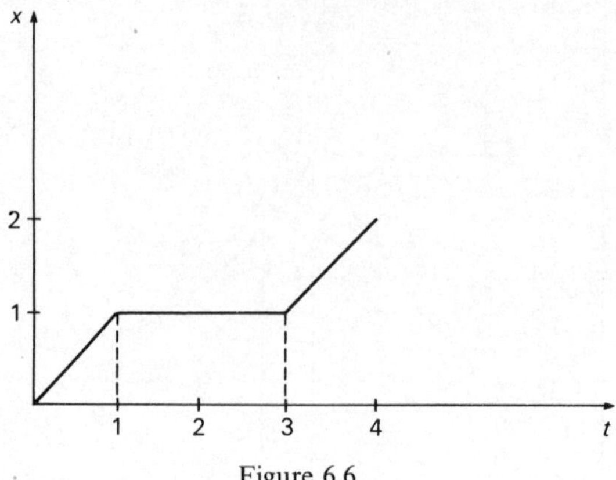

Figure 6.6

example). Generally, if a smooth extremal can be found which satisfies the boundary conditions, there is no need to look for other solutions (see Hadley and Kemp, 1971, p. 40).

EXAMPLE 6.6

$$\text{Maximize} -\int_0^4 (x-1)^2(\dot{x}-1)^2 \mathrm{d}t$$

subject to $x(0)=0$, $x(4)=2$.

The Euler equation is satisfied if either $x=1$ or $\dot{x}=1$ (implying $x(t)=t+c$). Neither on their own will satisfy both boundary conditions – but the extremal defined by

$$\begin{aligned} x &= t & 0 \leqslant t \leqslant 1 \\ x &= 1 & 1 \leqslant t \leqslant 3 \\ x &= t-2 & 3 \leqslant t \leqslant 4 \end{aligned}$$

will do so (see Figure 6.6). This has corners at $t=1, 3$. The corner conditions are easily checked; f^0 and $f^0_{\dot{x}}=0$ for all $t, 0 \leqslant t \leqslant 4$.

6.8 SUMMARY OF RESULTS FOR $x(t) \in \mathbb{R}^n$

The extension of results to the case where $x(t), \dot{x}(t)$ are vectors is routine, and we merely state these results.

6.8.1 Fixed endpoint problem

PROBLEM 6.5

Maximize $J(x)=\int_{t_0}^{t_1} f^0(x,\dot{x},t)\mathrm{d}t$

Subject to $x(t_0)=x_0$,

$x(t_1)=x_1$,

t_0, t_1 fixed.

THEOREM 6.9 If x is optimal for Problem 6.5, then the following conditions hold on $[t_0, t_1]$:

(a) The Euler equation

$$f_x^0-\frac{\mathrm{d}}{\mathrm{d}t}f_{\dot{x}}^0=\theta,$$

where $f_x^0=(f_{x_1}^0,\ldots,f_{x_n}^0)$,

$$\frac{\mathrm{d}}{\mathrm{d}t}f_{\dot{x}}^0=\left(\frac{\mathrm{d}}{\mathrm{d}t}f_{\dot{x}_1}^0,\ldots,\frac{\mathrm{d}}{\mathrm{d}t}f_{\dot{x}_n}^0\right),$$

where, at corners, $\mathrm{d}/\mathrm{d}t$ is interpreted as the appropriate left- or right-hand limit (if $x(t)\in\mathbb{R}^n$, there are n Euler equations).

(b) The Legendre condition

$f_{\dot{x}\dot{x}}^0$ negative semi-definite,

where $f_{\dot{x}\dot{x}}^0$ is the matrix $\left\{\dfrac{\partial^2 f^0}{\partial\dot{x}_i\partial\dot{x}_j}\right\}$.

(c) Corner conditions

$f_{\dot{x}}^0$ and $f^0-f_{\dot{x}}^0\dot{x}$ continuous across corners.

(d) Endpoint conditions

$x(t_0)=x_0$,

$x(t_1)=x_1$.

6.8.2 Variable endpoint problem

PROBLEM 6.6

$$\text{Maximize } J(x)=\int_{t_0}^{t_1} f^0(x,\dot{x},t)\mathrm{d}t$$

subject to $x(t_0)=x_0$,
t_0 fixed,
$x(t_1)$, t_1; conditions to be specified.

THEOREM 6.10 If x is optimal for Problem 6.6, then conditions (a), (b), (c) of Theorem 6.9 must hold, and, in addition,

$$x(t_0)=x_0,$$

and the transversality conditions;

(i) If t_1 is free, then
$$\{f^0-f^0_{\dot{x}}\dot{x}\}=0,$$
(ii) If x^i_1 is free for $i=1,\ldots,a$, then
$$\{f^0_{\dot{x}_i}\}|_{t_1}=0 \text{ for } i=1,\ldots,a,$$
(iii) If x^i_1 must satisfy $x^i_1 \geqslant \bar{x}_i$, $i=a+1,\ldots,b$, then
$$x^i_1 \geqslant \bar{x}_i,\ f^0_{\dot{x}_i}|_{t_1} \leqslant 0,\ [x^i_1-\bar{x}_i]f^0_{\dot{x}_i}|_{t_1}=0,\ i=a+1,\ldots,b.$$

Alternatively, if
(iv) the endpoint is constrained to lie on a terminal surface $G(x(t_1),t_1)=0$, then

$$[f^0-f^0_{\dot{x}}\dot{x}]_{t_1}+[f^0_{\dot{x}}]_{t_1}\left.\frac{\mathrm{d}x}{\mathrm{d}t}\right|_{G(x_1,t_1)=0}=0, \tag{6.47}$$

where $\mathrm{d}x/\mathrm{d}t|_{G=0}$ is a column vector normal to the terminal surface, defined by $G_x(x_1,t_1)\mathrm{d}x/\mathrm{d}t|_{G=0}-G_t(x_1,t_1)=0$. (6.47) follows from the vector equivalent of (6.32). Thus

$$(f^0-f_{\dot{x}}\dot{x})|_{t_1}\delta t_1+f^0_{\dot{x}}|_{t_1}\delta x_1 \leqslant 0.$$

Dividing by δt_1, letting $\delta t_1 \to 0$, and noting that

$$G_x(x_1,t_1)\left.\frac{\mathrm{d}x}{\mathrm{d}t}\right|_{G=0}+G_t(x_1,t_1)=0, \text{ we obtain (6.47).}$$

At this stage, it is worth pointing out some of the difficulties involved in analyzing these necessary conditions. The principal difficulty in economic applications is that the Euler equation when $x(t)\in\mathbb{R}$ gives rise to a non-linear second order differential equation for which typically only qualitative properties are known. With $x(t)\in\mathbb{R}^n$, there are n such equations. There are also in many cases additional differential equations involved in the economic problem. If the total number of such equations is n, then, for $n\leqslant 2$, analysis is fairly straightforward (see Chapter 5), but, if $n>2$, the problem is usually much less tractable. Occasionally, transformations can reduce this 'dimensionality' problem (see Chapter 5 in Pitchford and Turnovsky, 1977, for a discussion of the problems and possibilities of treating 2 state variable problems). The examples in Section 6.10 and 6.11 illustrate the qualitative analysis of the above necessary conditions in some detail.

6.9 SUFFICIENT CONDITIONS FOR FINITE HORIZON PROBLEMS

Necessary conditions cannot tell us whether an optimal trajectory exists (they only give us some properties which such optimal trajectories must possess *if* they exist). There may be none, or several, paths which satisfy the necessary conditions. In the latter case, the problem of picking out the maximizing extremal(s) remains. If the following sufficient conditions are satisfied, then the trajectory is optimal. We present only certain sufficient conditions based upon concavity properties; these are the most useful in economic applications (for other, classical, sufficiency conditions, see Hadley and Kemp, 1971, pp. 105–27, Hestenes, 1966, Chapter 3). The following exposition is based on Hadley and Kemp (1971).

THEOREM 6.11 (Sufficient conditions) For the fixed endpoint Problem 6.5 or the variable endpoint Problem 6.6, if $f^0(z,\dot{z},t)$ is concave in $(z,\dot{z})$ for all admissible z and a trajectory x satisfies the Euler equation, corner conditions, and transversality conditions

(Problem 6.6), then $J(x) \geqslant J(z)$ for all admissible z; x yields a strong global maximum.

Proof. For the fixed endpoint problem, let $z = x + \eta$ (the concern is with strong extremals). Then, from concavity

$$f^0(z, \dot{z}, t) \leqslant f^0(x, \dot{x}, t) + f_x^0 \eta + f_{\dot{x}}^0 \dot{\eta}, \tag{6.48}$$

where $\dot{x}$, $\dot{\eta}$ are piecewise continuous. Integration yields

$$J(z) \leqslant J(x) + \int_{t_0}^{t_1} [f_x^0 \eta + f_{\dot{x}}^0 \dot{\eta}] \mathrm{d}t. \tag{6.49}$$

The latter integral is zero if the Euler and corner conditions are satisfied. A detailed proof for Problem 6.6 is omitted; clearly, depending upon the endpoint conditions, additional endpoint terms arise in (6.49) which are equal to zero by the relevant transversality conditions.

THEOREM 6.12 (Uniqueness) If the conditions of Theorem 6.11 are satisfied and $f^0(z, \dot{z}, t)$ is strictly concave in $(z, \dot{z})$, then, if x maximizes J, x is unique; $J(x) > J(z)$ for all admissible $z, z \neq x$.

Proof. Suppose $x_1, x_2, (x_1 \neq x_2)$ both maximize J. Then $J(x_1) = J(x_2)$. Now consider a trajectory $y = \alpha x_1 + (1-\alpha)x_2$, for any α, $0 < \alpha < 1$. y naturally satisfies the necessary conditions and strict concavity implies $f^0(y, \dot{y}, t) > \alpha f^0(x_1, \dot{x}_1, t) + (1-\alpha) f^0(x_2, \dot{x}_2, t)$. Integrating between t_0 and t_1, $J(y) > \alpha J(x_1) + (1-\alpha) J(x_2)$ for all $\alpha, 0 < \alpha 01$. This contradicts x_1, x_2 being maxima.

EXAMPLE 6.7

In Example 6.1 (the mine), $f^0 = \dot{y} g(\dot{y}) e^{-rt}$, where $g'(\dot{y}) < 0$. This is concave in $(y, \dot{y})$ if

$$\begin{pmatrix} f_{yy}^0 & f_{y\dot{y}}^0 \\ f_{\dot{y}y}^0 & f_{\dot{y}\dot{y}}^0 \end{pmatrix} = \begin{pmatrix} 0 & 0 \\ 0 & 2g' + \dot{y} g'' \end{pmatrix}$$

is negative semi-definite. Thus, if the Legendre necessary condition is satisfied ($f^0_{\dot{y}\dot{y}} \leqslant 0$), f^0 is also concave in $(y, \dot{y})$ and the extremal discussed in Example 6.1 will be optimal. For example, if demand is linear ($g'' = 0$), then f^0 is concave. Note also that, if $f^0_{\dot{y}\dot{y}} > 0$ at any point on the extremal, the Legendre condition rules out this extremal being optimal.

The extended examples in Sections 6.11, 6.12 give additional applications of the above theorems.

6.10 CONSTRAINTS IN THE CALCULUS OF VARIATIONS

Most constraints tend to be handled more easily via control theory (particularly constraints on state variables). The treatment here is accordingly brief. Constraints may be classified as either global or local.

6.10.1 Global, integral constraints

(Integral constraints can often be avoided by appropriate transformations, as in Example 6.1.)

PROBLEM 6.7

$$\text{Maximize } J = \int_{t_0}^{t_1} f^0(x, \dot{x}, t)\mathrm{d}t \tag{6.50}$$

$$\text{subject to } \int_{t_0}^{t_1} f^1(x, \dot{x}, t)\mathrm{d}t = S,$$

$$x(t_0) = x_0,$$

$$x(t_1) = x_1,$$

where f^0, $f^1 \in C_2$ and S is some given number.

A Lagrange multiplier method is adopted as follows (see Chapter 7).

Define

$$\mathscr{L}=\lambda_0\int_{t_0}^{t_1} f^0(x,\dot{x},t)\mathrm{d}t-\lambda_1\left\{S-\int_{t_0}^{t_1} f^1(x,\dot{x},t)\mathrm{d}t\right\}$$

$$=\int_{t_0}^{t_1}\{\lambda_0 f^0(x,\dot{x},t)+\lambda_1 f^1(x,\dot{x},t)\}\mathrm{d}t-\lambda_1 S.$$

Then, the following **Hamiltonian** function is defined as

$$H(x,\dot{x},t,\lambda_0,\lambda_1)=\lambda_0 f^0(x,\dot{x},t)+\lambda_1 f^1(x,\dot{x},t),$$

where λ_0, λ_1 are constants. The necessary conditions refer to this function.

THEOREM 6.13 (Necessary conditions) If x yields a maximum for the Problem 6.7, then there exist multipliers λ_0, λ_1, not both zero, such that, between corners,

$$H_x-\frac{\mathrm{d}}{\mathrm{d}t}H_{\dot{x}}=0.$$

Furthermore, $H_{\dot{x}}$, $H-H_{\dot{x}}\dot{x}$ are continuous on $[t_0,t_1]$.

Proof. See Hadley and Kemp (1971), pp. 177–83.

The two endpoint conditions and the integral constraint serve to determine the two constants of integration and the multipliers λ_0, λ_1, up to a scalar multiple. For meaningful problems, $\lambda_0\neq 0$ and is conventionally set to unity (an extended discussion of the multiplier λ_0 may be found in the next chapter). Note that the multipliers here are constants, not functions of time. Several such constraints may be dealt with by simply extending the Hamiltonian function to $H=\lambda_0 f^0+\sum_i\lambda_i f^i$.

EXAMPLE 6.8 (Example 6.1 revisited).

$$\text{Maximize}\int_0^T x(t)g[x(t)]\mathrm{e}^{-rt}\mathrm{d}t$$

$$\text{subject to}\int_0^T x(t)\mathrm{d}t=S\ \text{(resource stock).}$$

Thus

$$H=\lambda_0 xg(x)e^{-rt}+\lambda_1 x.$$

The Euler equation is

$$\lambda_0(xg'+g)e^{-rt}+\lambda_1=0.$$

If $\lambda_0=0$, then $\lambda_1=0$, contradicting the necessary condition that the multipliers are not both zero. Hence take $\lambda_0=1$.

$$MR=(xg'+g)=-\lambda_1 e^{rt}.$$

As in Example 6.1, assuming sufficient demand, the solution involves $-\lambda_1>0$; that is, marginal revenue increases exponentially over time.

With a constraint of the form $\int_0^T x(t)dt\leqslant S$, an alternative solution would be where this constraint is inactive; complementary slackness would then dictate that $\lambda_1=0$ (and hence $\lambda_0=1$) and so $MR=0$ at all times. Note also that, in the analytic problem of Example 6.1 where $p=a-bx$, we obtain

$$-\lambda_1=r(2bS-aT)/(1-e^{rT})$$

and the solution as before.

6.10.2 Local constraints

Constraints of the form $g^i(x,\dot{x},t)=0$ and $g^i(x,\dot{x},t)\geqslant 0$ are considered in much greater detail in Chapter 7; however, this problem of **Lagrange** may be treated by introducing time varying multipliers. In this section, we shall consider only the case of equality constraints. For inequality constraints, see Chapter 7.

PROBLEM 6.8

$$\text{Maximize } J(x)=\int_{t_0}^{t_1} f^0(x,\dot{x},t)dt$$

subject to $x(t_0)=x_0$,

$x(t_1)=x_1$,

$g^i(x,\dot{x},t)=0 \qquad i=1,\ldots,q(q<n).$

Assume $f^0, g^1, \ldots, g^q \in C_2$. Define

$$H(x, \dot{x}, t, \lambda) = \lambda_0 f^0 + \sum_{i=1}^{q} \lambda_i(t) g^i(x, \dot{x}, t).$$

THEOREM 6.14 (Necessary conditions) If x is an optimal solution to Problem 6.8 and if the Jacobian matrix $\{\partial g^i(x, \dot{x}, t)/\partial x_j\}$ has full rank q for all $t \in [t_0, t_1]$, then there exist multipliers $\lambda_0, \lambda_1(t), \ldots, \lambda_q(t)$, not all simultaneously zero, such that, writing $\lambda = (\lambda_0, \lambda_1(t), \ldots, \lambda_q(t))'$, the following conditions hold between corners

$$H_x(x, \dot{x}, t, \lambda) - \frac{\mathrm{d}}{\mathrm{d}t} H_{\dot{x}}(x, \dot{x}, t, \lambda) = 0,$$

$H_{\dot{x}\dot{x}}(x, \dot{x}, t, \lambda)$ negative semidefinite, and, furthermore,

$H_{\dot{x}}(x, \dot{x}, t, \lambda)$, $H(x, \dot{x}, t, \lambda) - H_{\dot{x}}(x, \dot{x}, t, \lambda)\dot{x}$ are continuous everywhere on $[t_0, t_1]$.

λ_0 may be taken as 0 or 1.

Proof. See Hadley and Kemp (1971) or Hestenes (1966).

With variable endpoints, transversality conditions arise as before (with H replacing f^0). Note that λ_0 is a constant but $\lambda_1(t), \ldots, \lambda_q(t)$ are generally time varying. The constraint qualification on the Jacobian matrix requires that $\dot{x}$ is involved in all the constraints (again, see Chapter 7). Just as with static optimization, it may be that $\lambda_0 = 0$. This might occur, for example, if x was the only trajectory satisfying the constraints – but many other kinds of 'abnormal' behaviour may lead to $\lambda_0 = 0$. Only if $\lambda_0 \neq 0$ is the problem non-trivial (see Chapter 7 and Hadley and Kemp, 1971, pp. 225–8). Problems where $\lambda_0 \neq 0$ are termed **Normal**.

The following example illustrates the use of equality constraints as a means of dealing with problems which involve higher order derivatives.

EXAMPLE 6.9

Consider the problem

$$\text{Maximize } J(x)=\int_{t_0}^{t_1} f^0(x,\dot{x},\ddot{x},t)\mathrm{d}t$$

subject to $x(t_0)=x_0$,

$$x(t_1)=x_1,$$

where $f^0 \in C_2$, $x(t) \in \mathbb{R}$ and t_0, t_1, x_0, x_1 are given contants. With suitable assumptions, the Euler equations for this problem may be derived in the usual way to give

$$f^0_x - \frac{\mathrm{d}}{\mathrm{d}t} f^0_{\dot{x}} + \frac{\mathrm{d}^2}{\mathrm{d}t^2} f^0_{\ddot{x}} = 0. \tag{i}$$

As an alternative, consider the substitution $y(t)=\dot{x}(t)$. The problem becomes

$$\text{Maximize } J(x,y)=\int_{t_0}^{t_1} f^0(x,\dot{x},\dot{y},t)\mathrm{d}t$$

subject to $x(t_0)=x_0, x(t_1)=x_1$,

$$\dot{x}(t)=y(t).$$

The Jacobian has rank 1 so the constraint qualification is satisfied; the Hamiltonian is

$$H=\lambda_0 f^0(x,\dot{x},\dot{y},t)+\lambda_1(y-\dot{x}).$$

There are now two state variables, x, y, hence two Euler equations $H_x - \mathrm{d}/\mathrm{d}t\, H_{\dot{x}}=0$, $H_y - \mathrm{d}/\mathrm{d}t\, H_{\dot{y}}=0$ which give

$$\lambda_0 f^0_x - \frac{\mathrm{d}}{\mathrm{d}t}(\lambda_0 f^0_{\dot{x}} - \lambda_1)=0, \tag{ii}$$

$$\lambda_1 - \frac{\mathrm{d}}{\mathrm{d}t}(\lambda_0 f^0_{\dot{y}})=0. \tag{iii}$$

Now suppose $\lambda_0=0$. Then, from (iii), $\lambda_1=0$ for all t. This contradicts the necessary condition that the multipliers are not simultaneously zero. Hence take $\lambda_0=1$.

Equation (i) may then be obtained by performing the differentiation in (ii) followed by differentiating (iii) and substituting this into (ii).

Example 6.9 is easily extended in exactly the same manner as in Section 5.2 and Example 5.4. Specifically, if $f^0 = f^0(x, \dot{x}, \ldots, x^{(n)}, t)$, then introduce variables $y_1, \ldots, y_{n-1}$, where $y_i = \dot{y}_{i-1}$, $i = 2, \ldots, n-1$ and $y_1 = \dot{x}$, giving a problem with n state variables and $n-1$ constraints.

6.11 AN EXTENDED EXAMPLE

In this section, an individual's consumption and bequest problem (studied in some detail in Atkinson, 1971) is examined. We show how the necessary conditions may be manipulated to derive qualitative properties for the optimal solution. The impact upon the optimal path of various wealth taxes is then considered. Since the problem involves a terminal function and transversality conditions, it nicely illustrates much of the foregoing theoretical discussion.

The individual wishes to determine an optimum consumption plan over a fixed finite time horizon $[0, T]$ so as to maximize discounted utility from consumption and the final bequest. The objective function is

$$\text{Maximize} \int_0^T U[C(t)]\mathrm{e}^{-qt}\mathrm{d}t + F[S(T)]e^{-qT}. \tag{6.51}$$

$C(t)$ denotes consumption and $S(t)$ assets at time t. $U[.]$ is the instantaneous utility function and $F[.]$, the utility from bequests function. It is assumed that $U, F \in C_2$, that $U'[.], F'[.] > 0$ and $U''[.], F''[.] < 0$. The individual's marginal time preference rate is q. He is assumed to be able to borrow/lend at an interest rate r. Whilst alive, he earns a constant wage of w and receives bequests from time to time which we shall assume to be discounted to his birth and represented by initial assets $S_0 > 0$. He makes no gifts whilst alive; only the bequest at time T. Asset income is $rS(t)$, so assets accumulate according to

$$\dot{S}(t)=rS(t)+w-C(t). \tag{6.52}$$

Assets at time t are thus given by

$$S(t)\mathrm{e}^{-rt}=S_0+\int_0^t [w-C(\xi)]\mathrm{e}^{-r\xi}\,\mathrm{d}\xi. \tag{6.53}$$

Equation (6.52) may be used in (6.51) to give an intermediate function with arguments $(S,\dot{S},t)$. The initial condition is that

$$S(0)=S_0 \tag{6.54}$$

and final assets $S(T)$ at T are given by (6.53) evaluated at T. $S(T)$ is the bequest. The Euler equation is $f_S^0-\mathrm{d}/\mathrm{d}t\, f_{\dot{S}}^0=0$, where $f_S^0=rU'[C]\mathrm{e}^{-qt}$ and $f_{\dot{S}}^0=-U'[C]\mathrm{e}^{-qt}$. Thus, we have

$$rU'[C]\mathrm{e}^{-qt}+\frac{\mathrm{d}}{\mathrm{d}t}\{U'[C]\mathrm{e}^{-qt}\}=0 \tag{6.55}$$

which gives

$$\varepsilon(C)\dot{C}(t)=(r-q)C(t), \tag{6.56}$$

where $\varepsilon(C)=-U''[C]C/U'[C]$ is the elasticity of marginal utility of consumption ($\varepsilon>0$). The transversality condition for free terminal time with a bequest function is that (Section 6.6) $f_{\dot{x}}^0+B_x=0$. Here this is:

$$-U'[C(T)]\mathrm{e}^{-qT}+F'[S(T)]\mathrm{e}^{-qT}=0$$

$$\text{so } U'[C(T)]=F'[S(T)]. \tag{6.57}$$

Thus, the marginal utility of bequeathing \$1 must equal the marginal utility of consuming it on the death bed. If it is assumed that $r>q$ (conventional), then the right-hand side of (6.56) is non-negative, indicating that if $C(0)>0$, consumption will increase over time. The trajectory for assets may, however, fall, rise, or be humped over time. Figure 6.7 illustrates the latter two possibilities; the isokine $\dot{S}=0$ would need to be positioned to the left of point A for the former trajectory to be illustrated.

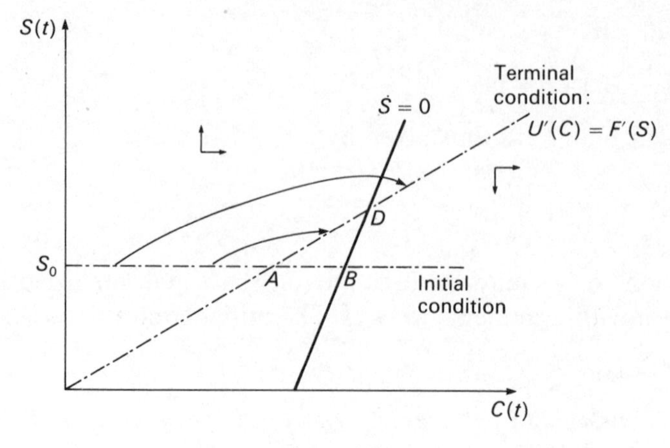

Figure 6.7

The terminal condition is given by (6.57). Thus, differentiating (6.57) gives

$$\frac{\mathrm{d}S(T)}{\mathrm{d}C(T)} = \frac{U''[C(T)]}{F''[S(T)]} > 0.$$

The isokine passes through the origin and has positive slope (if U, F have the same functional form, it is a straight line). There is no $\dot{C}=0$ isokine in the region where $C>0$. The $\dot{S}=0$ isokine is defined, from (6.52), by the equation $C=w+rS$. Establishing directions of motion and sketching trajectories is straightforward (and is discussed in Chapter 5, Section 5.6.5). Note that, if the optimal trajectory starts left of A, a humped trajectory may occur if, between A and B, it monotonically increases. The reader may care to verify that, if S_0 is set so that the initial condition lies above the point D, a trajectory on which S monotonically decreases becomes possible.

Consider now the impact of various types of wealth tax. If an unanticipated tax is introduced at time t, the individual will reoptimize his plan over the life remaining to him. The problem he faces is formally the same as that so far discussed; T may be reinterpreted as 'time remaining' and S_0 as assets at the time the tax is introduced. We consider (following Atkinson, 1971):

(1) a capital levy, a tax on net worth at time $0(\tau_1)$,

(2) a wealth transfer tax, on bequests (τ_2).

For simplicity, we shall assume the individual has already received his inheritances by the time the taxes are imposed. Introducing taxes transforms the problem in the following way:

$$\text{Maximize} \int_0^T U[C(t)]\mathrm{e}^{-qt}\,\mathrm{d}t + F[S(T)(1-\tau_2)]\mathrm{e}^{-qT}, \tag{6.58}$$

with budget equation

$$\dot{S}(t) = rS(t) + w - C(t), \tag{6.59}$$

and initial assets

$$S(0) = S_0(1-\tau_1). \tag{6.60}$$

The asset equation is

$$S(t)\mathrm{e}^{-rt} = S_0(1-\tau_1) + \int_0^t [(w - C(\xi)]\mathrm{e}^{-r\xi}\,\mathrm{d}\xi. \tag{6.61}$$

Case (1) The capital levy

Let $\tau_1 > 0$, $\tau_2 = 0$, and consider the impact of an increase in τ_1. This changes $S(0)$ in (6.60) and will lead to a new asset trajectory, via (6.61) and final bequest. The Euler equation (6.55) is unaffected, and this continues to hold as τ_1 is varied. Hence, if we differentiate (6.56) with respect to τ_1, this provides a means of studying the impact of τ_1 on the trajectory, as follows:

$$\frac{\partial}{\partial \tau_1}(\dot{C}) = \left(\frac{r-q}{\varepsilon}\right)(1 - C\varepsilon'/\varepsilon)\frac{\partial C}{\partial \tau_1}. \tag{6.62}$$

Note that this is legitimate if $C/\varepsilon(C)$ is C_1, in consequence of Theorem 5.3. Now, the order of differentiation may be interchanged

$\dfrac{\partial}{\partial \tau_1}\left(\dfrac{\partial C}{\partial t}\right) = \dfrac{\partial}{\partial t}\left(\dfrac{\partial C}{\partial \tau_1}\right)$, hence we may write

$$\frac{\partial}{\partial t}\left(\frac{\partial C}{\partial \tau_1}\right) = \left(\frac{r-q}{\varepsilon}\right)(1 - C\varepsilon'/\varepsilon)\frac{\partial C}{\partial \tau_1}. \tag{6.63}$$

Here, $r>q$, so if we make the reasonable assumption that ε does not vary too much with C (such that ε' is small and $1-C\varepsilon'/\varepsilon>0$) then this implies that sign $\partial/\partial t(\partial C/\partial\tau_1)=\text{sign}(\partial C/\partial\tau_1)$. Hence sign$(\partial C/\partial\tau_1)$ cannot change over time; if it starts positive/negative at time $t=0$, $dC/\partial\tau_1$ thereafter increases/decreases.

The transversality condition links $C(T)$ and $S(T)$; differentiating (6.57) with respect to τ_1 gives

$$U''[C(T)]\frac{\partial C(T)}{\partial\tau_1}=F''[S(T)]\frac{\partial S(T)}{\partial\tau_1}, \tag{6.64}$$

hence, since $U''[.],F''[.]<0$, $\partial C(T)/\partial\tau_1$ and $\partial S(T)/\partial\tau_1$ have the same sign. Thus, linking these facts, we have *either* (a) an increase in τ_1 increases consumption throughout the time interval *and* increases the bequest, $S(T)$, or (b) an increase in τ_1 decreases consumption throughout the time interval *and* decreases the bequest, $S(T)$.

Intuition indicates it must be the latter, and consideration of the lifetime budget constraint

$$S(T)\mathrm{e}^{-rT}=S_0(1-\tau_1)+\int_0^T[w-C(\xi)]\mathrm{e}^{-r\xi}\,\mathrm{d}\xi \tag{6.65}$$

confirms this. For, suppose, following an increase in τ_1, that $C(\xi)$ increased for all $\xi, 0\leqslant\xi\leqslant T$. Then the integral in (6.65) becomes smaller – but so too does $S_0(1-\tau_1)$ – hence $S(T)$ becomes smaller – which contradicts the assumption in (a) above.

Hence we conclude that the capital levy reduces the individuals' consumption $C(t)$, for all $t, 0\leqslant t\leqslant T$ *and* his bequest, $S(T)$.

To study the behaviour of assets and consumption under the capital levy, repeat the above differentiation process for (6.59), giving

$$\frac{\partial}{\partial\tau_1}\left(\frac{\partial S}{\partial t}\right)=\frac{\partial}{\partial t}\left(\frac{\partial S}{\partial\tau_1}\right)=r\frac{\partial S}{\partial\tau_1}-\frac{\partial C}{\partial\tau_1}. \tag{6.66}$$

The trick of changing the order of differentiation makes (6.63) and (6.66) into a pair of differential equations in $\partial C/\partial\tau_1$, $\partial S/\partial\tau_1$. To avoid notational clutter, define $C_\tau=\partial C/\partial\tau_1$ and $S_\tau=\partial S/\partial\tau_1$. Then

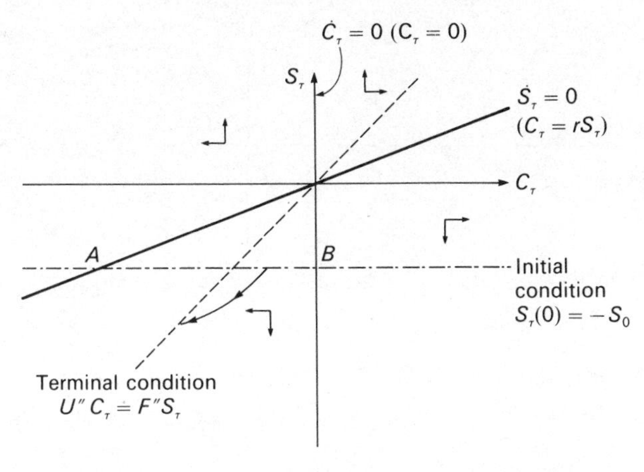

Figure 6.8

(6.63) and (6.66) become

$$\dot{C}_\tau = \left(\frac{r-q}{\varepsilon}\right)(1 - C\varepsilon'/\varepsilon)C_\tau, \tag{6.67}$$

$$\dot{S}_\tau = rS_\tau - C_\tau, \tag{6.68}$$

where $1 - C\varepsilon'/\varepsilon > 0$ by assumption. The phase diagram for this system is constructed in Figure 6.8.

The terminal condition is given by (6.64) (again, if U, F have the same functional form, this is a straight line). $\dot{S}_\tau = 0$ is given by $C_\tau = rS_\tau$ and $\dot{C}_\tau = 0$ by $C_\tau = 0$. The initial condition follows from differentiating (6.60) w.r.t. τ_1; $S_\tau(0) = -S_0$. In Figure 6.8, the diagram is constructed on the assumption that $U''[C(T)]/F''[S(T)] > 1/r$. Taking this case first, we know already that $C_\tau(t) \leqslant 0, 0 \leqslant t \leqslant T$ – thus, if the path starts to the left of A, the terminal condition cannot be attained. This is also true if it started to the right of B. Hence the path must start between A and B; hence S_τ, C_τ start negative and become more negative as time passes. Also, since $S_\tau < 0$ and $\partial/\partial t(\partial S/\partial \tau_1) = \partial/\partial \tau_1(\dot{S})$, in this case, assets *and* savings are diminished for all $t, 0 \leqslant t \leqslant T$. Thus, his reduction in consumption is not sufficient to offset the fall in interest income.

However, if $U''[C(T)]/F''[S(T)] < 1/r$, although assets are

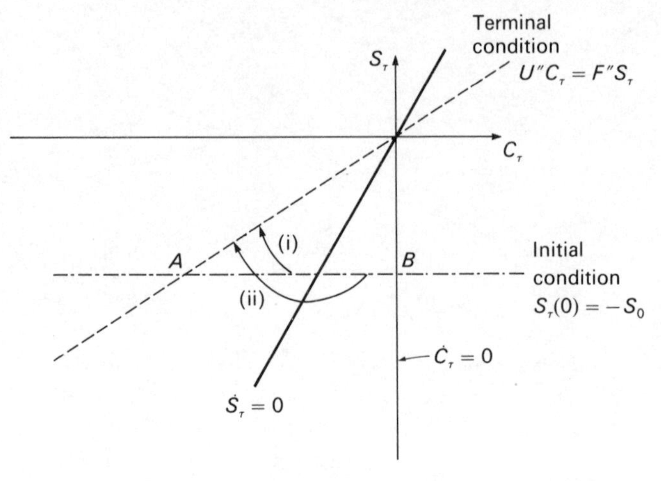

Figure 6.9

reduced for all $t, 0 \leqslant t \leqslant T$, savings ($S$) may rise for all (trajectory i), or part (trajectory ii) of his life (see Figure 6.9).

Case (2) *Wealth transfer tax*

Here, $\tau_2 > 0$, $\tau_1 = 0$. Let $S_\tau \equiv \partial S(t)/\partial \tau_2$ and $C_\tau \equiv \partial C(t)/\partial \tau_2$. Only the transversality condition is affected; it becomes

$$U'[C(T)] = F'[S(T)(1-\tau_2)](1-\tau_2). \tag{6.69}$$

First, we show that C_τ and S_τ have opposite signs. Thus, differentiating the Euler equation with respect to τ_2 gives a parallel result to (6.63) and the same conclusion that $C_\tau(t)$ has the same sign for all $t \in [0, T]$. With $t = T$, (6.59) defines a relation between $C(T)$ and $S(T)$; this may also be differentiated with respect to τ_2, giving

$$S_\tau(T)\mathrm{e}^{-rt} = \frac{\partial}{\partial \tau_2}\int_0^T [w - C(\xi)]\mathrm{e}^{-r\xi}\,\mathrm{d}\xi = -\int_0^T C_\tau(\xi)\mathrm{e}^{-r\xi}\,\mathrm{d}\xi.$$

Since $C_\tau(\xi)$ has the same sign on $[0, T]$, this implies $S_\tau(\xi)$ and $C_\tau(\xi)$ have opposite signs, and hence, in particular $S_\tau(T)$ and $C_\tau(T)$ have opposite signs.

Writing the elasticity of $F'[.]$ as $\eta = F''[.]S(1-\tau_2)/F'[.]$, then differentiating (6.69) w.r.t. τ_2 gives

$$U''[C(T)]C_\tau = -F'[.] + F''[.](1-\tau_2)[S_\tau(1-\tau_2) - S] \tag{6.70}$$

so

$$U''[C(T)]C_\tau(T) = -F'[S(T)](\eta+1) + (1-\tau_2)^2 F''[.]S_\tau. \tag{6.71}$$

Examining the signs of the various terms in (6.71), note that, if $\eta < -1$, then $S_\tau > 0$ (and so $C_\tau < 0$) and if $\eta > -1$, $S_\tau < 0$ and $C_\tau > 0$ (for example, in the latter case, if we assume $S_\tau > 0$, (6.71) requires that $C_\tau > 0$ which contradicts the earlier observation that S_τ, C_τ have opposite signs).

Hence, if $F'[.]$ is elastic, consumption is decreased and savings (S) are increased at all points (so that gross of tax bequests are increased) by an increase in the wealth tax. However, if $F'[.]$ is inelastic, consumption increases and savings fall (gross of tax bequests decrease). In either case, it remains true that net of tax bequests always fall. To see this, note that

$$\frac{\partial}{\partial \tau_2}[S(1-\tau_2)] = S_\tau(1-\tau_2) - S \tag{6.72}$$

Clearly this is negative if $S_\tau < 0$. Alternatively, if $S_\tau > 0$, then $C_\tau < 0$ and, from (6.70),

$$S_\tau(1-\tau_2) - \mathrm{S} = \{\mathrm{U}''[.]\mathrm{C}_\tau + F'[.]\}/\{F''[.](1-\tau_2)\} < 0.$$

Hence $\partial/\partial\tau_2[S(1-\tau_2)] < 0$; net of tax bequests fall in response to an increase in the wealth tax.

6.12 INFINITE HORIZON PROBLEMS

Many economic problems feature an unbounded horizon (particularly economic growth models). In this section, we present some basic theorems (see also Chapter 7). Consider the problem

$$\text{Maximize} \int_0^\infty f^0(x, \dot{x}, t)\mathrm{d}t$$

$$\begin{aligned} \text{subject to } & x(t_0) = x_0, \\ & x_0, t_0 \text{ fixed}, \\ & x_1 \text{ to be specified}, \\ & x(t) \in \mathbb{R}^n. \end{aligned}$$

With an upper limit of ∞, the integral may not converge. Hence, if the existing concept of optimality is to be retained, the set of admissible variations must be restricted to those for which the integral does converge (the alternative to this is to reformulate the concept of optimality – see later).

Clearly, the optimal solution must satisfy the endpoint conditions. Furthermore, since the optimal solution must also be optimal for all subproblems of type (6.4), the Euler and corner conditions must continue to be necessary conditions. In fact, the necessary conditions for infinite horizon problems are identical to those for finite horizon problems if the following points are recognized:

(1) the concept of admissibility must include the requirement that the integral converges, and
(2) the transversality conditions associated with the terminal endpoint do not straightforwardly carry over to the infinite horizon case.

Thus, suppose the problem requires the terminal condition

$$\lim_{t \to \infty} x_i(t) = \bar{x}_i \qquad i = 1, \ldots, n.$$

We might expect the infinite horizon transversality condition analogous to (6.34) to be

$$\lim_{t \to \infty} f^0_{\dot{x}}(x, \dot{x}, t)[z(t) - x(t)] = 0. \tag{6.73}$$

However, various counter examples exist which show that conditions such as this are not in general necessary for optimality

(Arrow and Kurz, 1970, p. 46, quote an example due to Halkin). Without imposing some endpoint condition, the Euler equation may only serve to identify a family of potential solutions. However, in many economic applications, it is often possible to establish from first principles that a solution exists and that all but one trajectory satisfying the necessary conditions are definitely non-optimal (this approach is illustrated in Example 6.11).

Sufficient conditions for infinite horizon problems naturally parallel closely those presented for finite horizon problems (cf. Theorem 6.11). Also, the transversality conditions do play a role in this case.

PROBLEM 6.9

$$\text{Maximize } J(x)=\int_{t_0}^{\infty} f^0(x,\dot{x},t)\mathrm{d}t$$

$$\text{subject to } x(t_0)=x_0,$$

$$t_0, x_0 \text{ fixed},$$

$$\lim_{t\to\infty} x_i(t)=\bar{x}_i \qquad i=1,\ldots,n.$$

THEOREM 6.15 For Problem 6.9, if $f^0(z,\dot{z},t)$ is concave in $(z,\dot{z})$ and if x is admissible and satisfies the Euler and corner conditions and, for all admissible z,

$$f_{\dot{x}}^0(x,\dot{x},t)(x-z)\geqslant 0 \quad \text{for all} \quad t\geqslant t' \quad \text{for some} \quad t'>0$$

(generally depending on z), then x is an optimal solution; $J(x)\geqslant J(z)$ for all admissible z and x is a strong global maximum.

Uniqueness follows from strict concavity without modification, as in Theorem 6.12.

A useful modification to Theorem 6.15 is as follows. Suppose we restrict admissible trajectories to those satisfying $(z(t), \dot{z}(t), t)\in A$, where A is some open subset of $\mathbb{R}^{2n+1}$. Then, if f^0 is concave in $(z,\dot{z})$ on A, $J(x)\geqslant J(z)$ for all admissible z. Furthermore, if A is a convex set, then strict concavity of f^0 entails that x is unique.

The problem of studying existence, boundedness and so on has been treated in some detail in the literature on infinite horizon optimal growth models; useful introductory discussions may be found in Hadley and Kemp (1971) and Takayama (1985).

The above approach rules out trajectories for which the integral does not converge. This is not always, or even usually, satisfactory. Often, it is precisely the trajectories for which the integral does not converge which are of principal economic interest. In such cases, an alternative concept of optimality is required. Section 7.10 of Chapter 7 considers this issue in greater detail; here we shall confine attention to the following **Overtaking Criterion**. Define

$$\Delta J = \int_{t_0}^{t} \{f^0(x(\xi), \dot{x}(\xi), \xi) - f^0(z(\xi), \dot{z}(\xi), \xi)\} \mathrm{d}\xi.$$

DEFINITION 6.7 If there exists a τ such that $\Delta J(t) \geqslant 0$ for all $t \geqslant \tau$, then we say $x(t)$ **overtakes** $z(t)$.

PROBLEM 6.10

$$\text{Optimize } J(x) = \int_{t_0}^{\infty} f^0(x, \dot{x}, t) \mathrm{d}t$$

subject to $x(t_0) = x_0$,

$$\lim_{t \to \infty} x(t) \geqslant \bar{x}.$$

THEOREM 6.16 If x is an admissible solution to Problem 6.10 and satisfies the Euler and corner conditions, if $f^0(z, \dot{z}, t)$ is concave in $(z, \dot{z})$ and if there exists a number t' (in general dependent on z) such that

$$f^0_{\dot{x}}(x(t), \dot{x}(t), t)[z(t) - x(t)] \geqslant 0 \quad \text{for all} \quad t \geqslant t',$$

then x is optimal with respect to the overtaking criterion.

Proof. This may be derived as a special case of Theorem 7.10.

In practice, many of the infinite horizon problems in economics involve an integral of the form

$$J(x)=\int_0^\infty g(x,\dot{x})e^{-rt}dt.$$

Clearly, in this case, the integral exists if g is bounded. The typical optimal solution to this kind of problem is that of an adjustment process toward a long-run steady state (in which $\dot{x}=\ddot{x}=\theta$). Denoting $\hat{x}$ as the steady state, its solution may be derived from the Euler equation

$$g_x e^{-rt}-\frac{d}{dt}\{g_{\dot{x}}e^{-rt}\}=\theta.$$

At the steady state, $d/dt\, g_{\dot{x}}=\theta$; hence the steady state may be determined as the solution to

$$g_x(\hat{x},\theta)+rg_{\dot{x}}(\hat{x},\theta)=\theta. \tag{6.74}$$

The following is an example of this type.

EXAMPLE 6.10 (Capital stock adjustment)

This example is a simplified version of a problem analyzed by Frank Milne (Chapter 5 in Pitchford and Turnovsky, 1977).

Consider a firm employing capital stock $K(t)$ which depreciates at a rate ψ. The firm earns revenue (or quasirents), $R[K(t)]$, undertakes investment $I(t)$ and experiences adjustment costs $C[I(t)]$. Discounted profits are to be maximized over an infinite horizon with discount rate $r>0$. The firm's problem is

$$\text{Optimize}\int_0^\infty [R(K)-C(I)]e^{-rt}\,dt \tag{i}$$

$$\text{subject to } \dot{K}=I-\psi K, \tag{ii}$$

$$K(0)=K_0\geqslant 0,$$

$$K(\infty) \text{ unrestricted.}$$

Equation (ii) is the stock adjustment equation. Non-negativity restrictions on K will be discussed in Chapter 7, Example 7.6.

$R(K)$, $C(I)$ are assumed to have the following properties:

$$R(K) \gtreqless 0 \quad \text{as } K \lesseqgtr \bar{K} \qquad (\bar{K} > 0),$$

$$R'(K) \gtreqless 0 \quad \text{as } K \lesseqgtr K^*, \qquad 0 < K^* < \bar{K},$$

$$R''(K) < 0,$$

$$C(I) > 0, \qquad I \neq 0,$$

$$C(I) = 0, \qquad I = 0,$$

$$C'(I) \gtreqless 0 \quad \text{as } I \gtreqless 0,$$

$$C''(I) > 0,$$

where $\bar{K}$, K^* are given constants. The functions are illustrated in Figure 6.10. The revenue function is thus concave, and positive or negative investment is costly (with increasing adjustment costs being incurred, the faster the adjustment, positive or negative). Only the case where $K(0) = 0$ will be considered in what follows. Equation (ii) allows I to be replaced in the objective function, leaving an intermediate function

$$f^0(K, \dot{K}, t) = \{R(K) - C[\dot{K} + \psi K]\}e^{-rt}.$$

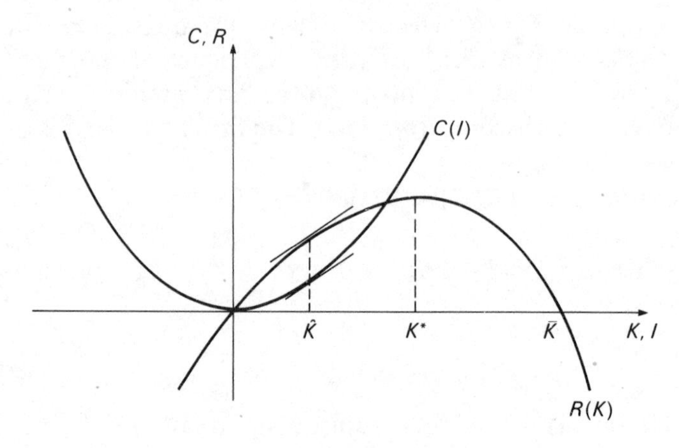

Figure 6.10

The Euler equation then becomes

$$\dot{I}=\{(r+\psi)C'(I)-R'(K)\}/C''(I), \qquad \text{(iii)}$$

where $I=\dot{K}+\psi K$. It proves convenient to retain I as a distinct variable; note that, if K had corners, I would have jump discontinuities.

The Weierstrass necessary condition holds (and, *a fortiori*, the Legendre condition); recall that $E\leqslant 0$ if f^0 is concave in $\dot{K}$. This is true since $C[.]$ is convex in $\dot{K}$; $C''(.)>0$.

Theorem 6.15 indicates the Euler equation will identify a maximizing extremal if $f^0(K,\dot{K},t)$ is concave in $(K,\dot{K})$ for all admissible trajectories and if $f^0_{\dot{K}}(K(t)-z(t))\geqslant 0$ for $t>t'$ for some $t'>0$ (K denotes the optimal trajectory and z is some other admissible trajectory). Now $f^0(K,\dot{K},t)$ is concave in $(K,\dot{K})$, since the Hessian matrix

$$A=\begin{pmatrix} f^0_{kk} & f^0_{k\dot{k}} \\ f^0_{\dot{k}k} & f^0_{\dot{k}\dot{k}} \end{pmatrix}=\mathrm{e}^{-rt}\begin{pmatrix} R''(K)-C''(I) & -C''(I) \\ -C''(I) & -C''(I) \end{pmatrix}$$

is negative definite: $R''(K)-C''(I)<0$ and

$$|A|=\mathrm{e}^{-rt}[-C''(I)R''(K)]>0.$$

If we wish to apply Theorems 6.15 or 6.16, we need to be able to show that

$$f^0_{\dot{K}}(K,\dot{K},t)[K(t)-z(t)]\geqslant 0 \text{ for sufficiently large } t.$$

The problem here is that $z(t)$ is unrestricted as $t\to\infty$ and so this condition need not hold. We therefore consider the trajectories which satisfy the Euler necessary condition and establish from first principles that one is admissible and that the others are definitely not optimal. To do this, we examine the phase diagram (Figure 6.11) for the system of differential equations (ii) and (iii); the trajectories sketched there satisfy the initial condition [$K(0)=0$], the capital accumulation equation and the Euler necessary condition.

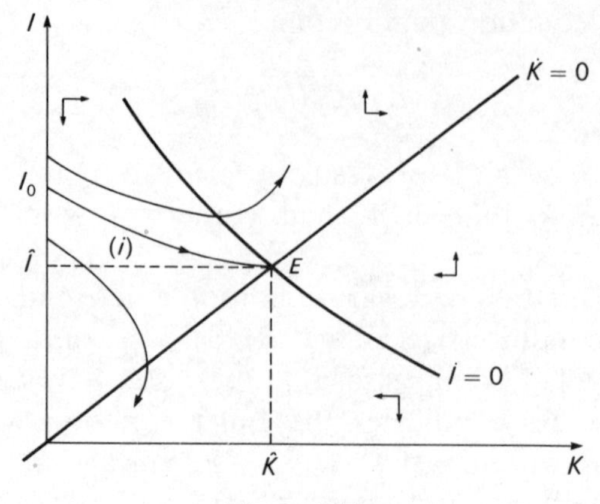

Figure 6.11

The isokines are

$$\dot{I}=0\colon\ (r+\psi)C'(I)-R'(K)=0, \qquad \text{(iv)}$$

$$\dot{K}=\colon\ I-\psi K=0, \qquad \text{(v)}$$

and the equilibrium denoted by E, is given by the simultaneous solution of (iv) and (v). The equilibrium values $\hat{I}$, $\hat{K}>0$. To see this, substitute (v) into (iv) and define the function $V(K)=(r+\psi)C'(\psi K)-R'(K)$. Now, $V(K)$ is continuous with $V(0)<0$ and $V(K^*)>0$ (given the assumptions on C, R).Continuity implies at least one value of K, denoted $\hat{K}$, for which $V(K)=0$ and $0<\hat{K}<K^*$. That $\hat{K}$ is unique follows from the strict monotonicity of $V(K)$; $V'(K)=(r+\psi)C''(\psi K)-R''(K)>0$. The phase diagram is clearly a saddlepoint (the reader may find it instructive to check the local approximation system properties). Figure 6.11 illustrates extremal trajectories; that is, trajectories which satisfy the capital accumulation equation and the Euler necessary condition. Trajectory i, indicates the stable arm with I_0, its corresponding initial level of investment. Our aim is to show that this is admissible and that the other trajectories are not optimal.

Trajectory i, is admissible since it satisfies the initial condition and the integral converges: $C(I_0) > C(I_i)$ so

$$-C(I_0)/r < \int_0^\infty \{R(K_i) - C(I_i)\}e^{-rt}dt < \int_0^\infty \{R(\hat{K}) - C(\hat{I})\}e^{-rt}dt$$

$$= [R(\hat{K}) - C(\hat{I})]/r,$$

since $K_i < \hat{K} \Rightarrow R(K_i) < R(\hat{K})$,

and $I_i > \hat{I} \Rightarrow C(I_i) > C(\hat{I})$.

Now, trajectories starting below i, may be ruled out since they eventually have both $I(t)$ and $K(t)$ becoming increasingly negative. Apart from the economic interpretation of negative capital stock, such trajectories cannot be optimal since it is possible to replace them with alternative admissible trajectories which yield an unambiguously greater value for the integral. The same holds true for trajectories starting above I_0 which have monotonically increasing $K(t)$ and $I(t)$. We shall examine only the latter (and leave it as an exercise for the reader to construct comparison trajectories for the former case).

Consider, then, trajectory starting above I_0. At some point in time t', it will reach a point $\bar{I}, \hat{K}$ where $\bar{I} > \hat{I}$. Thereafter, $K > \hat{K}$ and $I > \bar{I}$. Compare this trajectory on $[t', \infty)$ with the alternative admissible trajectory

$$K(t) = \hat{K}, I(t) = \hat{I} = \psi\hat{K} \quad \text{for} \quad t \geqslant t'$$

(thus, on this alternative trajectory, I jumps from $\bar{I}$ to $\psi\hat{K}$ at t') and clearly

$$\int_{t'}^{\tau} \{R(K) - C(I)\}e^{-rt}dt < \int_{t'}^{\tau} \{R(\hat{K}) - C(\hat{I})\}e^{-rt}dt$$

for all $\tau > t'$ (as the reader may care to verify, given the specification of R and C). Hence, trajectories starting above I_0 are not optimal.

So, having ruled out all other trajectories, assuming a solution exists, trajectory i, is in fact the optimal solution. Thus, investment starts at I_0 at time zero and monotonically declines as capital stock increases. The trajectory converges to a steady state in which $\hat{I}(>0)$ just suffices to maintain the capital stock $\hat{K}$.

CHAPTER 7

OPTIMAL CONTROL THEORY

Chapter 6 provides an introduction to Control theory and the Calculus of Variations. In particular, Sections 6.1–6.5 are important to an understanding of the material of this chapter. In view of the complexity of the derivation of necessary conditions, we first consider a variational approach to the 'simplest' control problem in Section 7.1. In Section 7.2, the necessary conditions are stated for various problems and Section 7.3 contains a sketch of the proof of Pontryagin's maximum principle. Section 7.4 deals with constraints, Section 7.5 with sufficient conditions and Section 7.6 with the particular problems which arise with bounded state variables. Section 7.7 contains an extended economic example which illustrates the application of necessary and sufficient conditions (as well as comparative statics). Finally Section 7.8 is concerned with the difficulties which arise in infinite horizon control problems.

7.1 INTRODUCTION

7.1.1 A variational approach to necessary conditions

The variational approach is not necessarily straightforward, but, if we consider the following 'simplest' control problem and make suitable assumptions about differentiability etc., it provides a useful heuristic introduction to Pontryagin's maximum principle. The problem is

PROBLEM 7.1

$$\text{Maximize } J(u)=\int_{t_0}^{t_1} f^0(x(t),u(t),t)\mathrm{d}t, \tag{7.1a}$$

$$\text{subject to } x(t_0)=x_0, \tag{7.1b}$$

t_0, x_0, t_1 fixed, x_1 free.

$$\dot{x}(t)=f(x(t),u(t),t), \tag{7.1c}$$

$$x(t)\in\mathbb{R}^n, \tag{7.1d}$$

$$u(t)\in U\subset\mathbb{R}^m, \tag{7.1e}$$

$u(t)$ piecewise continuous on $[t_0, t_1]$,

where $f=(f^1,\ldots,f^n)'$.

At this stage, we shall assume f, $f^0\in C_2$. Given the initial condition (7.1b) and a particular choice of control trajectory, u, a unique solution $\phi(t)$ to (7.1c) is assumed to exist.

In this section, it is assumed that there are no constraints on $u(t)$, so $U=\mathbb{R}^m$. The free endpoint problem is more straightforward than the fixed endpoint problem (in contrast to the calculus of variations) since in the fixed endpoint problem, $u(t)$ cannot be totally arbitrary since it has to be chosen to control $x(t)$ to its terminal endpoint (in Section 7.2, alternative assumptions are considered). The derivation of necessary conditions follows the usual procedure; a trajectory which maximizes the integral is assumed to exist and the variation induced in the integral by a variation in the control is calculated. This is followed by setting the first variation to zero. u^* denotes the optimal control which maximizes $J(u)$ and x^* is determined according to

$$\dot{x}^*(t)=f(x^*(t),u^*(t),t) \qquad t_0\leqslant t\leqslant t_1, \tag{7.2a}$$

$$x^*(t_0)=x_0. \tag{7.2b}$$

As in Chapter 6, Section 6.10, we make use of Lagrange multipliers to take account of the constraints. Let

$$\mathscr{L}(u)=J(u)+\int_{t_0}^{t_1}\lambda'(t)[f(x,u,t)-\dot{x}]\mathrm{d}t,$$

so that $$\mathscr{L}(u)=\int_{t_0}^{t_1}\{f^0(x,u,t)+\lambda'(t)[f(x,u,t)-\dot{x}]\}\mathrm{d}t, \tag{7.3}$$

where $\lambda(t)=(\lambda_1(t),\ldots,\lambda_n(t))'$ is the set of multipliers (or 'co-state variables') associated with (7.1c). It is convenient to define the function

$$G(x,\dot{x},u,\lambda,t)\equiv f^0(x,u,t)+\lambda'(t)[f(x,u,t)-\dot{x}]. \tag{7.4}$$

More generally, there is also a multiplier (λ_0) on f^0 as in Section 6.10. At this stage, it is convenient to assume a normal problem such that $\lambda_0=1$. The role of this multiplier is considered in more detail in Section 7.3, where it is shown that the necessary conditions will in fact entail that $\lambda_0\neq 0$ for the above problem and the assumption made here justified.

The optimal trajectory is assumed to terminate at a particular point x_1 at t_1. Of course, variations need not terminate at x_1. Accordingly, consider the variations $\delta x, \delta\dot{x}, \delta u$, and $\delta x(t_1)$. For simplicity, it is assumed that $\dot{x}$, u, $\delta\dot{x}$ and δu are continuous (this assumption is relaxed in Section 7.3). The method is exactly that of Chapter 6, Section 6.5 and the Taylor series expansion yields

$$\Delta\mathscr{L}(u)=\int_{t_0}^{t_1}\{G_x\delta_x+G_{\dot{x}}\delta\dot{x}+G_u\delta u+G_\lambda\delta\lambda\}\mathrm{d}t+\mathscr{O}(.), \tag{7.5}$$

where $\mathscr{O}(.)$ denotes the higher order terms. Note that $G=G(x^*(t), \dot{x}^*(t), u^*(t), \lambda(t), t)$; i.e. all functions are evaluated 'on the extremal'. If (x^*, u^*) is the trajectory which maximizes J, then $\Delta\mathscr{L}\leqslant 0$ for all admissible variations sufficiently small. In the terminology of the Calculus of Variations, we are restricting attention to weak variations (cf. Chapter 6, Section 6.2; Section 7.3 provides a more general approach). Here, $G_x=f_x^0+\lambda' f_x$; $G_{\dot{x}}=-\lambda'$; $G_u=f_u^0+\lambda' f_u$ and $G'_\lambda=f-\dot{x}=\theta$ (hence $G_\lambda\delta\lambda=0$). Substituting these into the above integral

$$\Delta\mathscr{L}=\int_{t_0}^{t_1}\{[f_x^0+\lambda' f_x]\delta x-\lambda'\delta\dot{x}+[f_u^0+\lambda' f_u]\delta u\}\mathrm{d}t+\mathcal{O}(.)\leqslant 0. \quad (7.6)$$

Now

$$\int_{t_0}^{t_1}-\lambda'(t)\delta\dot{x}(t)\mathrm{d}t=[-\lambda'(t)\delta x(t)]_{t_0}^{t_1}+\int_{t_0}^{t_1}\dot{\lambda}'(t)\delta x(t)\mathrm{d}t \quad (7.7)$$

and $\delta x(t_0)=\theta$, given the initial condition that $x(t_0)=x_0$. However, $\delta x(t_1)$ is arbitrary, hence, putting (7.7) into (7.6) yields

$$\int_{t_0}^{t_1}\{[f_x^0+\lambda' f_x+\dot{\lambda}']\delta x+[f_u^0+\lambda' f_u]\delta u\}\mathrm{d}t \\ -\lambda'(t_1)\delta x(t_1)+\mathcal{O}(.)\leqslant 0. \quad (7.8)$$

Now, δx, δu are arbitrary and variations *could* terminate at x_1, hence the integral must equal zero. Furthermore, the Lagrange multipliers are arbitrary (recall $G_\lambda\delta\lambda=0$ since $G_\lambda=\theta$) so may be selected to satisfy the equation

$$\dot{\lambda}'(t)=-f_x^0(x^*(t),u^*(t),t)-\lambda'(t)f_x(x^*(t),u^*(t),t). \quad (7.9)$$

This makes the first term equal to zero. Finally, we must have

$$f_u^0(x^*(t),u^*(t),t)+\lambda'(t)f_u(x^*(t),u^*(t),t)=\theta, \quad (7.10)$$

since δu is an arbitrary continuous function and $f_u^0+\lambda' f_u$ is continuous. This follows from the following theorem (which also holds if $Q(t)$, $\alpha(t)$ are merely piecewise continuous).

THEOREM 7.1 Suppose $Q(t)$ is a continuous function on $[t_0,t_1]$ and $\int_{t_0}^{t_1}Q(t)\alpha(t)\mathrm{d}t=0$ for all continuous functions $\alpha(t)$ on $[t_0,t_1]$. Then $Q(t)=0$ for $t\in[t_0,t_1]$.

Proof. Theorem 6.3 establishes that $Q(t)=k$, constant (when $\alpha(t)$ is constrained to satisfy $\int_{t_0}^{t_1}\alpha(t)\mathrm{d}t=0$). Now, if $k\neq 0$, we can choose $\alpha(t)=k$ on $[t_0,t_1]$ such that $\int_{t_0}^{t_1}Q(t)\alpha(t)\mathrm{d}t=k^2(t_1-t_0)\neq 0$, establishing a contradiction.

Finally, returning to (7.8), since the integral equals zero and $\delta x(t_1)$ is arbitrary, we must also require

$$\lambda(t_1)=\theta \tag{7.11}$$

as a transversality condition.

Equations (7.9), (7.10) and (7.11) in conjunction with (7.1c) are the equations the extremal must satisfy. It proves convenient to define the following **Hamiltonian** function

$$H[x(t),u(t),\lambda(t),t]=f^0(x(t),u(t),t)+\lambda'(t)f(x(t),u(t),t). \tag{7.12}$$

In terms of this function, the necessary conditions may be expressed in the easily remembered format:

$$\dot{x}^*(t)=H'_\lambda[x^*(t),u^*(t),t], \qquad t\in[t_0,t_1], \tag{7.13a}$$

$$\dot{\lambda}(t)=-H'_x[x^*(t),u^*(t),t], \qquad t\in[t_0,t_1], \tag{7.13b}$$

$$\theta=H'_u[x^*(t),u^*(t),t], \qquad t\in[t_0,t_1]. \tag{7.13c}$$

It turns out that even when there are constraints on $u(t)$, only (7.13c) need be modified; if the Hamiltonian is not differentiable and/or u is restricted in some way, (7.13c) is replaced by the intuitively appealing 'Maximum principle' that the Hamiltonian be maximized at each point in time:

$$H[x^*(t),u^*(t),\lambda(t),t]\geqslant H[x^*(t),u(t),\lambda(t),t],$$
$$t\in[t_0,t_1] \text{ for all } u(t)\in U. \tag{7.13c'}$$

In fact, the above conditions could have been derived directly from Theorem 6.13 of the Calculus of Variations, so nothing 'new' is involved in this unconstrained control problem.

Consider the equations (7.13) for the single state variable problem, where $x(t)\in\mathbb{R}$. Then $\lambda(t)\in\mathbb{R}$ and (7.13a), (7.13b) define a pair of first order differential equations of the form

$$\dot{x}=f(x,u,t), \tag{7.14a}$$

$$\dot{\lambda}=g(x,u,\lambda,t). \tag{7.14b}$$

Suppose condition (7.13c) allows u to be expressed in terms of x and

λ. Then (7.14) may be transformed to yield

$$\dot{x} = M(x, \lambda, t),$$

$$\dot{\lambda} = N(x, \lambda, t).$$

Furthermore, in many economic problems, the functions M and N will be autonomous (or a suitable change of variable will render them autonomous), so the problem often becomes one of studying the behaviour of a pair of non-linear autonomous differential equations. This may be done using the techniques developed in Chapter 5. Transversality conditions give the necessary boundary conditions (see Sections 7.1.2 and 7.2). Single state variable problems are relatively much more tractable than 2-state variable problems since the latter involve (at least) $2+2$ differential equations; in practice, the analysis of 2-state variable problems usually requires that they simplify in some particular way (often, that they 'collapse' to 1-state variable problems; for a discussion of 2-state variable problems in an economic context, see Chapter 5 in Pitchford and Turnovsky, 1977).

Economic problems often involve objective functions such as

$$f^0(x, u, t) = Q(x, u)e^{-rt},$$

where e^{-rt} denotes a discount factor. It has been a common practice in such cases to work with transformed multipliers of the type

$$\psi(t) = \lambda(t)e^{rt},$$

since this renders the resulting differential equations autonomous.

7.1.2 Transversality conditions

In this section, the objective function in (7.1) is extended to include a bequest function

$$J(u) = \int_{t_0}^{t_1} f^0(x, u, t)dt + B(x(t_1), t_1), \tag{7.15}$$

and variations in the terminal point (x_1, t_1) are considered.

In Section 7.1.1, with a free endpoint but fixed time, we obtained a term $-\lambda'(t_1)\delta x(t_1)$ (see equation (7.8)). Now, consider a variation in x_1 and t_1. The method is that of Chapter 6, Section 6.5 (equation (6.30); see also Figure 6.3) where we obtained the equation

$$\delta x(t_1) = \delta x_1 - \dot{x}\,\delta t_1 + \mathcal{O}(.). \tag{7.16}$$

The analysis here is identical, except that the function G replaces f^0 and there is an additional bequest function. As usual, if we apply the variational method as in Section 7.1.1, variations could terminate at (x_1, t_1) so we must require the integral terms to equal zero. We are then left with

$$\Delta\mathcal{L} = -\lambda'(t_1)\delta x(t_1) + G|_{t_1}\delta t_1 + B_x\delta x_1 + B_t\delta t_1 + \mathcal{O}(.) \leqslant 0. \tag{7.17}$$

Using (7.16) and noting that $G = f^0 + \lambda' f - \lambda'\dot{x}$, this yields the following necessary condition:

$$[B_x - \lambda']\delta x_1 + [f^0 + \lambda' f + B_t]\delta t_1 \leqslant 0, \tag{7.18}$$

where all the functions are evaluated at t_1 – thus $f^0 = f^0(x^*(t_1), u^*(t_1), t_1)$ etc. Equation (7.18) may be used to derive transversality conditions for various kinds of endpoints; these are considered in some detail in Section 7.2. However, as an example, consider the case where both x_1 and t_1 are free; in this case both δx_1 and δt_1 are arbitrary, hence necessary conditions are that

$$B_x - \lambda' = \theta,$$

$$f^0 + \lambda' f + B_t = 0,$$

where again, all functions are evaluated at the endpoint (x_1, t_1).

7.1.3 Interpretation of the multipliers

The analog here to the interpretation of the multiplier in static optimization is that, under certain conditions, $\lambda(t)$ measures the sensitivity of the maximum value to perturbations in the state vector. Consider the control problem (7.1) and let u^*, x^* denote the optimal control-response pair. Now consider the problem

$$J^*(x_\tau)=\text{Maximize}\int_\tau^{t_1} f^0(x,u,t)dt$$

$$\begin{aligned}\text{subject to } & x(\tau)=x_\tau=x^*(\tau),\\ & \tau, t \text{ fixed, } x_1 \text{ free.}\\ & \dot{x}=f(x,u,t),\end{aligned}$$

where we write $J^*(x_\tau)$ to denote that the maximum value depends on the initial condition. The initial condition for the subproblem is that the trajectory starts at a point on the optimal trajectory for the overall problem; $x_\tau=x^*(\tau)$. By the principle of optimality (Theorem 6.1), the subarcs $u^*_{[\tau,t_1]}$, $x^*_{[\tau,t_1]}$ are the optimal solutions to the subproblem. Furthermore, the Lagrange multiplier $\lambda(t)$ takes the same value for both the overall problem and the subproblem on $[\tau, t_1]$ since the transversality condition defining $\lambda(t_1)$ is the same for both problems and they both must satisfy the necessary condition that $\dot{\lambda}(t)=-H'_x(x^*,u^*,\lambda,t)$. Now consider a change in initial condition for the subproblem such that

$$x(\tau)=x_\tau+\delta x_\tau.$$

Referring to the variational analysis of Section 7.1.2, $J^*(x_\tau)$ and the Lagrangian $\mathscr{L}(x_\tau)$ are given by (7.1a) and (7.3) – in both cases with τ replacing t_0. A perturbation in the initial condition by δx_τ gives rise to the following consequences (note that $J^*(x_\tau)=\mathscr{L}(x_\tau)$)

$$\begin{aligned}\Delta J^*(x_\tau)&=J^*(x_\tau+\delta x_\tau)-J^*(x_\tau)=\Delta\mathscr{L}(x_\tau)\\ &=\int_\tau^{t_1}\{G_x\delta x+G_{\dot{x}}\delta\dot{x}+G_u\delta u+G_\lambda\delta\lambda\}dt\\ &\quad+G_{\dot{x}}|_{t_1}\delta x_1+[G-G_{\dot{x}}\dot{x}]\delta t_1-G_{\dot{x}}|_\tau\delta x_\tau+\mathcal{O}(.),\end{aligned}\qquad(7.19)$$

where $\mathcal{O}(.)$ represents the higher order terms. The necessary conditions take care of the first three terms so we have (since $G_{\dot{x}}|_\tau=\lambda'(\tau)$)

$$\Delta J^*(x_\tau)=-G_{\dot{x}}|_\tau\delta x_\tau+\mathcal{O}(.)=\lambda'(\tau)\delta x_\tau+\mathcal{O}(.).\qquad(7.20)$$

Hence, at points where the limit exists,

$$\partial J^*/\partial x_\tau = \lim_{\delta x_\tau \to 0} \Delta J^*/\delta x_\tau = \lambda'(\tau). \tag{7.21}$$

Furthermore, if alternative terminal endpoint conditions are specified, note that the endpoint transversality conditions will still remain identical for the overall and subproblems so that the multiplier trajectories remain identical for the two problems on $[\tau, t_1]$.

Finally, for the overall problem, the objective function is additive. So, if we denote the maximum value for the overall problem when subject to a perturbation at time τ as

$$J^{**} = J^*(x_0, x_\tau^-) + J^*(x_\tau^+), \tag{7.22}$$

where $x_\tau^- = x^*(\tau)$, $x_\tau^+ = x^*(\tau) + \delta x_\tau$ and $J^*(x_0, x_\tau^-)$ denotes the maximum value for the problem between endpoints x_0, x_τ^-, then since, by the principle of optimality, the subproblem up to time τ remains optimal (so $J^*(x_0, x_\tau^-)$ is a constant), we have

$$\partial J^{**}/\partial x_\tau = \lambda'(\tau). \tag{7.23}$$

Thus, the impact upon the maximum value of a perturbation in x at time τ is measured by the value of the Lagrange multiplier $\lambda(\tau)$.

For example, if J denoted a firm's discounted profits and x_i, its capital stock, then $\lambda_i(\tau)$ would be a measure of how a discrete change in capital stock at time τ would affect overall discounted profits; in this case, it is in the nature of a shadow price.

7.2 NECESSARY CONDITIONS

In this section, the assumptions of Section 7.1 are relaxed and necessary conditions are stated for the control problems of Mayer and Bolza. The latter has an objective function of the form

$$J(u) = \int_{t_0}^{t_1} f^0(x, u, t)\mathrm{d}t + B(x(t_1), t_1), \tag{7.24}$$

and the former

$$J(u)=c'x(t_1). \tag{7.25}$$

The following notation will be adopted: $\hat{f}=(f^0, f^1, \ldots, f^n)'$ and $f=(f^1, \ldots, f^n)'$. In fact, any vector with a ^ will have $n+1$ components. $\hat{f}$ and B are taken to be real valued, defined on $\mathbb{R}^n \otimes \mathbb{R}^m \otimes [T_0, T_1]$ and $\mathbb{R}^n \otimes [T_0, T_1]$ respectively. $[T_0, T_1]$ is some given time interval. f^i, f^i_x, f^i_t, B_x, B_t, $i=0, \ldots, n$ are assumed continuous on $\mathbb{R}^n \otimes \mathrm{cl}\, U \otimes [T_0, T_1]$, where cl U denotes the closure of U in $\mathbb{R}^m$ (U is assumed time invariant). Note that $\hat{f}$ is *not* required to have a continuous partial derivative with respect to u.

The Mayer form of objective function is adopted in the derivation of necessary conditions in Section 7.3 and so is the first problem considered here. It looks like a special case of that of Bolza, but, as we shall see, it is possible to transform the latter into the former (and vice versa). Consider then the following problem (see Takayama, 1985, and Hestenes, 1966, Chapter 7).

PROBLEM 7.2 Maximize $J=c'x(t_1)$

subject to $x(t_0)=x_0$

t_0, t_1, x_0 fixed, x_1 free.

$\dot{x}=f(x(t), u(t), t)$,

$u(t) \in U \subset \mathbb{R}^m$,

$x(t) \in \mathbb{R}^n$,

u piecewise continuous on $[t_0, t_1]$,

$t_0, t_1 \in [T_0, T_1]$,

where $c=(c_1, \ldots, c_n)'$ is a vector of constants (not all zero) and U is independent of $(x(t), t)$. The Hamiltonian for the problem is

$$H[x(t), u(t), \lambda(t), t]=\lambda'(t).f(x(t), u(t), t).$$

THEOREM 7.2 If u^* maximizes J, then there exists a non-zero continuous function $\lambda(t)$ such that, for all $t, t_0 \leqslant t \leqslant t_1$:

$$\dot{x}^*(t)=H'_\lambda[x^*(t), u^*(t), \lambda(t), t], \tag{7.26a}$$

$$\dot{\lambda}(t)=-H'_x[x^*(t), u^*(t), \lambda(t), t], \tag{7.26b}$$

$$H[x^*(t), u^*(t), \lambda(t), t] \geqslant H[x^*(t), u(t), \lambda(t), t]$$

for all $u(t) \in U$, (7.26c)

and

$$\lambda(t_1) = c. \quad (7.26d)$$

At points of discontinuity in $u^*(.)$, (i), (ii), (iii) hold for both $u^*(t^-)$, $u^*(t^+)$; $\dot{x}$, $\dot{\lambda}$ being interpreted as left, resp, right-hand derivatives.

This theorem is commonly called a **Maximum Principle**, or **Pontryagin's Maximum Principle**, in view of the fact that the Hamiltonian is maximized at each point in time for all $u(t) \in U$. Naturally, if f is differentiable with respect to u and $u(t)$ lies in the interior of U at time t, then (7.26c) implies that

$$\partial H[x^*(t), u^*(t), \lambda(t), t] / \partial u = \theta,$$

and, if H_{uu} exists, then H_{uu} must be negative semi-definite (the necessary conditions for an unconstrained maximum). If final time is open, there is an additional degree of freedom in the problem, but there is an additional transversality condition to be added to (7.26); that is (see Section 7.1)

$$(v) \quad \lambda'(t_1)\dot{x}(t_1) = H[x(t_1), u(t_1), \lambda(t_1), t_1] = 0. \quad (7.26e)$$

In an autonomous problem, where $f = f(x(t), u(t))$, $H = 0$ for all t, $t_0 \leqslant t \leqslant t_1$ (see Section 7.3). For alternative endpoint conditions, see Takayama (1985).

Theorems for problems with objective functions of type (7.1a), (7.24) follow from Theorem 7.2. Thus, suppose

$$J(u) = \int_{t_0}^{t_1} f^0(x(t), u(t), t)\mathrm{d}t.$$

Define a new variable, $x^0(t)$, by

$$\dot{x}^0(t) = f^0(x(t), u(t), t), \quad (7.27a)$$

$$x^0(t_0) = 0. \quad (7.27b)$$

Then $J(u)=x^0(t_1)$ and Problem 7.1 becomes

$$\text{Maximize } x_0(t_1) \tag{7.28}$$

subject to the constraints in (7.1) and, additionally, (7.27). Similarly, if

$$J(u)=B(x(t_1).t_1)$$

then define a new variable

$$x_{n+1}(t)=B(x(t),t). \tag{7.29}$$

The problem is to maximize $x_{n+1}(t_1)$, subject to the constraints as in (7.1) plus, in addition, the constraint

$$\dot{x}_{n+1}(t)=B_x\dot{x}(t)+B_t=B_xf+B_t. \tag{7.30}$$

Given these observations, the maximum principle for the following problem may be straightforwardly derived from Theorem 7.2;

PROBLEM 7.3 Maximize $J(u)=\int_{t_0}^{t_1} f^0(x,u,t)\mathrm{d}t+B[x(t_1),t_1]$

subject to $\dot{x}=f(x,u,t)$,
$x(t_0)=x_0$,
x_0,t_0,t_1 fixed, x_1 free.
$u(t)\in U\subset\mathbb{R}^m$, $x(t)\in\mathbb{R}^n$.

Equation (7.27) is used to define $x^0(t)$. The associated Hamiltonian is defined as $H[x(t),u(t),\hat{\lambda}(t),t]=\hat{\lambda}'(t)f(x,u,t)$ where $\hat{\lambda}'(t)=(\lambda_0(t),\ldots,\lambda_n(t))$.

THEOREM 7.3 If u^* maximizes J, then there exists a non-zero continuous function $\hat{\lambda}(t)=(\lambda_0(t),\ldots,\lambda_n(t))'$ such that, for all $t, t_0\leqslant t\leqslant t_1$:

(i) $\dot{\hat{x}}^*(t)=H'_{\hat{\lambda}}[x^*(t),u^*(t),\hat{\lambda}(t),t]$,
(ii) $\dot{\hat{\lambda}}(t)=-H'_{\hat{x}}[x^*(t),u^*(t),\hat{\lambda}(t),t]$,
(iii) $H[x^*(t),u^*(t),\hat{\lambda}(t),t]\geqslant H[x^*(t),u(t),\hat{\lambda}(t),t]$ for all $u(t)\in U$, and
(iv) $\lambda_0=1$, $\lambda'(t_1)=B_x[x(t_1),t_1]$.

To see this, define x_0, x_{n+1} as in (7.27), (7.29). Let

$\psi' = (\psi_0, \ldots, \psi_{n+1})$ and define the Hamiltonian as $H = [f^0, f^1, \ldots, f^{n+1}]\psi$. Apply Theorem 7.2 to this. Theorem 7.3 then follows if $\hat{\lambda}$ is defined by $\lambda_0 = \psi_0$, $\lambda_i = \psi_i + \partial B/\partial x_i$, $i = 1, \ldots, n$.

Problem 7.3 is in a form commonly encountered in economic analysis; Theorem 7.3 addresses a particular endpoint problem. We now consider various endpoint conditions. In Section 7.1 (equation (7.18)), the following necessary condition was obtained:

$$[B_x - \lambda'(t_1)]\delta x_1 + [f^0 + \lambda' f + B_t]\delta t_1 \leqslant 0, \tag{7.31}$$

where all the functions are evaluated at the endpoint (x_1, t_1). Equation (7.31) may be used to derive the transversality conditions for the following problem.

PROBLEM 7.4 Maximize $J(u) = \int_{t_0}^{t_1} f^0(x(t), u(t), t)\mathrm{d}t + B(x(t_1), t_1)$

subject to $\dot{x} = f(x, u, t)$,
$u(t) \in U \subset \mathbb{R}^m$, $x(t) \in \mathbb{R}^n$,
$x(t_0) = x_0$; x_0, t_0 fixed.

C (i) $x_i(t_1)$ free, $i = 1, \ldots, a$,
C (ii) $x_i(t_1) = \bar{x}_i$, fixed, $i = a+1, \ldots, b$,
C(iii) $x_i(t_1) \geqslant \bar{x}_i$, fixed, $i = b+1, \ldots, d$,
C(iv) $h(x_{d+1}, \ldots, x_n) \geqslant 0$, where $h \in C_1$
and where $1 \leqslant a \leqslant b \leqslant d \leqslant n$.

THEOREM 7.4 If u^* maximizes $J(u)$, then there exists a non-zero, continuous function $\hat{\lambda}(t)$ such that, for all $t_0 \leqslant t \leqslant t_1$:

(i) $\dot{\hat{x}}^*(t) = H'_{\hat{\lambda}}[x^*(t), u^*(t), \hat{\lambda}, t)]$,
(ii) $\dot{\hat{\lambda}}(t) = -H'_{\hat{x}}[x^*(t), u^*(t), \hat{\lambda}, t]$,
(iii) $H[x^*(t), u^*(t), \hat{\lambda}, t] \geqslant H[x^*(t), u(t), \hat{\lambda}, t]$ for all $u(t) \in U$,
(iv) $\lambda_0 \geqslant 0$,
(v) Certain transversality conditions as outlined below.

Case (A): t_1 fixed

$\delta t_1 = 0$, hence (7.31) becomes $[B_x - \lambda'(t_1)]\delta x_1 \leqslant 0$. Hence:
A (i) $\lambda_i(t_1) = \partial B[x(t_1), t_1]/\partial x_i$, $i = 1, \ldots, a$,

A (ii) $\lambda_i(t_1)$ free, $i=a+1,\ldots,b$,
A(iii) $x_i(t_1)\geqslant\bar{x}_i$, $\lambda_i(t_1)\geqslant\partial B[x(t_1), t_1]/\partial x_i$, and
$[x_i(t_1)-\bar{x}_i][\lambda_i(t_1)-\partial B/\partial x_i]=0$, $i=b+1,\ldots,d$.
A(iv) There exists a $\psi\geqslant 0$ such that $\psi h=0$ and
$\lambda_i(t_1)=\partial B/\partial x_i+\psi\,\partial h/\partial x_i$, $i=d+1,\ldots,n$.

Notes to transversality conditions for case (A):
C (i) δx_{i1} arbitrary $\Rightarrow \partial B/\partial x_1-\lambda_i(t_1)=0$.
C (ii) $\delta x_{i1}=0$.
C(iii) If $x_{i1}(t_1)>\bar{x}_{i1}$, condition as C(i).
If $x_{i1}(t_1)=\bar{x}_{i1}$, then $\delta x_{i1}\geqslant 0$ but otherwise arbitrary;
$\delta x_{i1}\geqslant 0\Rightarrow[\partial B/\partial x_i-\lambda_i(t_1)]\leqslant 0$.
A(iii) summarizes these conditions.
C(iv) If $h>0$, conditions as for C(i).
If $h=0$, then δx_1 is not completely arbitrary but must satisfy

$$dh=\sum_{i=d+1}^{n}\frac{\partial h}{\partial x_i}\delta x_{i1}\geqslant 0.$$

So, the necessary condition requires that, for any δx_1 satisfying the above,

$$\sum_{i=d+1}^{n}\{\partial B/\partial x_i-\lambda_i(t_1)\}\delta x_{i1}\leqslant 0.$$

Farkas' Lemma (chapter 1, page 14) states that

$\sum_i\{\lambda_i-\partial B/\partial x_i\}\delta x_{i1}\geqslant 0$ if, and only if, there exists a $\psi>0$ such that

$$\psi\,\partial h/\partial x_i=\{\lambda_i(t_1)-\partial B/\partial x_i\}\qquad i=d+1,\ldots,n.$$

The cases $h>0$, $h=0$ are put together under (A)(iv).

Case (B): t_1 free

Conditions are (A)(i)–(iv), but, in addition, since now δt_1 is arbitrary, from (7.31):

(v) $f^0+\lambda'(t_1)f+\partial B/\partial t=0$.

Again, all functions are evaluated at the endpoint.

Case (C): Constraints on t_1

Consider the case $t_1 \leqslant T$ where T is some given number. Then either $t_1 < T$, when case B conditions hold, or the constraint binds and $t_1 = T$; if so, $\delta t_1 \leqslant 0$. We require $(f^0 + \lambda' f + \partial B/\partial t)\delta t_1 \leqslant 0$ to hold. If $\delta t \leqslant 0$, then $f^0 + \lambda' f + \partial B/\partial t \geqslant 0$ is required.
Combining the two cases gives the condition

(v') $t_1 \leqslant T$, $(f^0 + \lambda' f + \partial B/\partial t) \geqslant 0$, $(t_1 - T)(f^0 + \lambda' f + \partial B/\partial t) = 0$,

where all functions are evaluated at the endpoint.

It is worth noting that it is sometimes possible to infer $\lambda_0 > 0$ from the necessary conditions by noting that $\hat{\lambda}(t)$ is required to be non-zero for all t. Thus, for example, if there is no bequest function and x_1 is free, $\lambda(t_1) = \theta \Rightarrow \lambda_0(t_1) \neq 0$ hence we can take $\lambda_0 > 0$ (conventionally, when non-zero, λ_0 is set equal to unity since it is simply a 'scaling factor'; its significance is discussed in Section 7.3).

The key to a tractable problem formulation often lies in the choice of control and state variables. This choice may not be self-evident, and a process of trial, error and inspiration (see Example 7.4) may be necessary. However, a useful rule of thumb is to note if differential equations appear in the basic model; if so, it will often be appropriate to choose as state variables those which appear in derivative form, as in the following two examples (for compactness, these have already been discussed in Chapter 6, Example 6.1 and Section 6.11, respectively).

EXAMPLE 7.1 (The mineowners depletion problem; see Example 6.1). $x(t)$ denotes the rate of extraction at time t and $y(t)$, the amount extracted so far. The inverse demand schedule is given by $p(t) = g[x(t)]$ where $g \in C_2$ and $g'[\,.\,] < 0$. The total stock of resource is S and there are no extraction costs. The problem is to maximize discounted revenues. The control formulation treats

$x(t)$ as the choice or control variable and $y(t)$ as the state variable:

$$\underset{x}{\text{Maximize}} \int_0^T x(t)g[x(t)]e^{-rt}dt$$

subject to $y(0)=0$,
$y(T) \leqslant S$,
$\dot{y}(t)=x(t)$.

The inequality $y(T) \leqslant S$ allows that not all the resource may be extracted within the planning period. The Hamiltonian is

$$H=\lambda_0 x(t)g[x(t)]e^{-rt}+\lambda(t)x(t),$$

and a necessary condition is that

$$\dot{\lambda}=-\partial H/\partial y=0 \Rightarrow \lambda(t)=\lambda, \text{ a constant.}$$

By assumption, there are no constraints on $x(t)$, the control variable. Hence, assuming an interior solution exists, we must have

$$\partial H/\partial x=\lambda_0\{g(x)+xg'(x)\}e^{-rt}+\lambda=0. \tag{i}$$

Hence, if $\lambda_0=0$, so does $\lambda(t)$, contradicting the necessary condition that not all multipliers are simultaneously zero. Hence we may take $\lambda_0=1$. Then (i) becomes, naturally enough, the same result as was obtained using the Calculus of Variations (Examples 6.1 and 6.8). Now, if $\lambda=0$, then $g(x)+xg'(x)=0$ (marginal revenue equals zero for all t). Let $\bar{x}$ denote the solution to this equation; then $\bar{x}$ must also satisfy the terminal condition that

$$y(T)=\int_0^T \bar{x}\,dt=T\bar{x} \leqslant S.$$

If satisfied, this would constitute an unconstrained solution. If violated, then we must have $\lambda<0$ (from transversality condition C(iii) of Theorem 7.4) and the constrained solution discussed in Example 6.1; in this case, (i) indicates that marginal revenue increases exponentially over time at the rate of interest (discounted marginal revenue is constant).

EXAMPLE 7.2 (The lifetime consumption problem) This problem

is analyzed in detail as a problem in the Calculus of Variations in Chapter 6, Section 6.11 (to which the reader is referred). The objective function is

$$\text{Maximize } J(c)=\int_0^T U[c(t)]\mathrm{e}^{-qt}\mathrm{d}t+F[S(T)]\mathrm{e}^{-qT}, \tag{i}$$

and the asset accumulation equation is

$$\dot{S}(t)=rS(t)+w-c(t). \tag{ii}$$

A natural control formulation of this problem takes consumption as the control variable and assets as the state variable. The Hamiltonian is

$$H=\lambda_0 U[c]\mathrm{e}^{-qt}+\lambda(rS+w-c). \tag{iii}$$

If we ignore possible non-negativity restrictions on c, S, the solution, if it exists, will be an interior solution (sufficient conditions and state variable constraints are discussed in Sections 7.5–7.6). The necessary conditions follow from Theorem 7.3:

$$\dot{S}=\partial H/\partial\lambda, \tag{iv}$$

$$\dot{\lambda}=-\partial H/\partial S, \tag{v}$$

$$\partial H/\partial c=0, \tag{vi}$$

$$\lambda_0\geqslant 0,$$

and the transversality condition

$$\lambda(T)=\partial B/\partial S=F'[S(T)]\mathrm{e}^{-qT} \tag{vii}$$

from A(i) above. Equation (iv) gives the asset equation (ii). Equation (v) gives

$$\dot{\lambda}=-r\lambda\Rightarrow\lambda(t)=\lambda(0)\mathrm{e}^{-rt}, \tag{viii}$$

(vi) gives

$$\lambda_0 U'(c)\mathrm{e}^{-qt}=\lambda(t)=\lambda(0)\mathrm{e}^{-rt}, \tag{ix}$$

(vii) implies that $\lambda(T)>0$, hence, from (viii), $\lambda(t)>0$ and, from (ix), $\lambda_0>0$. Take therefore $\lambda_0=1$.

$\lambda(0)$ can be eliminated from (ix) by differentiation, giving

$$U''(c)\dot{c}=(q-r)U'(c)). \tag{x}$$

Equation (x) is precisely the same as that derived in Section 6.11. The transversality condition is also naturally the same; from (vii) and (ix)

$$U'[c(T)]e^{-qT}=\lambda(T)=F'[S(T)]e^{-qT},$$

hence $U'[c(T)]=F'[S(T)]$ as in equation (6.57).

Equations (x) and (ii) form a pair of differential equations, the analysis of which is, of course, identical to that in Section 6.11 to which the reader is referred.

7.3 PONTRYAGIN'S MAXIMUM PRINCIPLE

7.3.1 The control problem

In this section, we present an outline of the proof for the autonomous problem. Our presentation follows the line of the original proof by Pontryagin *et al.* (1962) and, in spirit, that of Athans and Falb (1966). In contrast to the variational approach of Section 7.1, f^0, f are no longer required to be differentiable with respect to u and constraints on $u(t)$ are now permissible. This complicates the variational analysis, particularly for the fixed endpoint problem (the variation in the control trajectory in this case cannot be totally arbitrary since the control has to carry the system to a particular endpoint) and motivates the following approach. Space considerations preclude a detailed and rigorous presentation; our aim is to explain its nature and to make the full proof (Pontryagin *et al.*, 1962, Lee and Markus, 1967) more accessible to the reader. The problem under consideration is given as follows.

PROBLEM 7.5

$$\text{Maximize } J(u)=\int_{t_0}^{t_1} f^0(x(t),u(t))\mathrm{d}t \tag{7.32}$$

subject to $\dot{x}=f(x(t),u(t))$,
$x(t_0)=x^0$, fixed,
$x(t_1)=x^1$, fixed,
t_0 fixed, t_1 free.
$x(t)\in\mathbb{R}^n$,
$u(t)\in U\subset\mathbb{R}^{nm}$,
where u is a piecewise continuous function.

As before, we adopt the notation $f\equiv(f',\ldots,f^n)'$, $\hat{f}\equiv(f^0,\ldots,f^n)'$, $x\equiv(x_1,\ldots,x_n)'$, $\hat{x}\equiv(x_0,\ldots,x_n)'$ and so on; that is, all variables with a ^ are $n+1$-tuples. As before, $\hat{f}$ is assumed to be continuous in (x,u) and continuously differentiable in x.

The above problem is autonomous; t does not feature explicitly in $\hat{f}$. This is not unduly restrictive since non-autonomous problems may be readily transformed into this format. Thus, consider the objective function

$$J(u)=\int_{t_0}^{t_1} f^0(x(t),u(t),t)\mathrm{d}t,$$

and suppose

$$\dot{x}=f(x(t),u(t),t).$$

Define an additional variable x_{n+1} by

$$\dot{x}_{n+1}=1, \qquad x_{n+1}(t_0)=t_0.$$

By this means, time may be embedded in the state vector:

$$J(u)=\int_{t_0}^{t_1} f^0(x(t),u(t),x_{n+1}(t))\mathrm{d}t,$$

and

$$\dot{x}=f(x(t),u(t),x_{n+1}(t))$$

$$\dot{x}_{n+1}=1.$$

Returning to the Problem 7.5, we define the additional variable $x^0(t)$ as in Section 7.2 by

$$\dot{x}_0 = f^0(x(t), u(t)),$$

$$x_0(t_0) = 0,$$

so that the objective function becomes

$$\text{Maximize } J(u) = x^0(t_1).$$

The Hamiltonian function is defined as

$$H[x(t), u(t), \hat{\lambda}(t)] \equiv \hat{\lambda}'(t)\hat{f}(x(t), u(t)). \tag{7.33}$$

Note that the Hamiltonian does not involve $x^0(t)$ explicitly. Points at which $u(t)$ is continuous are termed regular points (there are thus only a finite number of non-regular points; almost all points are regular points). The theorem to be 'proved' is as follows.

THEOREM 7.5 If u^* is admissible and maximizes $J(u)$, then there exists a non-zero continuous function $\hat{\lambda}(t) \equiv (\lambda_0(t), \ldots, \lambda_n(t))'$ such that:

[i] $\dot{\hat{x}}^*(t) = H'_{\hat{\lambda}}[x^*(t), u^*(t), \hat{\lambda}(t)]$, $t_0 \leqslant t \leqslant t_1$,

[ii] $\dot{\hat{\lambda}}(t) = -H'_{\hat{x}}[x^*(t), u^*(t), \hat{\lambda}(t)]$, $t_0 \leqslant t \leqslant t_1$,

[iii] $H[x^*(t), u^*(t), \hat{\lambda}(t)] \geqslant H[x^*(t), u(t), \hat{\lambda}(t)]$, $t_0 \leqslant t \leqslant t_1, u(t) \in U$,

[iv] $\lambda_0(t) = \lambda_0$, constant, $\geqslant 0$,

In the autonomous case, it is also true that

[v] $H[x^*(t), u^*(t), \hat{\lambda}(t)] = k$, constant

although this is not generally true for non-autonomous problems. Furthermore, with t_1 free, $k = 0$.

Because H does not depend on x_0 explicitly, $H_{\hat{x}} = (0, \partial H/\partial x_1, \ldots, \partial H/\partial x_n)$. Thus (7.33) above implies that $\dot{\lambda}_0(t) = 0$ so we may write

$$\lambda_0(t) = \lambda_0 = \text{constant}.$$

Geometrically, the problem is one of moving from x^0 to x^1 so as to maximize the value of $x_0(t_1)$. This is illustrated in Figure 7.1 for $x(t) \in \mathbb{R}^2$, where $\hat{x}^*$ denotes the optimal trajectory in $\mathbb{R}^{2+1}$ and $\hat{x}$, a suboptimal trajectory.

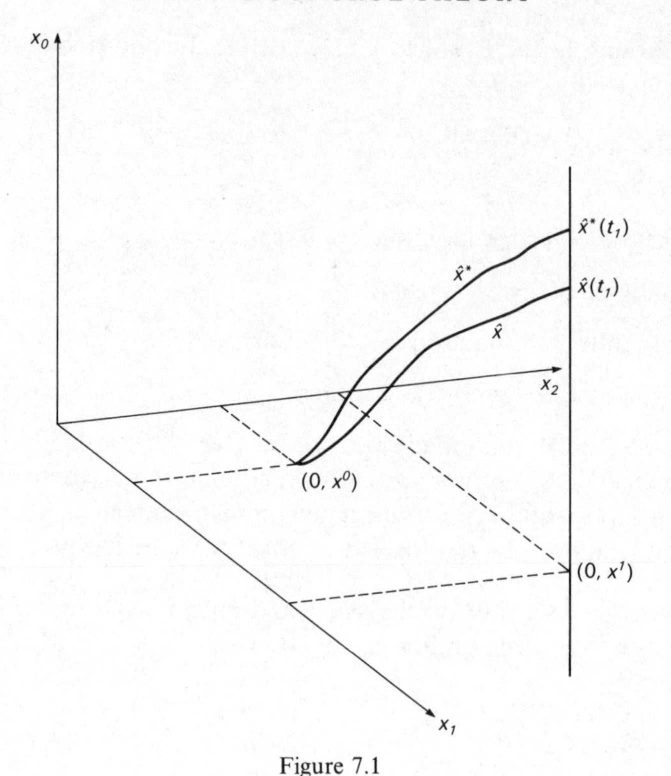

Figure 7.1

We shall refer to $\hat{x}(t)$ as a point in extended state space, and $x(t)$ as a point in state space. t_1 is the time at which the optimal trajectory reaches x^1.

7.3.2 Behaviour of the co-state variables

According to the theorem, the multipliers or co-state variables are governed by the differential equation [ii];

$$\dot{\hat{\lambda}} = -H'_{\hat{x}}[x(t), u(t), \hat{\lambda}(t)] = -\hat{f}'_{\hat{x}} \cdot \hat{\lambda}. \tag{7.34}$$

We first examine some properties of this system. For given $\hat{x}$, u, the matrix $\hat{f}_{\hat{x}}$ is clearly of the form $A(t)$ where $A(t)$ is an $n+1 \times n+1$ matrix of time varying coefficients; it is thus a linear system. $A(t)$ is piecewise continuous and hence the solution, $\lambda(t)$, is continuous (see Section 5.3). (7.34) is often referred to as an **adjoint** system to the linear system

$$\dot{\hat{w}} = \hat{f}_{\hat{x}} \, . \, \hat{w}. \tag{7.35}$$

Before discussing the interpretation and significance of (7.34) and (7.35), we consider some properties of adjoint systems. The systems

$$\dot{w}(t) = A(t) . w(t), \tag{7.36}$$

$$\dot{v}(t) = -A'(t) . v(t), \tag{7.37}$$

where $w(t)$, $v(t) \in \mathbb{R}^n$ and $A(t)$ is an $n \times n$ matrix of time varying elements, are said to be adjoint systems. Let $\theta(t, t_0)$ and $\psi(t, t_0)$ denote their respective fundamental matrices such that $\theta(t,t)$ and $\psi(t,t) = I$ for all t and such that $w(t_0) = w_0, v(t_0) = v_0$. Then the solutions to (7.36), (7.37) may be written as

$$w(t) = \theta(t, t_0) w_0, \tag{7.38}$$

$$v(t) = \psi(t, t_0) v_0. \tag{7.39}$$

An important property of these equations is that

$$\psi'(t, t_0) \theta(t, t_0) = I \tag{7.40}$$

(where I is the identity matrix), so that

$$\theta(t, t_0) = [\psi'(t, t_0)]^{-1}.$$

To see this, note that (7.40) holds for $t = t_0$ since $\psi(t_0, t_0) = \theta(t_0, t_0) = I$. Now, consider an arbitrary vector $\xi \in \mathbb{R}^n$ and let $y(t) \in \mathbb{R}^n$ be defined by

$$y(t) = \psi'(t, t_0) \theta(t, t_0) \xi. \tag{7.41}$$

Differentiating this gives

$$\dot{y}(t) = [\dot{\psi}'(t, t_0) \theta(t, t_0) + \psi'(t, t_0) \dot{\theta}(t, t_0)] \xi. \tag{7.42}$$

Now

$$\dot{\psi}(t, t_0) = -A'(t) \psi(t, t_0)$$

(since $\dot{v} = -A'v$, $v = \psi v_0$ hence $\dot{\psi} v_0 = \dot{v} = -A'v = -A' \psi v_0$).

Transposing

$$\dot{\psi}'(t,t_0) = -\psi'(t,t_0)A(t). \tag{7.43}$$

Also, similarly,

$$\dot{\phi}(t,t_0) = A(t)\Theta(t,t_0), \tag{7.44}$$

Substituting (7.43), (7.44) into (7.42) gives

$$\dot{y}(t) = \{-\psi'(t,t_0)A(t)\Theta(t,t_0) + \psi'(t,t_0)A(t)\Theta(t,t_0)\}\xi$$

so $\dot{y}(t) = \theta$ and, since $y(t_0) = \xi$, we have $y(t) = \xi$ for all t. Hence

$$y(t) = \psi'(t,t_0)\Theta(t,t_0)\xi = \xi$$

and hence (7.40). Furthermore,

$$v'(t)w(t) = v_0'\psi'(t,t_0)\Theta(t,t_0)w_0 = v_0'w_0 = \text{constant} \tag{7.45}$$

for all t. That is, geometrically, the angle between $v(t)$ and $w(t)$ remains constant over time.

In what follows, we shall use a system of type (7.35) to describe how perturbations to x at a point in time propagate over time. The interpretation of the adjoint co-state system is deferred until this has been done.

7.3.3 Spatial variations

The consequences of pulse variations in the optimal control u^* are now considered (Figure 7.2 illustrates such a pulse and the consequences for the state variable for the case where $x(t), u(t) \in \mathbb{R}$). The interval over which the pulse is defined is $T = (b - \varepsilon a, b] \subset [t_0, t_1]$ for some $\varepsilon > 0$. It is assumed that b is a regular point and a is some positive constant. The variation is defined as

$$\begin{aligned} u_\varepsilon(t) &= u^*(t) \qquad t \notin T \\ &= v \qquad\qquad t \in T, \qquad v \in U, \end{aligned} \tag{7.46}$$

where v is an admissible vector of constants. The pulse in u^* will influence the state trajectory for all $t > b - \varepsilon a$. Denoting the

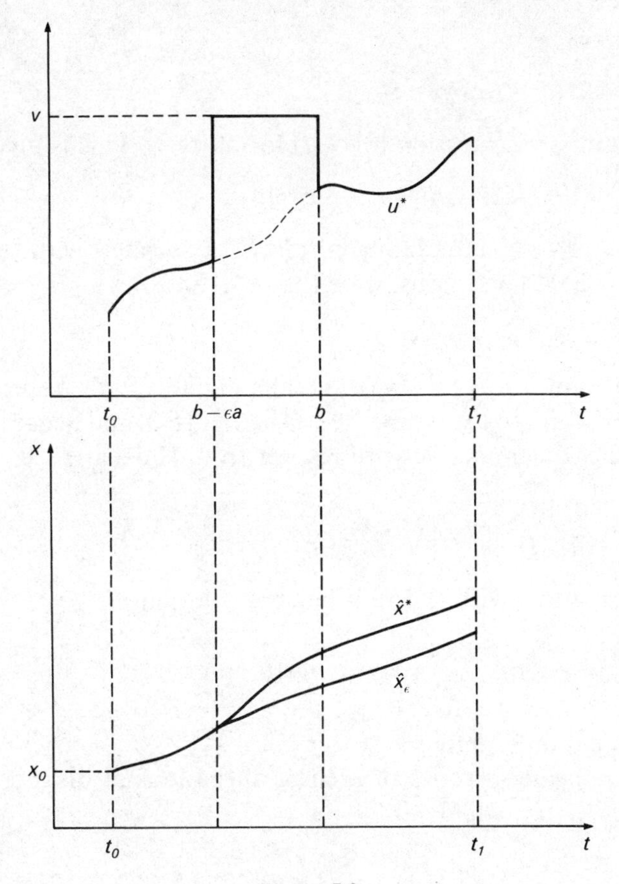

Figure 7.2

associated state trajectory as $\hat{x}_\varepsilon(t)$, then

$$\hat{x}_\varepsilon(t)=\hat{x}^*(t), \qquad t_0 \leqslant t \leqslant b-\varepsilon a.$$

The divergence induced by time b may be expressed as

$$\begin{aligned}\hat{x}_\varepsilon(b) &= \hat{x}^*(b)+\varepsilon a[\dot{\hat{x}}(b)-\dot{\hat{x}}^*(b)]+\mathcal{O}(\varepsilon)\\ &= \hat{x}^*(b)+\varepsilon a\{\hat{f}(x^*(b),v)-\hat{f}(x^*(b),u^*(b))\}+\mathcal{O}(\varepsilon).\end{aligned} \tag{7.47}$$

For $t>b$, this deviation propagates over time. Defining

$$\hat{w}(t,v)\equiv \hat{f}(x^*(t),v)-\hat{f}(x^*(t),u^*(t)), \tag{7.48}$$

(7.47) becomes

$$\hat{x}_\varepsilon(b) = \hat{x}^*(b) + \varepsilon a \hat{w}(b, v) + \mathcal{O}(\varepsilon). \tag{7.49}$$

We require an expression for $\hat{x}_\varepsilon(t)$ for all $t > b$. This is given by

$$\hat{x}_\varepsilon(t) = \hat{x}^*(t) + \varepsilon a \Theta(t, b) \hat{w}(b, v) + \mathcal{O}(\varepsilon), \tag{7.50}$$

where $\Theta(t, b)$ is the fundamental matrix associated with the linear system (7.35). To see this, note that $\hat{x}_\varepsilon(t)$ satisfies

$$\dot{\hat{x}}_\varepsilon = \hat{f}(x_\varepsilon, u^*), \qquad b \leqslant t \leqslant t_1 \tag{7.51}$$

and the solution to this, $\hat{x}_\varepsilon(t)$, is differentiable with respect to the initial condition $x_\varepsilon(b)$ at time b (see Theorem 5.3) and hence is, by the chain rule, differentiable with respect to ε. Thus, at $\varepsilon = 0$,

$$\frac{\partial}{\partial \varepsilon} \dot{\hat{x}}_\varepsilon = \frac{\mathrm{d}}{\mathrm{d}t} \left\{ \frac{\partial \hat{x}_\varepsilon}{\partial \varepsilon} \right\} = \hat{f}_{\hat{x}}(\hat{x}^*, u^*) \frac{\partial \hat{x}_\varepsilon}{\partial \varepsilon}. \tag{7.52}$$

Comparing this with (7.35), it has the solution

$$\frac{\partial}{\partial \varepsilon} \hat{x}_\varepsilon(t) = \Theta(t, b) \frac{\partial \hat{x}_\varepsilon(b)}{\partial \varepsilon} \; (= a \Theta(t, b) \hat{w}(b, v))$$

hence equation (7.50).

If we consider a series of such perturbations at distinct regular points $b_1, \ldots, b_s$, where

$$t_0 < b_1 < \ldots < b_s < t_1$$

and s is any finite positive integer such that the pulses are defined as constant vectors v_i on intervals $(b_i - \varepsilon a_i, b_i)$, $i = 1, \ldots, s$, then the resulting response may be written as

$$\hat{x}_\varepsilon(t) = \hat{x}^*(t) + \varepsilon \sum_{i=1}^{s} a_i \Theta(t, b_i) \hat{w}_i(b_i, v_i) + \mathcal{O}(\varepsilon), \tag{7.53}$$

where

$$\hat{w}_i(b_i, v_i) \equiv [\hat{f}(x^*(b_i), v_i) - \hat{f}(x^*(b_i), u^*(b_i))].$$

Equation (7.53) evaluated at $t = t_1$ gives the overall result of these perturbations. The vectors $\hat{w}_i \in \mathbb{R}^{n+1}$ are commonly referred to as

state perturbation vectors. Examining (7.50) and (7.53), the effect of all such perturbations, at t_1, is, to first order in ε, a convex cone. Let us therefore define the following cones

$$k(\tau)=\left\{\sum_{i=1}^{s} a_i\hat{w}_i(\tau, v_i)\,\middle|\,a_i\geqslant 0, v_i\in U, s \text{ any positive integer}\right\}, \quad (7.54)$$

$$K=\left\{\sum_{i=1}^{n} a_i z_i\,\middle|\,a_i\geqslant 0, z_i=\theta(t_1,\tau_i)\hat{w}_i, \tau_i\in(t_0,t_1), \hat{w}_i\in k(\tau_i)\right\}. \quad (7.55)$$

Thus, by construction, the terminal state can be deflected in any direction pointing into, or along, the cone K, through the application of a pulse or series of pulses. However, this cannot include the direction $(1,0,0,0,\ldots,0)\equiv\hat{d}$ since this would imply the attainability of a higher value of x^0, contradicting the optimality of u^*. Therefore, since K is both convex and does not contain $\hat{d}$, it follows that there must exist a plane $\mathscr{P}$ separating K from the direction $\hat{d}$; that is to say, all rays of K must point to one side of (or along) $\mathscr{P}$, whilst $\hat{d}$ points in a direction to the other side of $\mathscr{P}$ (such a $\mathscr{P}$ need not be unique, but, if the set of attainability is smooth at $\hat{x}^*(t_1)$, then naturally, $\mathscr{P}$ is the tangent plane to this set at this point). The hyperplane $\mathscr{P}$ may be defined by the equation

$$\hat{\psi}'.\hat{z}=0, \quad (7.56)$$

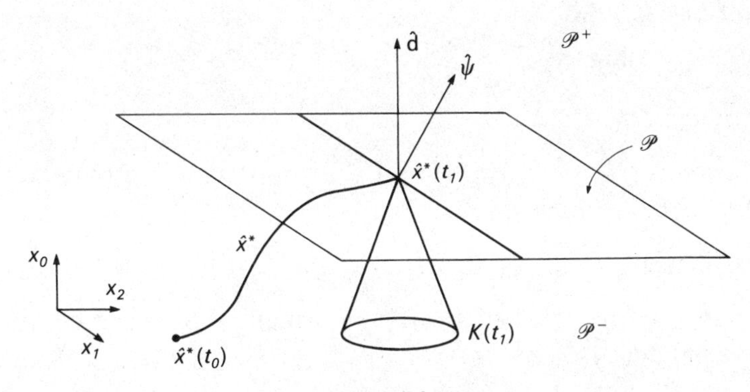

Figure 7.3

where $\hat{z}\in\mathbb{R}^{n+1}$ and $\hat{\psi}\in\mathbb{R}^{n+1}$ is the normal to the hyperplane such that

$$\hat{\psi}'.\hat{z}\leqslant 0 \quad \text{if } \hat{z}\in K, \text{ and} \tag{7.57}$$

$$\hat{\psi}.\hat{d}\geqslant 0. \tag{7.58}$$

Identifying the co-state vector with the normal to this hyperplane by defining $\hat{\lambda}(t_1)\equiv\hat{\psi}$, then (7.56) becomes

$$\hat{\lambda}'(t_1).\hat{z}=0. \tag{7.59}$$

The adjoint system

$$\dot{\hat{\lambda}}=-\hat{f}_{\hat{x}}'\hat{\lambda} \tag{7.60}$$

is used to define how this hyperplane evolves over time. Letting

$$\hat{\lambda}(t)=\psi(t,t_0)\hat{\lambda}(t_0) \tag{7.61}$$

denote the solution to (7.60), where $\psi(t,t_0)$ is the fundamental matrix for (7.36), $\hat{\lambda}(t_0)$ is thus defined via

$$\hat{\lambda}(t_0)=\psi^{-1}(t_1,t_0)\hat{\lambda}(t_1). \tag{7.62}$$

We have already noted that λ_0 is constant over time; (7.58) implies that $\lambda_0\geqslant 0$ (part (iv) of the theorem) since $\hat{d}=(1,0,\ldots,0)'$.

Now consider a regular $t\in(t_0,t_1]$ and state perturbation vector

$$\hat{w}(t,v)=[\hat{f}(x^*(t),v)-\hat{f}(x^*(t),u^*(t))] \qquad (v\in U) \tag{7.63}$$

(see equation (7.48)); then, for all $\varepsilon>0$, sufficiently small,

$$\varepsilon\theta(t_1,t)\hat{w}(t,v)\in K$$

and hence, from (7.57)

$$\hat{\lambda}'(t_1)\theta(t_1,t)\hat{w}(t,v)\leqslant 0. \tag{7.64}$$

However, $\hat{\lambda}'(t_1)=\hat{\lambda}'(t)\psi(t_1,t)$ and the property of adjoint systems that $\psi'(a,b)\theta(a,b)=I$ implies that

$$\hat{\lambda}'(t)\hat{w}(t,v)\leqslant 0 \qquad [v\in U, t\in(t_0,t_1)]. \tag{7.65}$$

Written out in terms of the Hamiltonian, this becomes, for all regular t

$$H[x^*(t),u^*(t),\hat{\lambda}(t)]\geqslant H[x^*(t),v,\hat{\lambda}(t)] \quad \text{for all } v\in U. \tag{7.66}$$

This completes our sketch of the proof; we have shown that there exists a continuous vector $\hat{\lambda}(t)$ such that conditions (i)–(iv) hold if u^* maximizes $J(u)$. Additionally, it can be shown that H is a constant in the autonomous problem and equals zero in the free terminal time problem (see Athans and Falb, 1966, Chapter 5). The result is obvious if sufficient differentiability and an interior solution ($u^*(t) \in \operatorname{Int} U$ for all t) is assumed, since it then follows that

$$\frac{\mathrm{d}}{\mathrm{d}t} H = H_{\hat{x}}\dot{\hat{x}} + H_u \dot{u} + H_{\hat{\lambda}}\dot{\hat{\lambda}}') + H_t = (H_{\hat{x}} + \dot{\hat{\lambda}}')f + H_u \dot{u} + H_t = H_t$$

(since $H_{\hat{x}} + \dot{\hat{\lambda}} = \theta$ and $H_u = \theta$ by the maximum principle). If the problem is autonomous, $H_t = 0$, hence $\mathrm{d}H/\mathrm{d}t = 0$ so the Hamiltonian is constant over time. With free terminal time, the terminal condition is that

$$H(x^*(t_1), u^*(t_1), \hat{\lambda}(t_1)) = 0,$$

and so H equals zero for all t.

The reason why λ_0 is conventionally taken as 0 or 1 follows from the observation that, if $\lambda_0 \neq 0$, then clearly

$$\frac{1}{\lambda_0}\{\hat{\lambda}'(t_1)\hat{z}\} = \left\{\frac{1}{\lambda_0}\hat{\lambda}'(t_1)\right\}\hat{z} = 0$$

defines the same hyperplane as (7.59), hence we may as well take $\lambda_0 = 1$. If $\lambda_0 = 0$, the hyperplane includes the direction of increasing pay-off but the problem does not depend explicitly on the pay-off.

7.4 CONSTRAINTS ON STATE AND CONTROL VARIABLES

So far, only restrictions on $u(t)$ and the endpoints have been considered. In this section, global (iso-perimetric) and local equality and inequality constraints are considered. In fact, iso-perimetric constraints do not usually require a separate analysis since (as in the calculus of variations) they may be accounted for by introducing suitably defined additional state variables. Thus consider the

Problem 7.1 and suppose it to be subject to an additional constraint of the type

$$\int_{t_0}^{t_1} g(x,u,t)\mathrm{d}t \geqslant S, \tag{7.67}$$

where S is some given constant and $g \in C_1$. Define a new variable $x_{n+1}(t)$ by

$$\dot{x}_{n+1} = g(x,u,t),$$
$$x_{n+1}(t_0) = 0.$$

Problem 7.1 then becomes

$$\underset{u}{\text{Maximize}}\; J(u) = \int_{t_0}^{t_1} f^0(x,u,t)\mathrm{d}t$$

$$\begin{aligned}\text{subject to } & \dot{x} = f(x,u,t),\\ & \dot{x}_{n+1} = g(x,u,t),\\ & x(t_0) = x_0,\\ & x_{n+1}(t_0) = 0,\\ & x(t_1), t_1 \text{ – conditions to be specified},\\ & x_{n+1}(t_1) \geqslant S.\end{aligned}$$

This is a problem of a type already considered.

Local constraints may be inequalities or equalities:

$$g^j(x,u,t) \geqslant 0,$$

or

$$g^j(x,u,t) = 0.$$

These include constraints such as $g^j(x,u) = (\geqslant)0$, $g^j(u,t) = (\geqslant)0$ and $g^j(u) = (\geqslant)0$ as special cases – but constraints not involving the control vector $u(t)$ (the so-called bounded state variable case), $g^j(x,t) = (\geqslant)0$, will be treated separately in Section 7.6. Hestenes (1966, Chapter 8) considers the above problem and also includes, in addition, a vector of control parameters in the problem structure. Control parameters are constants over $[t_0, t_1]$ which may be chosen so as to maximize the objective function.

PROBLEM 7.6

$\underset{u,b}{\text{Maximize}}\, I^0$

$$\text{subject to } I^k \geqslant 0 \qquad k=1,\ldots,p', \tag{i}$$

$$I^k=0 \qquad k=p'+1,\ldots,p, \tag{ii}$$

$$\dot{x}=f(x,u,b,t), \tag{iii}$$

$$g^j(x,u,b,t)\geqslant 0 \qquad j=1,\ldots,q', \tag{iv}$$

$$g^j(x,u,b,t)=0 \qquad j=q'+1,\ldots,q, \tag{v}$$

where

$$I^0=S^0(b)+\int_{t_0}^{t_1} f^0(x,u,b,t)\mathrm{d}t,$$

$$I^k=S^k(b)+\int_{t_0}^{t_1} h^k(x,u,b,t)\mathrm{d}t,$$

$$t_0=t_0(b), \qquad t_1=t_1(b),$$

$$x(t_0)=x_0(b), \qquad x(t_1)=x_1(b),$$

$$S^i, f^i, g^i, h^i, x_0, x_1 \in C_1,$$

$$x(t)\in\mathbb{R}^n,$$

$u(t)\in\mathbb{R}^m$, where u is a piecewise continuous function,

$$b\in B\subset\mathbb{R}^r.$$

Here, $p, p', q, q' \in \mathbb{R}$ are given non-negative contants, $I^k, k=1,\ldots,p$ are a set of integral constraints and b, an r-dimensional control parameter. B is a given open set. The set W of admissible elements (x,u,b,t) is defined as

$$W\equiv\{(x,u,b,t); b\in B, t\in[t_0,t_1], g^j(x,u,b,t)\geqslant(=)0\, j=1,..,q'(q'+1,..,q)\}.$$

Any constraints on $u(t)$ must be defined in terms of the g-constraints in this case. It is possible to eliminate control parameters from the f, g and h functions by definition of suitable additional state variables, although this is not considered at this stage (but is discussed in Section 7.7).

THEOREM 7.6 If u^*, b^* are admissible, $\{(x^*(t), u^*(t), b^*, t) \in W$ for all $t \in [t_0, t_1]\}$, and maximize I^0, and the Constraint Qualification CQ (see below) holds at all points $(x^*(t), u^*(t), b^*, t)$, then there exist multipliers $\lambda_0 \in \mathbb{R}$, $\lambda(t) \in \mathbb{R}^n$, $\mu(t) \in \mathbb{R}^q$ and $w \in \mathbb{R}^p$, not all vanishing simultaneously, such that, given the functions defined by

$$L[x(t), u(t), b, \lambda(t), \mu(t), w, t] = H[x(t), u(t), b, \lambda(t), w, t] + \sum_{j=1}^{q} \mu_j(t) g^j(x(t), u(t), b, t)$$

$$H[x(t), u(t), b, \lambda(t), w, t] = \lambda_0 f^0(x(t), u(t), b, t) + \sum_{i=1}^{n} \lambda_i(t) f^i(x(t), u(t), b, t) + \sum_{k=1}^{p} w_k h^k(x(t), u(t), b, t)$$

$$J(b) = \lambda_0 S^0(b) + \sum_{k=1}^{p} w_k S^k(b),$$

the following must hold for $t \in [t_0, t_1]$:

N[1] λ is continuous with piecewise continuous derivative such that, where u^* is continuous

$$\dot{\lambda} = -L_x^{*\prime},$$

N[2] $L_u^* = \theta$,

N[3] $H[x^*(t), u^*(t), b^*, \lambda(t), w, t] \geqslant H[x^*(t), u(t), b^*, \lambda(t), w, t]$ for all $(x^*(t), u(t), b^*, t) \in W$,

N[4] $d/dt\, L^* = \partial/\partial t\, L^*$ at all points at which $u^*(t)$ is continuous (and the function L^* is continuous on $[t_0, t_1]$),

N[5] $\lambda_0 \geqslant 0$, $w_k \geqslant 0, k = 1, \ldots, p'$ and w_k, $k = p'+1, \ldots, p$ are constant and $w_k I^k = 0$, $k = 1, \ldots, p$,

N[6] $\mu_j(t), j = 1, \ldots, q$ are piecewise continuous (and continuous where u^* is continuous), and $\mu_j(t) \geqslant 0$, $\mu_j(t) g^i(x, u, b, t) = 0$, $j = 1, \ldots, q'$,

N[7] the transversality condition

$$-\frac{\partial J}{\partial b_j}+\left[-L_s\frac{\partial t_s}{\partial b_j}+\sum_{i=1}^{n}\lambda_i(t_s)\frac{\partial x_{is}}{\partial b_j}\right]_{s=0}^{s=1}=\int_{t_0}^{t_1}\frac{\partial L^*}{\partial b_j}\,\mathrm{d}t$$

$j=1,\ldots,q,$

where $L_s=L[x^*(t_s),u^*(t_s),b^*,\lambda(t_s),\mu(t_s),w,t_s]$ and x_{is}, $s=0,1$ denotes the ith component of endpoint x_0,x_1 respectively.

The constraint qualification CQ is as follows: Let $i_1,\ldots,i_\beta$ index the constraints which are active at any given time t; then

CQ: the matrix G below has rank q.

$$G=\begin{pmatrix} \partial g^1/\partial u_1 & \partial g^1/\partial u_2 & \ldots & \partial g^1/\partial u_m & g^1 & 0 & \ldots & 0 \\ \vdots & \vdots & & \vdots & 0 & g^2 & \ldots & 0 \\ \vdots & \vdots & & \vdots & \vdots & \vdots & & \\ \partial g^q/\partial u_1 & \partial g^q/\partial u_2 & \ldots & \partial g^q/\partial u_m & 0 & 0 & \ldots & g^q \end{pmatrix}$$

Equivalently, in terms of the active constraints, the matrix

$$\left\{\frac{\partial g^p}{\partial u_k}\right\}\begin{matrix} p=i_1,\ldots,i_\beta \\ k=1,\ldots,m \end{matrix}$$

has rank β.

CQ must hold for all $t\in[t_0,t_1]$. Some economics texts suggest that the above rank contraint qualification may be replaced by weaker constraint qualifications; this is not the case (Seierstad and Sydsaeter, 1977 discuss the problem and give a counter example).

Many problems will be special cases of the above general formulation. For example, if $x(t_1)$ is free, we can set $b_j=x_j(t_1)$, $j=1,\ldots,n$; if $x_k(t_1)\geqslant\bar{x}_k$, then set $I^k=S^k(b)=b_k-\bar{x}_k\geqslant 0$ and $x_k(t_1)=b_k$; if terminal time, t_1, is free, we can set another $b_j=t_1$. Terminal surface conditions such as

$$G^i(x(t_1))\geqslant 0 \qquad i=1,\ldots,z$$

may be handled by setting $x_j(t_1)=b_j$, $j=1,\ldots,n$ and setting $S^k(b)=G^k(b)\geqslant 0$, $k=1,\ldots,z$ and $S^0(b)=0$ etc.

The following peak load pricing problem (discussed in more detail in Takayama, 1985, pp. 671–83) provides a simple example of a control problem involving a control parameter.

EXAMPLE 7.3 (Peak load pricing) We shall assume familiarity with the basic ideas of peak load pricing (see Example 2.8 in Chapter 2). There is a willingness-to-pay function $W[q(t), t]$ which is a time varying function of the quantity of the (non-storable) good consumed, $q(t)$. Thus, if $\mathrm{p} = p(q, t)$ denotes the inverse demand function, then $W(q, t) = \int_0^q p(\xi, t)\mathrm{d}\xi$ and, of course, $W_q(q, t) = p(q, t)$. Production requires investment in capacity, Q; this limits feasible output ($q(t) \leqslant Q$) but, once installed, output at time t depends only upon variable inputs, $x(t)$ according to a linear production function

$$aq(t) = x(t) \qquad \text{for } q(t) \leqslant Q.$$

Only the non-trivial case where $Q > 0$ will be considered. Total costs at time t are given by

$$C(t) = \beta Q + wx(t),$$

where β, w are the factor prices. Given a fixed time horizon T (the length of the demand cycle) and no discounting, the objective function is

$$\underset{q(t), Q}{\text{Maximize}}\ V = \int_0^T [W(q(t), t) - \beta Q - wx(t)]\mathrm{d}t$$

$$= \int_0^T \{W(q(t), t) - \beta Q - waq(t)\}\mathrm{d}t$$

subject to $q(t) \leqslant Q$

$q(t) \geqslant 0.$

Choosing $q(t)$ as control variable and Q as control parameter (there is no state variable in this problem), in the notation of Theorem 7.6,

$$J(b) = 0,$$

$$H = \lambda_0\{W(q(t), t) - \beta Q - waq(t)\},$$

$$L = H + \mu_1(Q - q) + \mu_2 q.$$

The constraint qualification requires that

$$\begin{pmatrix} -1 & (Q-q) & 0 \\ 1 & 0 & q \end{pmatrix}$$

has rank 1 and is clearly satisfied. Working through the conditions of Theorem 7.6 gives

$$\lambda_0 \geqslant 0,\ \mu_1(t) \geqslant 0,\ \mu_2(t) \geqslant 0 \text{ (not all zero simultaneously)}, \tag{i}$$

$$\mu_2(t)q(t) = 0, \tag{ii}$$

$$\mu_1(t)(Q - q(t)) = 0, \quad \text{and} \quad q(t) \leqslant Q, \tag{iii}$$

$$\frac{\partial L^*}{\partial q} = \lambda_0\{W_q(q,t) - wa\} - \mu_1 + \mu_2 = 0. \tag{iv}$$

The left-hand side of the transversality condition is equal to zero, hence

$$\int_0^T \frac{\partial L^*}{\partial Q}\,\mathrm{d}t = 0 \quad \text{so} \quad \int_0^T \mu_1(t)\mathrm{d}t = \lambda_0 \beta T. \tag{v}$$

From (v), if $\lambda_0 = 0$, $\mu_1(t) = 0$ all t (since $\mu_1(t) \geqslant 0$). From (iv), if $\lambda_0 = 0$, $\mu_1 = \mu_2$, but if $\lambda_0 = 0$, $\mu_1 = 0$ so $\mu_2 = 0$. This contradicts (i). Hence $\lambda_0 \neq 0$ so we can take $\lambda_0 = 1$. (iv) then implies

$$p(q,t) = wa + \mu_1(t) - \mu_2(t). \tag{vi}$$

Clearly, if $0 < q < Q$, $p = wa$ (price must be set equal to marginal operating cost, which is a constant). Condition N[3] implies that

$$W(q^*, t) - W(q, t) \geqslant wa(q^* - q).$$

Thus, if $q^*(t) = 0$ at any t, this requires $\{W(q,t) - W(0,t)\}/q \leqslant wa$ for all q, $0 \leqslant q \leqslant Q$ which implies that $W_q(0,t) = p(0,t) \leqslant wa$; the top of the demand curve must fall below marginal operating cost before shut down is desirable. Consider now the case where $q^*(t) = Q$. Suppose this happens on a subset S of $[0, T]$; i.e. for $t \in S, q(t) = Q$. The optimal price is given by $\mathrm{p} = p(Q, t)$ and the value of Q is determined from (vi) and (v)

$$\int_0^T \mu_1(t)\mathrm{d}t = \int_S \{p(Q,t) - wa\}\mathrm{d}t = \beta T.$$

This states that the average excess of price over marginal cost equals the marginal capacity cost. Since Q is a constant, the price, for $t \in S$, varies so as to ration demand to available capacity.

The following example is included firstly as an economic problem which does not feature time as the independent variable; secondly, because it involves two state variables and, thirdly, since it nicely illustrates the fact that problems may require (considerable) manipulation in order to make use of the control theory format.

EXAMPLE 7.4 (Optimal income taxation) This example is based upon Cooter (1978) to which the reader is referred for additional results. Individuals are assumed to have a concave utility function $U(c, -l) \in C_2$ where c is consumption (with price set to unity) and l, labour supply. It is assumed that U_1, $U_2 > 0$ for all $c, l > 0$. Individuals differ only in respect of skill, s, which is assumed equal to their gross wage rate. An individual with skill s is called an s-person. An individual's before tax total wage is given by $W = ls$ and he/she pays taxes $T(W)$. $T(W)$ is assumed to be piecewise smooth. The s-person's problem is

$$\underset{W,c}{\text{Maximize}}\; U(c, -W/s)$$

subject to $W - T(W) - c \geqslant 0$.

An interior solution to this problem is assumed, for which necessary conditions are

$$U_2 = s[1 - T'(W)]\psi, \tag{i}$$

$$U_1 = \psi, \tag{ii}$$

$$W - T(W) - c = 0, \tag{iii}$$

where ψ is the Lagrange multiplier. Equations (i) and (ii) imply

$$U_2 = s[1 - T'(W)]U_1. \tag{iv}$$

The distribution of skills in the population is given by a continuous density function f with $f(s) > 0$ on $[s_1, s_2]$, where

$s_1, s_2 > 0$ denote the lower and upper limits to the skill distribution. The government is assumed to maximize additive social welfare according to

$$\int_{s_1}^{s_2} G[U(c, -W/s)]f(s)\mathrm{d}s \tag{v}$$

subject to raising a revenue (R) of at least $\bar{R}$ through taxation

$$R = \int_{s_1}^{s_2} T(W)f(s)\mathrm{d}s \geqslant \bar{R}. \tag{vi}$$

From an economic viewpoint, $\bar{R}$, $G[.]$, $f(.)$ and $U(.)$ are given. The government chooses the function T and s-persons their optimal responses, $c(s)$, $W(s)$ according to (i)–(iii). Mathematically, this may be viewed as a single optimization problem; one of choosing $T(W)$, $c(s)$, and $W(s)$ so as to maximize (v), subject to (iii), (iv) and (vi). As it stands, this problem is not in an appropriate format. However, a judicious choice of state and control variables can resolve this difficulty.

First, differentiating (vi) with respect to s and using (iii) gives

$$DR = T'(W)f(s) = (W - c)f(s), \tag{vii}$$

where $D \equiv \mathrm{d}/\mathrm{d}s$. W, c will in fact be chosen as controls; if there are any points of discontinuity in W, c, derivatives are interpreted as the appropriate left or right-hand side derivatives. Equation (vii) in conjunction with the endpoint conditions

$$R_{s_1} = 0, \qquad R_{s_2} \geqslant \bar{R} \tag{viii}$$

may be used to replace (vi). We shall therefore take R as a state variable.

At this stage, the reader may find it instructive to try to define state and control variables (possibly introducing new variables) so as to obtain a problem in the correct format which also takes account of the constraints (iii) and (iv). In fact, the maximum value function proves to be a useful choice for a state variable

$$V = \mathrm{Max}\, U(c, -W/s), \tag{ix}$$

where W, c are optimally chosen (so satisfying (iii) and (iv)). V thus denotes maximum utility obtained by an s-person. Differentiation of (ix) yields

$$DV = \{U_1 Dc - U_2 DW/s\} + U_2 W/s^2 \tag{x}$$

(again $D = d/ds$). This can be simplified by differentiating (iii)

$$[1 - T'(W)]DW - Dc = 0, \tag{xi}$$

so that $1 - T'(W) = Dc/DW$ (so long as $DW \neq 0$). Substituting this into (iv) gives

$$U_1 DW - U_2 sDc = 0. \tag{xii}$$

(Note that $DW \neq 0 \Rightarrow Dc \neq 0$.) Hence (x) becomes

$$DV = U_2 W/s^2. \tag{xiii}$$

So (xiii) follows from (iv) and (iv) follows from (xiii) so long as DW, $Dc \neq 0$. We shall assume that W, c are not constant with s, so that constraints (iii), (iv) are equivalent to and may be replaced by (xiii).

The problem is now stated as

$$\text{Maximize} \int_{s_1}^{s_2} G(V) f(s) \mathrm{d}s \tag{xiv}$$

subject to $DV = U_2(c, -W/s)W/s^2$, (xv)

$$DR = [W - c] f(s), \tag{xvi}$$

$$V - U(c, -W/s) = 0, \tag{xvii}$$

$$R_{s_1} = 0, \qquad R_{s_2} \geqslant \bar{R}, \tag{xviii}$$

$$V_{s_1}, V_{s_2} \text{ free.} \tag{xix}$$

In terms of Theorem 7.6, we have

$$f^0 = G(V) f(s),$$

$$f^1 = U_2(c, -W/s)W/s^2,$$

$$f^2 = (W - c) f(s),$$

$$g^1 = V - U(c, -W/s).$$

The Hamiltonian is $H=\sum_{i=0}^{2}\lambda^i f^i$ and the Lagrangean $L=H+\mu_1 g^1$. Notice that the constraint qualification is satisfied; $[-U_1, U_2/s]$ has rank 1. The necessary conditions are, in addition to (xv)–(xix),

$$\partial L/\partial c=\lambda_1 U_{21} W/s^2-\lambda_2 f(s)-\mu_1 U_1=0, \tag{xx}$$

$$\partial L/\partial W=\frac{\lambda_1}{s^2}\{U_2-U_{22}W/s\}+\lambda_2 f(s)+\mu_1 U_2/s=0, \tag{xxi}$$

$$D\lambda_1=-\lambda_0 G'(V)f(s)-\mu_1, \tag{xxii}$$

$$D\lambda_2=0, \tag{xxiii}$$

$$\lambda_1(s_1)=\lambda_1(s_2)=0, \tag{xxiv}$$

$$\lambda_2(s_0) \text{ free, } \lambda_2(s_1)\geqslant 0. \tag{xxv}$$

We shall now interpret the conditions (xv)–(xxv) in order to prove the following propositions:

[P1] A person with higher productive skill enjoys at least as high a utility as someone with lower skill.

[P2] The marginal tax rate is less than unity.

[P3] The most highly skilled and least highly skilled individuals have a zero marginal tax rate.

[P4] The marginal tax rate is non-negative if
- (a) $U_{12}\geqslant 0$, and
- (b) the marginal social value of leisure decreases with skill; $D[G'U_2]\leqslant 0$ ((a), (b) can be relaxed somewhat but have the merit of easy interpretation).

[P1] follows from (xv) since $U_2, W, s\geqslant 0$ hence $DV\geqslant 0$. Utility does not decrease with skill. In fact, if $W>0$, then $DV>0$ and more skill implies more utility. The intuition is that a more highly skilled person could always choose to earn the same income as a less skilled person – and take more leisure.

[P2] follows from (iv) which implies

$$T'(W)=1-U_2/sU_1<1 \qquad (U_1, U_2, s>0).$$

Hence an increase in gross wage leads to an increase in net wage (there is no 'poverty trap').

[P3] follows from (xxiv) which can be used in (xx) and (xxi) to yield (for $s_i = s_1$ or s_2)

$$\lambda_2 f(s_i) + \mu_1 U_1 = 0,$$

$$\lambda_2 f(s_i) + \mu_1 U_2/s = 0.$$

Hence $U_1 - U_2/s_i = 0$ (since $\mu_1 > 0, \lambda_2 > 0$ – as will be proved later). From (iv), this in turn implies

$$1 - T'(W)|_{s_i} = U_2/U_1 s_i = 1 \qquad \text{for} \quad i = 1, 2. \tag{xxvi}$$

Hence $T'(W)|_{s_1} = T'(W)|_{s_2} = 0$; zero marginal taxes at s_1 and s_2.

In spite of its trivial nature, [P4] is not easily established. We seek to show that $T'(W) \geqslant 0$ on $[s_1, s_2]$. Adding (xviii) to (xix) gives

$$\frac{\lambda_1}{s^2}[U_{21}W - U_{22}W/s + U_2] + \mu_1[U_2/s - U_1] = 0. \tag{xxvii}$$

Using (iv), this implies

$$T'(W) = \frac{\lambda_1}{s^2 \mu_1 U_1}\left(\frac{1}{U_{21}W - U_{22}(W/s) + U_2}\right). \tag{xxviii}$$

The term in [ ] > 0 given that $U_{12} \geqslant 0$. Hence, if we obtain the signs of λ_1, μ_1, the sign of $T'(W)$ will be determined. Now, substituting for μ_1 in (xxvii) using (xxi), we obtain

$$D\lambda_1 = [-\lambda_0 G' + \lambda_2 s/U_2]f(s) + \frac{\lambda_1}{sU_2}[U_2 - U_{22}W/s] \tag{xxix}$$

or

$$D\lambda_1 + \alpha(s)\lambda_1 = \beta(s), \tag{xxx}$$

where

$$\alpha(s) \equiv \{U_{22}W/s - U_2\}/sU_2 \tag{xxxi}$$

(note $\alpha(s) < 0$) and

$$\beta(s) \equiv \{\lambda_2 S - \lambda_0 G' U_2\} f(s)/U_2. \tag{xxxii}$$

Equation (xxx) is a linear differential equation. Multiplying through

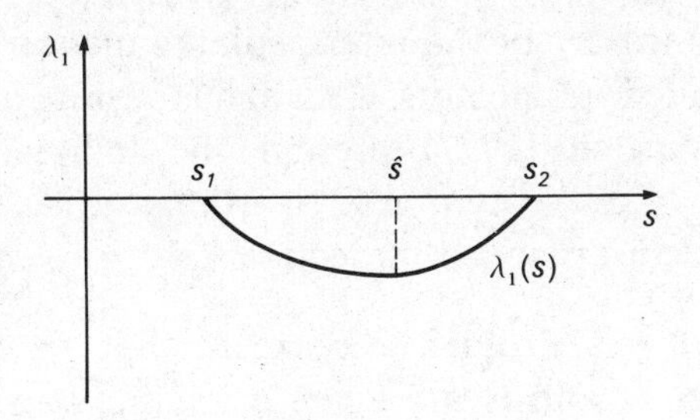

Figure 7.4

by the integrating factor $\exp[\int_{s_1}^{s} \alpha(\tau)\mathrm{d}\tau]$, we obtain the solution

$$\lambda_1 \exp\left[\int_{s_1}^{s} \alpha(\tau)\mathrm{d}\tau\right] = \int_{s_1}^{s} \beta(\tau)\exp\left[\int_{s_1}^{\tau} \alpha(\xi)\mathrm{d}\xi\right]\mathrm{d}\tau. \qquad \text{(xxxiii)}$$

Now, $\exp\int_a^b \alpha(\xi)\mathrm{d}\xi > 0$ and so the sign of λ_1 depends upon the behaviour of $\beta(\tau)$ which we now examine.

First we show that $\lambda_0 > 0$ (so $\lambda_0 = 1$) and $\lambda_2 > 0$. To see this, consider the alternatives: (a) If $\lambda_0 = \lambda_2 = 0$ then $\beta(s) = 0$ and $\lambda_1 = 0$ for all s from (xxxii) and (xxxiii). But then, by (xx), $\mu_1 = 0$ – this contradicts the necessary condition that not all the multipliers be simultaneously zero. (b) If $\lambda_0 = 0$ but $\lambda_2 > 0$ then $\beta(s) > 0$ for all s – and if $\lambda_0 = 1$ and $\lambda_2 = 0$ then $\beta(s) < 0$ for all s. If $\beta(s) > 0$ (resp. < 0) for all $s \in [s_1, s_2]$ and $\lambda_1(s_1) = 0$, then, by (xxxiii), $\lambda_1(s_2) > 0$ (resp. < 0) violating (xxiv). Hence we must have $\lambda_0 = 1$, $\lambda_2 > 0$ *and* $\beta(s)$ changing sign on (s_1, s_2). The sign is determined in (xxxii) by $\lambda_2 s - \lambda_0 G' U_2$ which is increasing in s(since $D[G' U_2] \leqslant 0$). Hence $\beta(s_1) < 0$, $\beta(s_2) > 0$ and there is some unique $\hat{s} \in (s_1, s_2)$ such that $\beta(\hat{s}) = 0$. Referring to (xxxiii), λ_1 is decreasing on $[s_1, \hat{s})$ and increasing on $(\hat{s}, s_2]$. Since $\lambda_1(s_1) = \lambda_1(s_2) = 0$, this implies $\lambda_1 \leqslant 0$ on $[s_1, s_2]$ (see Figure 7.4). Now, from (xx), $\mu_1(s) = \{\lambda_1 U_{21} W/s^2 - \lambda_2 f(s)\}/U_1 < 0$, hence, from (xxviii), $T'(W) > 0$ as was to be proved.

7.5 SUFFICIENT CONDITIONS: FINITE HORIZON PROBLEMS

In the Calculus of Variations, sufficient conditions were presented based upon concavity of the intermediate function in $(x, \dot{x})$. Strict concavity entailed uniqueness. Analogous results apply here. Van Long and Vousden (1977) present the following theorem, a generalization of a theorem due to Mangasarian (1966), for the Hestenes problem discussed in Section 7.4.

PROBLEM 7.7 As Problem 7.6 except that f^0, f, g, h do not involve b. Also, t_0 and t_1 are given constants and not dependent upon b. In fact, the absence of b from these functions is no loss of generality since, if present, they can be eliminated by defining appropriate new state variables x_{n+i} such that $x_{n+i}(t_0)=b_i$ and $\dot{x}_{n+i}=0$.

THEOREM 7.7 (Sufficient conditions for Problem 7.7) If there exists an admissible solution (x^*, u^*, b^*) together with multipliers $\lambda_0, \lambda(t), \mu(t), w$ satisfying the necessary conditions of Theorem 7.6, then the solution is optimal, provided that

S[1] $\lambda_0 > 0$,
S[2] $S^k(b)$, $k=0,\ldots,p$ are concave in b,
S[3] $\lambda'(t_0)x_0(b) - \lambda'(t_1)x_1(b)$ is concave in b,
S[4] Either

(i) L is differentiable and concave in (x, u) or

(ii) $\hat{H}(x, \lambda, w, t) = \operatorname*{Max}_{u \in V} H(x, u, \lambda, w, t)$
is concave in x,
where $V \equiv \{g^j(x, u, t) \geqslant 0, j=1,\ldots,q'$ and
$g^j(x, u, t)) = 0, j = q'+1, \ldots, q\}$.

If concavity is strict, the solution is unique. A sketch of the proof of Theorem 7.7 is as follows. Let (x, u, b) denote any other admissible solution and denote f^{0*}, f^* for $f^0(x^*, u^*, t)$, $f(x^*, u^*, t)$ etc. We require to show that

$$S^0(b^*) - S^0(b) + \int_{t_0}^{t_1} \{f^{0*} - f^0\} \mathrm{d}t \geqslant 0$$

for all admissible solutions (x, u, b). To do this, we use conditions S[1]–[4] above and the necessary conditions N[1]–[7] of Theorem 7.6.

By S[4](i) we have

$$L^* - L + L_x^*(x - x^*) + L_u^*(u - u^*) \geqslant 0. \tag{7.68}$$

By N[2], $L_u^* = \theta$, and by N[1], $\dot{\lambda}' = -L_x^*$. Writing (7.68) out and using these conditions, and substituting $\dot{x}^*$ for f^*, $\dot{x}$ for f, gives

$$\lambda_0(f^{0*} - f^0) + \lambda'(\dot{x}^* - \dot{x}) + w'(h^* - h) + \mu'(g^* - g) + \dot{\lambda}'(x^* - x) \geqslant 0. \tag{7.69}$$

Now, by N[6], $\mu' g = 0$ and $\mu \geqslant \theta, g \geqslant \theta$, so

$$\lambda_0(f^{0*} - f^0) + \frac{\mathrm{d}}{\mathrm{d}t}\{\lambda'(x^* - x)\} + w'(h^* - h) \geqslant \mu' g \geqslant 0. \tag{7.70}$$

Integrating (7.70) between t_0 and t_1 and adding $\lambda_0\{S^0(b^*) - S^0(b)\}$ to both sides gives

$$\lambda_0\left[S^0(b^*) - S^0(b) + \int_{t_0}^{t_1} (f^{0*} - f^0)\mathrm{d}t\right] \geqslant$$

$$\int_{t_0}^{t_1} \left\{w'(h - h^*) + \frac{\mathrm{d}}{\mathrm{d}t}[\lambda'(x - x^*)]\right\}\mathrm{d}t + \lambda_0[S^0(b^*) - S^0(b)]. \tag{7.71}$$

By S[1], $\lambda_0 > 0$, so the theorem is proved if the right-hand side of (7.71)$\geqslant 0$.

$$\lambda_0[S^0(b^*) - S^0(b)] + \int_{t_0}^{t_1} \left\{w'(h - h^*) + \mu' g + \frac{\mathrm{d}}{\mathrm{d}t}[\lambda'(x - x^*)]\right\}\mathrm{d}t$$
$$\geqslant \lambda_0[S^0(b^*) - S^0(b)] + w'[S(b^*) - S(b)]$$
$$-\lambda'(t_1)[x^*(t_1) - x(t_1)] + \lambda'(t_0)[x^*(t_0) - x(t_0)] \tag{7.72}$$

since $\int_{t_0}^{t_1} h^k \mathrm{d}t \geqslant -S^k(b)$ (note $S = (S^1, \ldots, S^p)'$). Now, from N[7], the transversality condition reduces to

$$-\lambda_0 \frac{\partial S^0}{\partial b} - w' \frac{\partial S}{\partial b} + \lambda'(t_1)\frac{\partial x_1(b)}{\partial b} - \lambda'(t_0)\frac{\partial x_0(b)}{\partial b} = 0 \tag{7.73}$$

since $\partial L^*/\partial b = \theta$ and $\partial t^s/\partial b = \theta$. Now, by S[2], S[3], concavity implies that the RHS of (7.72)$\geqslant$ the LHS of (7.73). Hence the right-hand side of (7.72)$\geqslant 0$.

It is possible to use S[4](ii) instead of S[4](i) by noting that $\hat{H}(x,.)\geqslant H(x,u,.)$, $L^*=\hat{H}(x^*,.)$, $L^*_x=\hat{H}_x(x^*,.)$ (by the Envelope Theorem) and that concavity in x implies

$$\hat{H}(x^*,.)+\hat{H}_x(x^*,.)(x-x^*)\geqslant\hat{H}(x,.)\geqslant H(x,u,.). \tag{7.74}$$

The Hestenes problem is very general and subsumes many special cases which, accordingly, we do not discuss. The following is a very simple application (see also Section 7.7).

EXAMPLE 7.5 The solution to Example 7.3 above clearly satisfies the sufficient conditions since $\lambda_0>0$, $S^k=0$ all k, there is no state variable so S[3] is satisfied and S[4] is satisfied since L is linear in the only control variable.

7.6 BOUNDED STATE VARIABLE CONSTRAINTS

The approach adopted in Section 7.4 fails if the constraints do not involve control variables since the constraint qualification no longer holds (since $\partial g^i/\partial u_j=0$ all j). However, equality constraints of the form

$$\theta^i(x,t)=0 \qquad i=1,\ldots,\alpha \tag{7.75}$$

($\theta\in C_2$) may be translated into the format of Section 7.4 as follows. Set

$$\dot{\theta}=\theta'_x\dot{x}+\theta_t=\theta_x f+\theta_t=\theta, \tag{7.76}$$

and

$$\theta(x_0,t_0)=\theta. \tag{7.77}$$

Equation (7.76) constrains θ to be constant for all t and (7.77) sets this to zero. Equations (7.76) and (7.77) are constraints of a type already discussed in Sections 7.1, 7.2 and 7.4. Naturally, if the rank condition is to hold, we require $\theta_x f_u=[\theta^i_{x_j}][f^i_{u_j}]$ to have rank α.

Inequality constraints of the type

$$\theta^i(x,t)\geqslant 0 \qquad i=1,\ldots,\alpha \tag{7.78}$$

are of much greater importance in economic models than equality constraints (which can often be analytically avoided by direct substitution). However, inequality constraints may be handled in similar fashion to that above, although certain difficulties do arise: in particular, co-state variables need no longer be continuous, and the size of jumps may be difficult to determine from the necessary conditions. Indeed, the solution to the problem in the absence of such constraints may be worth examining; it may be that the constraints remain inactive and can be legitimately ignored – or, if this is not the case, the solution may give some insight into the likely nature of the constrained solution (although note that introducing a constraint, even if it only binds on a subinterval of $[t_0, t_1]$, will usually have ramifications for the whole of the optimal trajectory).

Guinn (1965) provides a statement and proof of necessary conditions to the Hestenes' Problem 7.5 (discussed in Section 7.4) with added constraints of type (7.78). For simplicity, we shall confine discussion to the following special case.

PROBLEM 7.8

$$\underset{u}{\text{Maximize}} \int_{t_0}^{t_1} f^0(x, u, t) dt,$$

$$\begin{aligned}
\text{subject to } & \dot{x} = f(x, u, t), \\
& x(t) \in \mathbb{R}^n, \qquad u(t) \in \mathbb{R}^m, \\
& g^i(x, u, t) \geqslant 0 \qquad i = 1, \ldots, q, \\
& \theta^i(x, t) \geqslant 0 \qquad i = 1, \ldots, \alpha, \\
& t_0, t_1 \text{ fixed}, \\
& x(t_0) = x_0 \text{ fixed}, \\
& x^i(t_1) \text{ free} \qquad i = 1, \ldots, a, \\
& x^i(t_1) = x^{i1} \qquad i = a+1, \ldots, b, \\
& x^i(t_1) \geqslant x^{i1} \qquad i = b+1, \ldots, n, \\
& \text{where } x^{i1} \text{ denotes a fixed constant.}
\end{aligned}$$

Naturally, the initial condition must be compatible with the state variable constraint at t_0 in this case. Note that restrictions on $u(t)$ are dealt with in the above via g-constraints (rather than requiring

$u(t) \in U$). All functions of (x, u, t) are of class C_1 and $\theta^i \in C_2$. Define a set S by

$$S = \{(x, t);\ \theta^i(x, t) \geqslant 0, \text{ all } i\}$$

and denote $\bar{S}$ as its boundary: $\bar{S} = \{(x, t): \theta^i(x, t) = 0, \text{ some } i\}$. It is assumed that

[A] $\nabla\theta^i(x, t) \equiv (\partial\theta^i/\partial x^i, \ldots, \partial\theta^i/\partial x_n, d\theta^i/\partial t)$ is non-zero and, when $(x, t) \in \bar{S}$, the vectors $\nabla\theta^i(x, t)$ are linearly independent for all i for which $\theta^i(x, t) = 0$.

The treatment follows that for equality state variable constraints; define additional g-functions by setting

$$g^{q+i}(x, u, t) = [\theta^i_{x_j}][f^i] + \partial\theta^i/\partial t \qquad i = 1, \ldots, \alpha. \tag{7.79}$$

Clearly $g^{q+i} \in C_1$ since $f^i \in C_1$. Now, if $(x, t) \in \bar{S}$, we must require

$$g^{q+i}(x, u, t) \geqslant 0 \text{ for all } i \text{ such that } \theta^i(x, t) = 0. \tag{7.80}$$

This ensures the constraints $\theta^i(x, t) \geqslant 0$ will not be violated. For a given point (x, t), let $i_1, \ldots, i_{\beta_1} \in [1, \ldots, q]$ denote the indexes of active g-constraints such that $g^i(x, u, t) = 0$ and, when $(x, t) \in \bar{S}$, let $i_{\beta_1+1}, \ldots, i_{\beta_2} \in \{q+1, \ldots, q+\alpha\}$ denote the indexes of the added g-constraints which are binding, i.e. for which $g^{q+1}(x, u, t) = 0$ and $\theta^i(x, t) = 0$.

[CQ] The constraint qualification for this problem is that, if x^* denotes the optimal trajectory, the matrix $\{\partial g^i(x^*, u^*, t)/\partial u_k\}$:
(i) has rank β_1 if $(x^*, t) \in \text{Int } S$, and
(ii) has rank β_2 if $(x^*, t) \in \bar{S}$
(where, if u^* is discontinuous at t, the conditions hold for t^- and t^+).

DEFINITION 7.1 A point $(x^*(\tau), \tau) \in \bar{S}$ is termed a **junction point** of the trajectory x if there exists an $\varepsilon > 0$ such that the set of active constraints on $(\tau - \varepsilon, \tau)$ differs from that on $(\tau, \tau + \varepsilon)$.

Sometimes $x^*(\tau)$ is referred to as the junction point and τ as the junction time (Figure 7.5 illustrates for the case of a single constraint).

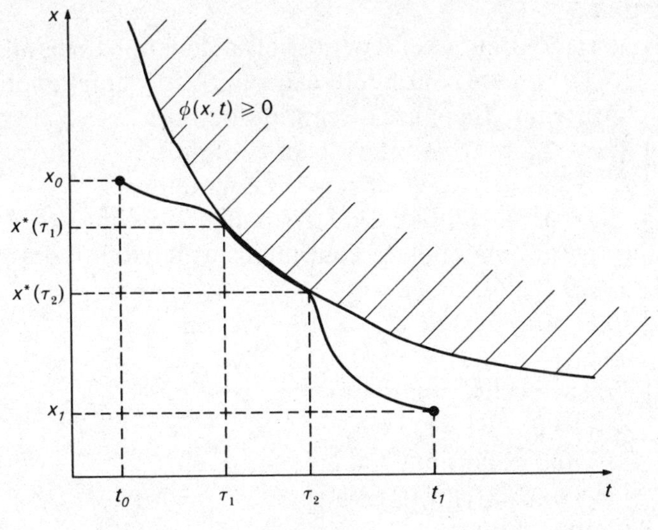

Figure 7.5

The Hamiltonian function is defined as

$$H(x,u,\lambda,t)=\sum_{i=0}^{n}\lambda_i(t)f^i(x,u,t)$$

and the Lagrange function is

$$L(x,u,\lambda,\mu,t)=H(x,u,\lambda,t)+\sum_{j=1}^{q+\alpha}\mu_j(t)g^j(x,u,t),$$

In general, the costate variables will be discontinuous at a junction point; the control variables may also be discontinuous at such points (as well as the usual discontinuities which arise in the controls at boundaries of the control constraint set). Let W denote the set of elements (x,u,t) which satisfy all the constraints in Problem 7.8.

THEOREM 7.8 (Necessary conditions, bounded state variable Problem 7.8) If (x^*,u^*) satisfies the constraints and maximizes $J(u)$ and if [A] and [CQ] above hold, then there exist multipliers $\lambda_0, \lambda(t)\in\mathbb{R}^n$, $\mu(t)\in\mathbb{R}^{q+\alpha}$ not vanishing simultaneously on $[t_0,t_1]$, such that

N[1] $\lambda(t)$ is continuous except possibly at junction times and $\mu_i(t)$, $i=1,\ldots,q$ are continuous except possibly at junction times – or where there are discontinuities in u^*.

N[2] $\lambda' = -L_x^*$ over its intervals of continuity.

N[3] $L_{u_i}^*=0$, $i=1,\ldots,m$ where u^* is continuous.

N[4] $H(x^*,u^*,\lambda,t) \geqslant H(x^*,u,\lambda,t)$ for all u satisfying $(x^*,u,t)\in W$. (Note: H may not be continuous at junction points.)

N[5] $\lambda_0 \geqslant 0$ ($\lambda_0=0$ or 1).

N[6] $\mu_j(t)\geqslant 0$; $\mu_j(t)g^j(x,u,t)=0$, $j=1,\ldots,q$, $\mu_{q+j}(t)=0$ when $\Theta^j(x,t)>0$, $j=1,\ldots,\alpha$.

N[7] Transversality conditions

$\lambda_i(t_1)=0 \qquad i=1,\ldots,a,$

$\lambda_i(t_1)$ no conditions $\qquad i=a+1,\ldots,b,$

$\lambda_i(t_1)\geqslant 0$; $\lambda_i(t_1)[x_i(t_1)-x^{i1}]=0, \qquad i=b+1,\ldots,n.$

N[8] Jump conditions:

$$\text{(i)}\quad \lambda(\tau_i^-)-\lambda(\tau_i^+)=\sum_{j=1}^{\alpha}\beta_{ij}\frac{\partial\Theta^j(x(\tau_i),\tau_i)}{\partial x}, \qquad i=1,\ldots,h,$$

where $t_0<\tau_1<\ldots<\tau_h<t_1$ are the junction times at which $\lambda(t)$ is discontinuous (if $\tau_h=t_1$, let $\lambda(\tau_h^+)=\lambda(t_1)$),

(ii) $\beta_{ij}\geqslant 0$ are non-negative constants such that

$$\beta_{ij}\Theta^j(x^*(\tau_i),\tau_i)=0, \quad i=1,\ldots,h, \quad j=1,\ldots,\alpha.$$

A detailed application of the above necessary conditions and the following sufficient conditions for the bounded state variable problem may be found in Section 7.7.

Define the function

$$H^*(x,\lambda,t)=\operatorname*{Max}_{u\in A(t)} H(x,u,\lambda,t),$$

where $A(t)=\{u{:}g^i(x,u,t)\geqslant 0,\ i=1,\ldots,q\}$

THEOREM 7.9 (Sufficient conditions, bounded state variable Problem 7.8) If (x^*,u^*) satisfies the assumptions and conditions of Theorem 7.8 and in addition:

S[1] $\theta^i(x,t)$ is quasiconcave in x and differentiable in x at $x^*(t)$, $i=1,\ldots,\alpha$.

S[2] Either:

(i) $H(x,u,\lambda,t)$ is jointly concave in (x,u), or

(ii) $H^*(x,\lambda,t)$ is concave in x on the set

$$\tilde{A}(t)=\{x; g^i(x,u,t)\geqslant 0,\ i=1,\ldots,q \text{ for } \textit{some } u\in\mathbb{R}^m\},$$

where $\tilde{A}(t)$ is convex (or, if $\tilde{A}(t)$ is not convex, $H^*(x,\lambda,t)$ has an extension to conv $\tilde{A}(t)$ which is concave in x).

S[3] $\lambda_0>0(\lambda_0=1)$

then (x^*,u^*) is an optimal solution.

For a proof of the above theorem (and an excellent exposition of sufficient conditions) see Seierstad and Sydsaeter (1977). The requirement S[2](ii) in Theorem 7.9 on H^* is slightly less restrictive than the requirement that the Hamiltonian be jointly concave in x and u; S[2](i) $\Rightarrow$ S[2](ii) but S[2](ii) $\not\Rightarrow$ S[2](i). This point is illustrated by an example in Section 7.7.

Clearly, these theorems may be difficult to apply – although most economic problems will involve only one or two junction points, so the piecing together of solutions is often possible (as in Section 7.7).

7.7 AN EXTENDED EXAMPLE

The following example examines a firm's demand for labour over the business cycle. The model is due to Nickel (1978). Nickell is concerned more with the economic aspects and the 'routine' mathematics is not discussed in great detail. However, the model is an excellent example of a control problem involving both constraints on the control variables and also a bounded state variable. In this section, we show how the necessary conditions (Theorem 7.8) and constraints can be used to identify a potential extremal and how the sufficient conditions (Theorem 7.9) can be applied to show the solution is in fact optimal.

The firm charges a constant price p for its product. Demand, $x(t)$, is cyclical (Figure 7.6). It has a fixed stock of machines $\bar{M}$ (we shall

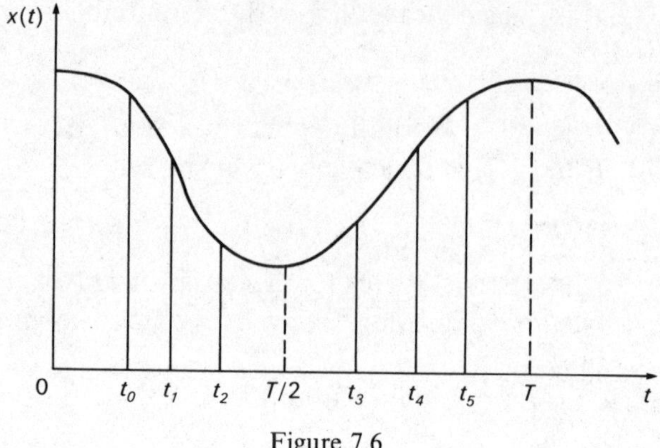

Figure 7.6

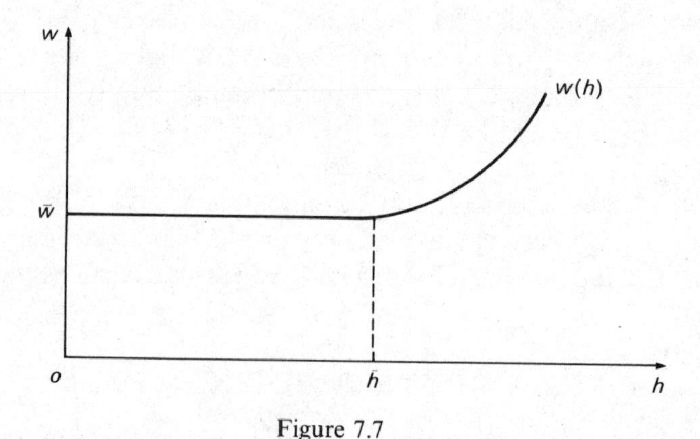

Figure 7.7

consider the choice of $\bar{M}$ as a second stage problem). Each machine requires one worker; with M workers, M machines can operate. Output per week is hM, where h is the hours worked/worker per week. Overtime rates of pay increase with hours worked, so the wage paid per week of h hours, $w(h)$, has $w'(h)>0$, $w''(h)>0$ (Figure 7.7). Typically $w'(h)\simeq 0$ over the normal working week of around 40 hours but then increases as in Figure 7.7. There may be a physical upper limit to hours worked, $\bar{h}$, but we shall assume $w(h)$ rises sufficiently, as $\bar{h}$ is approached, to guarantee that a solution involving $h=\bar{h}$ is never optimal. We shall also assume that the minimum wage paid per week, w, is sufficiently low that it always pays the firm to operate with $h>0$. There are no inventories in the model, and price is held constant so supply is limited by demand and demand is exogenous (as in Figure 7.6).

$$h(t)M(t) \leqslant x(t). \tag{7.81}$$

Also

$$0 \leqslant h(t) \leqslant \bar{h}, \tag{7.82}$$

$$0 \leqslant M(t) \leqslant \bar{M}. \tag{7.83}$$

The rate of accessions, $A(t)$, and discharges, $D(t)$, from employment, $M(t)$, implies the relation

$$\dot{M}(t) = A(t) - D(t), \tag{7.84}$$

where

$$A(t) \geqslant 0, \tag{7.85}$$

$$D(t) \geqslant 0. \tag{7.86}$$

Letting a, d, denote constant per worker hiring/firing costs, then, for given M, the objective function is

$$\underset{A,D,h}{\text{Max}} \int_0^T \{pMh - w(h)M - aA - dD\} e^{-rt} dt \tag{7.87}$$

subject to the constraints (7.81)–(7.86). The firm's discount rate is denoted by r and $[0, T]$ is the period of the assumed stationary business cycle. $\bar{M}$ is taken as a given constant in the first instance. Clearly, M is the state variable and A, D, h are controls. For simplicity we shall only consider the case where the parameters x, a, p etc. are such that the firm never shuts down completely – hence $M>0$ for all t. Our assumptions about the wage rate $w(h)$ will also entail an interior solution for $h(t)$ for all t, so we may ignore constraint (7.82). Define

$$H = \lambda_0 f^o + \lambda_1 f^1, \tag{7.88}$$

$$L = H + \sum_{i=1}^{4} \mu_i g^i, \tag{7.89}$$

where

$$f^0 = \{pMh - w(h)M - aA - dD\}e^{-rt}, \tag{7.90}$$

$$f^1 = A - D, \tag{7.91}$$

$$g^1 = A \geqslant 0; \qquad \mu_1 \geqslant 0, \qquad \mu_1 A = 0, \tag{7.92}$$

$$g^2 = D \geqslant 0; \qquad \mu_2 \geqslant 0, \qquad \mu_2 D = 0, \tag{7.93}$$

$$g^3 = x - hM \geqslant 0; \qquad \mu_3 \geqslant 0, \qquad \mu_3(x - hM) = 0. \tag{7.94}$$

The state variable constraint

$$\theta^1 = \bar{M} - M \geqslant 0$$

provides the additional g-constraint

$$\begin{aligned} &g^4 = -(A - D) \geqslant 0 \quad \text{when} \quad M = \bar{M}, \\ &\text{hence } \mu_4 = 0 \quad \text{when} \quad M < \bar{M} \end{aligned} \tag{7.95}$$

(including the complementary slackness conditions where relevant). We shall assume that $\bar{M}$, the stock of machines, is set at a level such that the firm's chice of $M(t)$ is not everywhere constrained by $\bar{M}$; for some time, there is idle capacity; $M(t) < \bar{M}$. Again, this is the economically most interesting case.

The necessary conditions are

$$\dot{\lambda}_1 = -\lambda_0[ph^* - w(h^*)]e^{-rt} + \mu_3 h, \tag{7.96}$$

$$\partial L/\partial h = \lambda_0[pM^* - w'(h^*)M^*]e^{-rt} - \mu_3 M^* = 0, \tag{7.97}$$

$$\partial L/\partial A = -\lambda_0 a\, e^{-rt} + \lambda_1 + \mu_1 - \mu_4 = 0, \tag{7.98}$$

$$\partial L/\partial D = -\lambda_0 d\, e^{-rt} - \lambda_1 + \mu_2 + \mu_4 = 0, \tag{7.99}$$

$$\lambda_1(t_1) = 0 \text{ since } M(t_1) \text{ is free.} \tag{7.100}$$

The Hamiltonian condition requires that

$$\begin{aligned} &\lambda_0[ph^*M^* - w(h^*)M^*]e^{-rt} + \lambda_1(A^* - D^*) \\ &\quad \geqslant \lambda_0[phM^* - w(h)M^*]e^{-rt} + \lambda_1(A - D) \\ &\quad \text{for all } M^*, A, D, h \text{ satisfying the constraints.} \end{aligned} \tag{7.101}$$

We also require $\lambda_0 \geqslant 0$ and that, at junction times, N[8] of Theorem

7.8 is satisfied. Now $\dot{M}=A-D$ must hold everywhere, hence $A^*-D^*=A-D$ in (7.101). If we write $Q(h)=ph-w(h)$, (7.101) becomes

$$Q(h^*)-Q(h)\geqslant 0 \tag{7.101'}$$

if $\lambda_0=1$ (which we show below) and $M^*>0$ (by assumption). Note that $Q(h)$ is a strictly concave function which attains its maximum value at some point h_0, where $p=w'(h_0)$. Hence it is strictly increasing for $h<h_0$ and strictly decreasing for $h>h_0$.

Clearly, at times of close to maximum demand, $M^*(t)=\bar{M}$ (otherwise the firm holds idle capacity for all t). We now show that $\lambda_0\neq 0$, so we may take $\lambda_0=1$. To see this, suppose $\lambda_0=0$. Now $M(t)>0$ implies $\mu_3=0$ by (7.97). Adding (7.98) to (7.99) gives

$$\mu_1+\mu_2=\lambda_0(a+d)e^{-rt} \tag{7.102}$$

so, if $\lambda_0=0$, $\mu_1+\mu_2=0$ which implies $\mu_1=\mu_2=0$ since $\mu_1,\mu_2\geqslant 0$. From (7.100), $\lambda_1(t_1)=0$, thus, at t_1, all multipliers are zero, contradicting the necessary condition that not all multipliers can be simultaneously zero. Hence we take $\lambda_0=1$.

Note that, given $\lambda_0=1$, then from (7.102), $\mu_1+\mu_2>0$ which implies $\mu_1>0$ or $\mu_2>0$ or both – and rules out $\mu_1=\mu_2=0$. This rules out the possibility that A^* and $D^*>0$. Thus if $D^*>0$, then $A^*=0$, and if $A^*>0$, then $D^*=0$. This is intuitively obvious; hiring and firing are costly, it is futile to do both at the same time.

We now piece together a solution which satisfies the necessary conditions. Referring to Figure 7.6, let $[0,t_0)$, $[t_0,t_1),\ldots,[t_4,t_5)$, $[t_5,T]$ define the set of regimes (where different constraints are binding). For each interval, we deduce a solution (M^*,A^*,D^*,h^*). In each case we make no use of the Hamiltonian condition (7.101') but it is trivial to verify that this is satisfied in each case.

R[1] $[0,t_0)$ **Excess demand**

Here $M^*(t)=\bar{M}$ and $x(t)-h^*(t)M^*(t)>0\Rightarrow\mu_3=0$. That is, all machines are in action but there is excess demand,

$$M^*=\bar{M}\Rightarrow\dot{M}^*=A-D^*=0\Rightarrow A^*=D^*.$$

From (7.97) $h^*(t)=h_0$, where h_0 is given by

$$p-w'(h_0)=0. \tag{7.103}$$

Since $\lambda_0=1$, (7.102) implies either $\mu_1>0$ or $\mu_2>0$ or both. $\mu_1>0 \Rightarrow A^*=0$ and $\mu_2>0 \Rightarrow D^*=0$ hence, either way, $A^*=D^*=0$. Note that (7.101′) is satisfied since $Q(h)$ attains its maximum value at h_0.

Solution: $(M, 0, 0, h_0)$.

R[2] $[t_0, t_1)$ **Hours adjustment**
We assume $x(t)$ declines far enough to reach the time t_0 when

$$\bar{M}h_0=x(t_0).$$

As $x(t)$ falls below this level, the firm must reduce hours – or lay off workers (reduce M) or both. Certainly, the constraint

$$M(t)h(t)=x(t)$$

binds. Suppose hours are held at h_0 and $M^*(t)$ declines. Then $M^*(t)<\bar{M} \Rightarrow \mu_4=0$ and $\dot{M}^*=\dot{x}(t)/h_0<0$ which implies $A^*=0$ and $D^*>0$ (hence $\mu_2=0$). Now (7.99) gives

$$\lambda_1=-d\,\mathrm{e}^{-rt} \text{ (so } \dot{\lambda}_1=dr\,\mathrm{e}^{-rt}),$$

(7.97) gives

$$\mu_3=[p-w'(h_0)]\mathrm{e}^{-rt}.$$

Substituting these into (7.96), we get

$$dr\,\mathrm{e}^{-rt}=-[ph_0-w(h_0)]\mathrm{e}^{-rt}+h_0[p-w'(h_0)]\mathrm{e}^{-rt},$$

or

$$w(h_0)-h_0w'(h_0)=dr. \tag{7.104}$$

This is unlikely to hold (see Figure 7.8) and, empirically, we might expect the value h_d which does solve the equation

$$w(h_d)-h_dw'(h_d)=dr \tag{7.105}$$

to be less than h_0 (note that $p=w'(h_0)$ so, if (7.104) held, the firm would be earning negative profits; $ph_0-w(h_0)<0$).

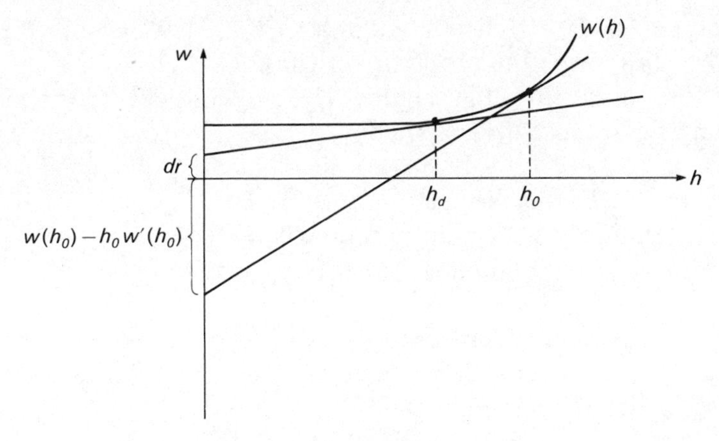

Figure 7.8

Consider therefore the alternative that hours adjust. Then

$$h^*(t)\bar{M} = x(t) \quad \text{or} \quad h^*(t) = x(t)/\bar{M}$$

and $M^*(t) = \bar{M} \Rightarrow A^* = D^* = 0$ as in R[1].

Solution: $[M, 0, 0, x(t)/M]$.

Equation (7.101′) is satisfied here. To see this, note that $h(t) \leqslant h^*(t) = x(t)/\bar{M} \leqslant h_0$ and $Q'(h) > 0$ for $h < h_0$. Hence $Q(h) \leqslant Q(h^*)$ if $h \leqslant h^*$.

R[3] $[t_1, t_2)$ **Shedding labour**

As $x(t)$ continues to decline, $h^*(t) = x(t)/\bar{M} \to h_d$. Thus, t_1 is determined by $x(t_1) = \bar{M}h_d$. For $t > t_1$, $h^*(t) = h^d(> x(t)/\bar{M})$ and adjustment takes place by shedding labour. Thus t_1 is a junction time and λ_1 may be discontinuous at this point. We have, for $t > t_1$,

$$x(t) - h_d M^*(t) = 0,$$

$$M^*(t) < \bar{M},$$

$$\dot{M}^*(t) = -D^*(t) = \dot{x}(t)/h_d < 0 \Rightarrow D^* > 0 \ (A^* = 0).$$

Solution: $[x(t)/h_d, 0, -x(t)/h_d, h_d]$.

(7.101′) is again satisfied as in R[2] because an admissible $h(t)$ must satisfy $x - hM^* \geqslant 0$, hence $h \leqslant x/M^* = h_d$. Then $h \leqslant h_d \leqslant h_0$, so $Q(h) \leqslant Q(h_d)$, given the properties of $Q(h)$.

Before discussing the regime $[t_2, t_3)$, consider the regimes symmetric to those discussed above.

R[4] $[t_5, T]$ Excess demand

t_5 is determined as the point where $\bar{M}h_0 = x(t_5)$ as in R[1].

Solution: $[\bar{M}, 0, 0, h_0]$.

R[5] $[t_4, t_5)$ Hours adjustment

As in R[2], the firm satisfies demand by adjusting hours worked: $\bar{M}h^*(t) = x(t)$.

Solution: $[\bar{M}, 0, 0, x(t)/\bar{M}]$.

R[6] $[t_3, t_4)$ Hiring of labour

The equivalent point to h_d on the upswing is denoted by h_a and is defined by (cf. (7.105)),

$$w(h_a) - h_a w'(h_a) = ar. \tag{7.106}$$

For the same reasons, we assume $h_a < h_0$. Thus t_4 is defined by (7.106) (which gives h_a) and the equation

$$\bar{M}h_a = x(t_4).$$

In this case, however, $A^* > 0$, $D^* = 0$, $M^*(t) < \bar{M}$ and $\dot{M}^* = A^* = \dot{x}(t)/h_a$.

Solution: $[x(t)/h_a, \dot{x}(t)/h_a, 0, h_a]$.

R[7] $[t_2, t_3)$ The slump

$\dot{x}$ changes sign at $T/2$ hence R[3] can last at most until $T/2$ and R[6] can start at earliest at $T/2$. However, there may be an intervening period in which both $A^* = 0$ and $D^* = 0$. We now consider the possible extent of this period. Since $A^* = D^* = 0$, $\dot{M}^* = 0$ so $M^*(t) = M_s$, a constant, over this period. Output is constrained by demand, so hours worked must bear the adjustment required to satisfy

$$x(t)h^*(t) = M_s.$$

(Notice that, in $[t_1, t_2]$, $D^*>0$, $A^*=0 \Rightarrow h^*(t)=h_d$ and in $[t_3, t_4]$, $A^*>0$, $D^*=0$, and this implied $h^*(t)=h_a$. With both A^* and $D^*=0$, h^* can vary). Now t_2 and t_3 must satisfy

$$M_s = h_d x(t_2) = h_a x(t_1). \tag{7.107}$$

Secondly, λ_1 is continuous through t_2 and t_3. At the end of R[3], from (7.99), since $\mu_4=\mu_2=0$ (since $M^*<\bar{M}, D^*>0$)

$$\lambda_1(t_2) = -d\,\mathrm{e}^{-rt}, \tag{7.108}$$

and at the beginning of R[6], $\mu_4=\mu_3=0$, hence

$$\lambda_1(t_3) = a\,\mathrm{e}^{-rt}. \tag{7.109}$$

Between t_2 and t_3, defining $Z(h)=w(h)-hw'(h)$, we obtain

$$\dot{\lambda}_1 = -[ph^* - w(h^*)]\mathrm{e}^{-rt} + h^*[p - w'(h^*)]\mathrm{e}^{-rt} = Z(h^*)\mathrm{e}^{-rt}$$

so

$$\dot{\lambda}_1 = Z\left(\frac{x(t)}{M_s}\right)\mathrm{e}^{-rt}, \tag{7.110}$$

where $x(t)/M_s$ is the argument of the function Z. Integrating and using (7.108), (7.109), we obtain

$$a\,\mathrm{e}^{-rt_3} + d\,\mathrm{e}^{-rt_2} = \lambda_1(t_3) - \lambda_1(t_2) = \int_{t_2}^{t_3} \dot{\lambda}_1 \mathrm{d}t = \int_{t_2}^{t_3} Z\left(\frac{x(t)}{M_s}\right) e^{-rt}\mathrm{d}t. \tag{7.111}$$

(7.107) and (7.111) serve to determine t_2, t_3 and M_s. It is easy to check that (7.101′) is satisfied in R[4]–R[7]. The other conditions requiring mention are the jump conditions N[8]. At junction times, t_1, t_3, we require

$$\lambda_1(t_1^-) - \lambda_1(t_1^+) = \beta_1 \frac{\partial \theta^1}{\partial M} = -\beta_1, \tag{7.112}$$

and

$$\lambda_1(t_3^-) - \lambda_1(t_3^+) = -\beta_2, \tag{7.113}$$

where $\beta_1, \beta_2 \geqslant 0$ are constants. It is unnecessary for us to calculate the size of β_1, β_2. (7.112), (7.113) indicate that λ_1 may jump upwards (and certainly not downwards) at the junction times t_1 and t_3.

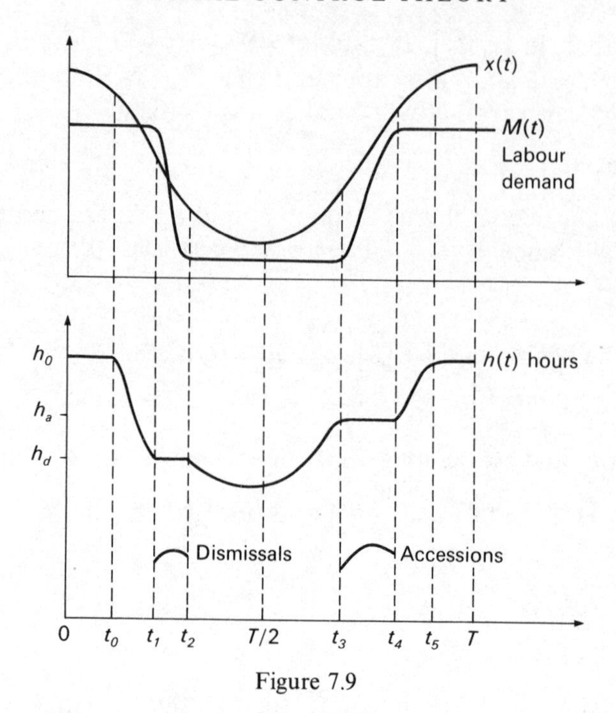

Figure 7.9

This completes the analysis of the full cycle. The nature of the solution is depicted in Figure 7.9. The solution satisfies the necessary conditions of Theorem 7.8. We now show that the sufficient conditions of Theorem 7.9 are also satisfied, and hence that our solution is in fact optimal.

Clearly S[1] is satisfied; $\bar{M}-M \geqslant 0$ is linear. For S[2], let us consider whether H is concave in (M, h, A, D). The Hessian matrix is

$$\begin{pmatrix} 0 & p-w'(h) & 0 & 0 \\ p-w'(h) & -w''(h)M & 0 & 0 \\ 0 & 0 & 0 & 0 \\ 0 & 0 & 0 & 0 \end{pmatrix}$$

This is not negative semidefinite since the principal minors of order 1, $|\tilde{D}_1|$ have values$=0$ or $-w''(h)M \leqslant 0$, those of order $|\tilde{D}_2|=0$ or

$-(p-w'(h))^2$ and the latter is <0 for some values of h on the optimal path. Hence H is not concave.

However, S[2](ii) is satisfied; the set $\tilde{A}(t) \equiv \{M : A \geqslant 0, D \geqslant 0, x-hM \geqslant 0$ for some $A, D, h\}$ is clearly convex and H^* is linear in M, hence concave in M.

Finally, we have already shown that $\lambda_0 > 0$. Hence the sufficient conditions are satisfied.

Optimal capacity

The second stage decision of choosing optimal capacity is now considered. Let q denote the rental price associated with capacity. Capacity costs are thus

$$C(\bar{M}) = \int_0^T q\bar{M}\,\mathrm{e}^{-rt}\mathrm{d}t.$$

Hence profits are given by $\pi(M) = \int_0^T [pM^*h^* - w(h^*)M^* - aA^* - dD^* - q\bar{M}]\mathrm{e}^{-rt}\mathrm{d}t$. We now calculate $\partial\pi(M)/\partial\bar{M}$ since a necessary condition for profit maximization, given the assumption that $\bar{M} > 0$, is that $\partial\pi(M)/\partial\bar{M} = 0$. The calculation is best done for each regime. Where $M^*(t) < \bar{M}$ (between t_1 and t_4), $\partial[.]/\partial\bar{M} = -q$ and where $M^*(t) = \bar{M}$, $\partial[.]/\partial\bar{M} = [ph^* - w(h^*) - q]$. So, referring to the optimal solutions, we obtain

$$\begin{aligned}\partial\pi/\partial\bar{M} = \int_0^{t_0} Z(h_0)\mathrm{e}^{-rt}\mathrm{d}t + \int_{t_0}^{t_1} Z(h^*)\mathrm{e}^{-rt}\mathrm{d}t + \int_{t_4}^{t_5} Z(h^*)\mathrm{e}^{-rt}\mathrm{d}t \\ + \int_{t_5}^{T} Z(h_0)\mathrm{e}^{-rt}\mathrm{d}t - a\,\mathrm{e}^{-rt_4} - d\,\mathrm{e}^{-rt_1} - \int_0^T q\,\mathrm{e}^{-rt}\mathrm{d}t = 0,\end{aligned} \tag{7.114}$$

where $Z(h) = hw'(h) - w(h)$. (7.114) implicitly determined the optimal choice of $\bar{M}$ (note that $t_0, \ldots, t_5$ all depend on $\bar{M}$). It is possible to examine comparative statics on the optimal solution in the manner of the example discussed in Chapter 6, Section 6.11. To illustrate, we calculate $\partial\bar{M}/\partial a$ and $\partial t_0/\partial a$.

Differentiating (7.114) with respect to a gives (note that $t_0, \ldots, t_5$ depend upon $\bar{M}$ and a. Also that the terms arising from differentiation with respect to upper and lower limits all cancel – h is continuous on $[0, T]$:

$$\int_{t_0}^{t_1} Z'(h^*)\frac{\partial h^*}{\partial a}\mathrm{e}^{-rt}\mathrm{d}t + \int_{t_4}^{t_5} Z'(h^*)\frac{\partial h^*}{\partial a}\mathrm{e}^{-rt}\mathrm{d}t - \mathrm{e}^{-rt_4} = 0.$$

Now $h^* = x/\bar{M}$ and

$$\frac{\partial h^*}{\partial a} = -\frac{x}{\bar{M}^2}\frac{\partial \bar{M}}{\partial a} = -\frac{h}{\bar{M}}\frac{\partial \bar{M}}{\partial a},$$

so

$$\frac{\partial \bar{M}}{\partial a} = -\bar{M}\,\mathrm{e}^{-rt_4}\left[\int_{t_0}^{t_1} h^*Z'(h^*)\mathrm{e}^{-rt}\mathrm{d}t + \int_{t_4}^{t_5} h^*Z'(h^*)\mathrm{e}^{-rt}\mathrm{d}t\right]^{-1} < 0,$$

since $Z'(h^*) > 0$. Turning to $\partial t_0/\partial a$, since $h^*(t_0)\bar{M} = x(t_0)$, we have

$$\bar{M}\frac{\partial h^*(t_0)}{\partial t_0}\frac{\partial t_0}{\partial a} + h^*(t_0)\frac{\partial \bar{M}}{\partial a} = \dot{x}(t_0)\frac{\partial t_0}{\partial a}.$$

Now $\dfrac{\partial h^*(t_0)}{\partial a} = 0$, so $\dfrac{\partial t^0}{\partial a} = \dfrac{h^*(t_0)}{\dot{x}(t_0)}\dfrac{\partial \bar{M}}{\partial a} > 0$, since $\dot{x}$, $\dfrac{\partial \bar{M}}{\partial a} < 0$.

Similar results may be developed for $t_1, \ldots, t_5$ and for other parameters (see Nickell, 1978).

7.8 INFINITE HORIZON PROBLEMS

Consider the problem of optimal economic growth which involves consumption and capital accumulation over time. For a given planning period, as a terminal condition, one might specify target capital stocks. However, any particular choice of time horizon (and stocks) is essentially arbitrary; there is no obviously natural termination point to the problem. One way to avoid this difficulty is to assume an infinite horizon. This too is clearly a fiction, and it only really makes sense if the optimal path is not too sensitive to what happens in the distant future. Still, it often proves a convenient simplification and has been widely adopted in the economic growth literature. The infinite horizon problem involves certain technical complications which lead to modifications in the statement of both necessary and sufficient conditions.

PROBLEM 7.9

$$\underset{u}{\text{Maximize}} \int_{t_0}^{\infty} f^0(x,u,t)\mathrm{d}t$$

$$\begin{aligned} \text{subject to } & \dot{x} = f(x,u,t), \\ & g^i(x,u,t) \geqslant 0 \qquad i=1,\ldots,q, \\ & \theta^i(x,t) \geqslant 0 \qquad i=1,\ldots,\alpha, \\ & t_0 \text{ fixed}, \\ & x(t_0)=x_0, \text{ fixed}, \\ & x_1: \text{conditions to be defined}, \\ & x(t)\in\mathbb{R}^n, \qquad u(t)\in\mathbb{R}^m. \end{aligned}$$

u is piecewise continuous on $[t_0,\infty)$. There may be terminal conditions such as $\lim_{t\to\infty}x(t)=x_1$ (fixed). The objective function, an improper integral, is sensible only if, for all admissible control pairs (x,u), the integral converges. A common form of economic objective involves $f^0(x,u,t)=g(x,u)e^{-rt}(r>0)$. The integral will clearly converge in this case if, for all admissible x,u, the function g is bounded.

If the integral converges for all admissible (x,u), then the necessary conditions are precisely those stated earlier for the equivalent finite horizon problem, except that the conditions are now required to hold on $[t_0,\infty)$ and the transversality conditions no longer hold. Thus, with an endpoint condition such as $\lim_{t\to\infty}x(t)=x_1$, one might expect $\lim_{t\to\infty}\lambda(t)=\theta$ to be the associated transversality condition; this is not the case. Halkin (1974) provides a counter-example.

In practice, it is desirable to admit as admissible, trajectories which do not converge to limits and for which the integral may not converge. If the integral does not converge, a reformulation of the concept of optimality is necessary. We deal first with the terminal conditions, and then with the criterion of optimality. Define, for any function $\theta(t)$

$$\underline{\lim}_{t\to\infty}\,\theta(t)=\lim_{t\to\infty}\,\inf\{\theta(t):t\in[\tau,\infty)\},$$

$$\overline{\lim}_{t\to\infty}\,\theta(t)=\lim_{\tau\to\infty}\,\sup\{\theta(t):t\in[\tau,\infty)\}.$$

Thus $\underline{\lim}_{t\to\infty}\,\theta(t)\geqslant 0$ means that, for every $\varepsilon>0$, there exists a t' such that, if $t\geqslant t'$, then $\theta(t)\geqslant -\varepsilon$. $\overline{\lim}_{t\to\infty}\,\theta(t)\geqslant 0$ means that this holds for *every* t'. Clearly, $\underline{\lim}_{t\to\infty}\,\theta(t)\leqslant \overline{\lim}_{t\to\infty}\,\theta(t)$.

The following endpoint conditions will be considered:

(i) There exists a τ such that $t\geqslant\tau$ implies $x_i(t)\geqslant x_{i1}$ $i=1,\ldots,n$,

(ii) $\underline{\lim}_{t\to\infty}\,x_i(t)\geqslant x_{i1}$, $i=1,\ldots,n$,

(iii) $\overline{\lim}_{t\to\infty}\,x_i(t)\geqslant x_{i1}$, $i=1,\ldots,n$.

Each is a further relaxation which increases the set of admissible solutions (since we wish to consider as admissible, paths for which $\lim_{t\to\infty} x(t)$ does not exist). Define

$$\Delta J(t)=\int_{t_0}^{t}\{f^0(x^*(\xi),u^*(\xi),\xi)-f^0(x(\xi),u(\xi),\xi)\}\mathrm{d}\xi$$

and consider the following concepts of optimality:

(i) If there exists a τ such that $\Delta J(t)\geqslant 0$ for all $t\geqslant\tau$, then we say $x^*(t)$ **overtakes** $x(t)$.

(ii) If $\underline{\lim}_{t\to\infty}\,\Delta J(t)\geqslant 0$, we say $x^*(t)$ **catches up** with $x(t)$.

(iii) If $\overline{\lim}_{t\to\infty}\,\Delta J(t)\geqslant 0$, we say $x^*(t)$ **sporadically catches up** with $x(t)$.

Clearly (i) $\Rightarrow$ (ii) $\Rightarrow$ (iii). The case where the integral converges for every admissible (x,u) is, of course, a special case of (ii) or (iii) since, with convergence, as $t\to\infty$, $\lim \Delta J=\overline{\lim}\,\Delta J=\underline{\lim}\,\Delta J$. We are now in a position to state the following sufficiency theorem for Problem 7.9.

THEOREM 7.10 If, in Theorems 7.8, 7.9, we put $t_1=\infty$, and if the conditions specified there are satisfied on $[t_0, \infty)$ (except for the transversality conditions) then the following are sufficient for optimality:

S(1) if the terminal condition is (i), then, according to the overtaking criterion, if, for all admissible $x(t)$, there exists a number τ (in general dependent on x) such that

$$\lambda'(t)[x(t)-x^*(t)]\geqslant 0 \quad \text{for} \quad t>\tau,$$

S(2) if the terminal condition is (ii), then, according to the catching up criterion, if, for all admissible $x(t)$

$$\underline{\lim}_{t\to\infty} \lambda'(t)[x(t)-x^*(t)]\geqslant 0,$$

S(3) if the terminal condition is (iii), then, according to the sporadically catching up criterion, if, for all admissible $x(t)$

$$\overline{\lim}_{t\to\infty} \lambda'(t)[x(t)-x^*(t)]\geqslant 0.$$

Proof. See Seierstad and Sydsaeter (1977).

Some care is required in checking whether the limits such as $\lim_{t\to\infty} \lambda'(t)(x(t)-x^*(t))\geqslant 0$ are in fact satisfied. For example, it would be incorrect to conclude that establishing $\lim_{t\to\infty} \lambda(t)\geqslant \theta$, $\lim_{t\to\infty} \lambda'(t)x^*(t)=0$ and $x(t)\geqslant \theta$ for all t, would suffice to establish that $\lim_{t\to\infty} \lambda'(t)(x(t)-x^*(t))\geqslant 0$, as consideration of the following example (see Seierstad and Sydsaeter, 1987) makes clear. Let $\lambda(t)=-\exp(-t)$, $x(t)=\exp(t)$, $x^*(t)=1$. Then the former conditions are satisfied, but not the latter. Some sufficient conditions for the above limits to hold are discussed in Seierstad and Sydsaeter (1977), also (1987).

Work on infinite horizon problems (particularly concave problems) in the context of economic growth continues – and space precludes an adequate discussion here. See, for example, the Journal of Economic Theory Symposium proceedings (1976).

Example 7.7 at the end of this chapter provides an application of

Theorem 7.10 (sufficient conditions). However, there are many situations in which the theorem cannot be applied. Still, the necessary conditions usually serve to identify a set of candidate trajectories, and often one of these suggests itself as the optimal trajectory. It may then be possible to eliminate the other trajectories from first principles as definitely non-optimal. Example 7.6 illustrates this approach.

In many infinite horizon economic applications, there is a single state variable and the integrand takes the quasiautonomous form

$$f^0(x,u)\mathrm{e}^{-rt}.$$

In these problems, the optimal solution is often an adjustment toward a long-run steady state (along the stable arm of a saddlepoint equilibrium).

$$\text{Maximize} \int_0^\infty f^0(x,u)\mathrm{e}^{-rt}\mathrm{d}t$$

$$\begin{aligned}\text{subject to } \dot{x} &= f(x,u),\\ x(0) &= x_0,\\ x(t), u(t) &\in \mathbb{R}.\end{aligned}$$

The Hamiltonian can be written as

$$H = \lambda_0 f^0(x,u)\mathrm{e}^{-rt} + \lambda f(x,u),$$

and, if we assume an interior solution, $H_u = 0$ is a necessary condition;

$$\lambda_0 \mathrm{e}^{-rt} f_u^0 + \lambda f_u = 0. \tag{7.115}$$

The multipliers should not be simultaneously zero, hence $\lambda_0 \neq 0$, so we may take $\lambda_0 = 1$. Now

$$\dot{\lambda} = -H_x = -f_x^0\,\mathrm{e}^{-rt} - \lambda f_x. \tag{7.116}$$

As with the Calculus of Variations, an autonomous system may be obtained by letting $q(t) = \lambda(t)\mathrm{e}^{-rt}$. (7.115) then becomes

$$f_u^0 + q f_u = 0.$$

Often it is possible to write this implicit equation involving u, x, and

q in the form

$$u = g(x, q).$$

(7.116) becomes

$$\dot{q} = -f_x^0(x, g(x, q)) - qf_x(x, g(x, q)) + rq. \tag{7.117}$$

(7.117) in conjunction with

$$\dot{x} = f(x, g(x, q)) \tag{7.118}$$

forms a pair of autonomous differential equations. The long-run steady state $(\hat{x}, \hat{q})$ is then obtained as the solution to

$$-f_x^0(\hat{x}, g(\hat{x}, \hat{q})) - \hat{q}f_x(\hat{x}, g(\hat{x}, \hat{q})) + r\hat{q} = 0,$$

$$f(\hat{x}, g(\hat{x}, \hat{q})) = 0.$$

The nature of the optimal trajectory in the region of the steady state can then be analyzed using the phase plane or linear approximation techniques, as in the following example.

EXAMPLE 7.6 (Capacity adjustment)

This problem has been analyzed in a slightly simpler form in Chapter 6, Example 6.10) to which the reader is referred for introductory discussion. The problem we address here is

$$\text{Maximize} \int_0^\infty \{R(K) - C(I)\}\mathrm{e}^{-rt}\mathrm{d}t$$

$$\text{Subject to } \dot{K} = I - \psi K, \tag{i}$$

$$K(0) = 0, \tag{ii}$$

$$K(t) \geqslant 0 \text{ for all } t \geqslant \tau \text{ (for some } \tau). \tag{iii}$$

I is chosen as the control and K as state variable. The Hamiltonian is

$$H = \lambda_0[R(K) - C(I)]\mathrm{e}^{-rt} + \lambda[I - \psi K],$$

and the Lagrange function $L = H$, since there are no g^i-constraints.

The necessary conditions of Theorem 7.8 for this problem are

$$\dot{\lambda} = -L_k = -\lambda_0 R'(K)e^{-rt} + \psi\lambda, \tag{v}$$

$$L_I = -\lambda_0 C'(I)e^{-rt} + \lambda = 0. \tag{vi}$$

The Hamiltonian condition is

$$\lambda_0[C(I) - C(I^*)]e^{-rt} + \lambda[I^* - I] \geqslant 0 \text{ for all } I. \tag{vii}$$

Referring to the sufficient conditions of Theorem 7.9, S[1] does not apply; S[2](i) requires the Hamiltonian to be jointly concave in (K, I) which it is; i.e. the Hessian matrix is

$$\begin{bmatrix} R''(K)e^{-rt} & 0 \\ 0 & -C''(I)e^{-rt} \end{bmatrix}$$

which is negative semidefinite. S[3] requires $\lambda_0 = 1$, which is straightforward to establish. Unfortunately, Theorem 7.10 also requires

$$\lambda(t)[K(t) - K^*(t)] \geqslant 0 \quad \text{for } t \geqslant t' \quad \text{for some } t' \tag{viii}$$

and it is not possible to establish this condition. We therefore consider the set of trajectories which satisfy the necessary conditions and show from first principles that only one of these can be optimal.

If $\lambda_0 = 0$, then (vi) implies $\lambda = 0$, violating the necessary condition that the multipliers do not simultaneously vanish. Hence take $\lambda_0 = 1$. From (i), (v) and (vi), and making the substitution $q(t) = \lambda(t)e^{rt}$, we obtain the system

$$\dot{K} = I - \psi K, \tag{ix}$$

$$\dot{q} = (r + \psi)q - R'(K), \tag{x}$$

$$q = C'(I). \tag{xi}$$

Setting $\dot{q} = \dot{K} = 0$, the equilibrium point $(\hat{K}, \hat{q})$ may be determined and, as in the Example 6.10, $\hat{K} > 0$, $\hat{q} > 0$ established. The phase diagram (Figure 7.10) is a saddlepoint, and the stable arm trajectory (K^*, q^*) starting from q_0 on the vertical axis is in fact the optimal trajectory. The analysis is precisely that used in Example 6.10, to which the reader is referred.

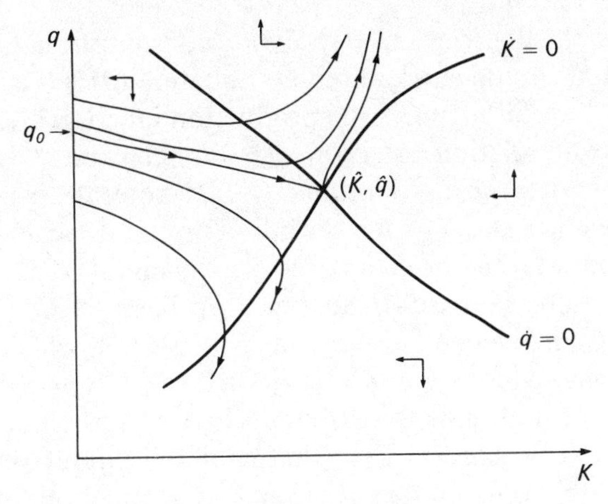

Figure 7.10

The optimal trajectory (K^*, q^*) converges to a steady state $(\hat{K}, \hat{q})$ with K^* monotonically increasing and q^* decreasing (and hence also I^* decreasing, since $q^* = C'(I^*)$ and $C''(.) > 0$). In the steady state solution, investment just suffices to maintain the capital stock. Notice that the optimal solution involves $K^*(t), I^*(t) \geqslant 0$ for all t, so, if there had been non-negativity constraints in the problem, the above solution would still constitute the optimal solution (for extensions, see Milne, 1977).

EXAMPLE 7.7 (Optimal growth) Consider the following growth problem:

$$\text{Optimize} \int_0^\infty U[C(t)]e^{-rt}dt \tag{i}$$

$$\text{subject to } Y(t) = C(t) + I(t), \tag{ii}$$

$$Y(t) = f(K(t)), \tag{iii}$$

$$\dot{K}(t) = I(t) - \psi K(t), \tag{iv}$$

$$C(t) \geqslant 0, \tag{v}$$

$$K(0) = K_0 > 0, \tag{vi}$$

where $C(t)$, $I(t)$, $Y(t)$, $K(t)$ denote consumption, investment, output and capital stock at time t. Equation (ii) is the accounting identity; f in (iii) denotes the production function and (iv) is the capital accumulation equation. $\psi>0$ is the capital depreciation rate and $r>0$ is the discount rate. $U[.]$ denotes the instantaneous utility function and $K_0>0$ is the initial capital stock. For a general discussion of this problem, see Takayama (1985). Here, we consider a special case (treated as a problem in the Calculus of Variations in Seierstad and Sydsaeter (1987)) in which the utility function has a constant elasticity of 1−v: $U(C)=C^{1-v}/(1-v)$, $0<v<1$. In addition, we assume a linear production function; $Y(t)=\alpha K(t)$, where $\alpha>\psi$ (so that capital accumulation is possible) and $(\alpha-\psi-r)/v<\alpha-\psi$. Using these functional forms and eliminating $C(t)$ from (i)–(iv), the problem becomes

$$\text{Optimize} \int_0^{\infty} [\alpha K(t)-I(t)]^{1-v} e^{-rt} dt \tag{vii}$$

$$\text{subject to } \dot{K}(t)=I(t)-\psi K(t), \tag{viii}$$

$$\alpha K(t)-I(t)\geqslant 0, \tag{ix}$$

$$K(0)=K_0>0, \tag{x}$$

$$\underline{\lim}_{t\to\infty} K(t)\geqslant 0, \tag{xi}$$

where we have included the 'non-negativity' transversality condition on capital stock such that Theorem 7.10 may be applied to obtain an optimal solution with respect to the catching up criterion. $K(t)$ denotes the state variable and $I(t)$, the control. The Hamiltonian is

$$H(K(t),I(t),\lambda(t),t)=$$
$$\lambda_0\left\{\frac{\alpha K(t)-I(t)}{1-v}\right\}^{1-v} e^{-rt}+\lambda_1(t)\{I(t)-\psi K(t)\},$$

and the Lagrange function

$$L(K(t),I(t),\lambda(t),\mu(t),t)$$
$$=H(K(t),I(t),\lambda(t),t)+\mu(t)\{\alpha K(t)-I(t)\},$$

Referring to Theorem 7.8, the necessary conditions are (suppressing arguments and denoting the optimal trajectory by K^*, I^*):

$$\dot{\lambda}_1 = -\alpha\lambda_0[\alpha K^* - I^*]^{-v}\mathrm{e}^{-rt} + \psi\lambda_1 - \mu\alpha, \tag{xii}$$

$$\partial L/\partial I = -\lambda_0[\alpha K^* - I^*]^{-v}\mathrm{e}^{-rt} + \lambda_1 - \mu = 0, \tag{xiii}$$

$$\mu \geqslant 0, \alpha K^* - I^* \geqslant 0, \qquad \mu(\alpha K^* - I^*) = 0. \tag{xiv}$$

Intuition suggests that constraint (ix) does not bind in the optimal solution; accordingly we examine whether a solution for which $\alpha K^* - I^* > 0$ can be found which satisfies both the necessary and the sufficient conditions. If $\alpha K^* - I^* > 0$, then $\mu(t) = 0$ for all t. From (xiii), if $\lambda_0 = 0$, so does λ_1, contradicting the necessary condition that the multipliers should be non-zero. Hence take $\lambda_0 = 1$. (xiii) gives

$$\lambda_1 = [\alpha K^* - I^*]^{-v}\mathrm{e}^{-rt} \tag{xv}$$

which, applied to (xii), yields

$$\dot{\lambda}_1 = (\psi - \alpha)\lambda_1. \tag{xvi}$$

Integrating (xvi) and using (xv),

$$\lambda_1(t) = c_1\mathrm{e}^{(\psi-\alpha)t} = [\alpha K^* - I^*]^{-v}\mathrm{e}^{-rt}$$

$$\Rightarrow [\alpha K^* - I^*]^{-v} = c_1\mathrm{e}^{(\psi-\alpha+r)t}$$

$$\Rightarrow \alpha K^* - I^* = c_2\exp\left\{\left(\frac{\alpha-\psi-r}{v}\right)t\right\}. \tag{xvii}$$

From (viii), $I^* = \dot{K}^* + \psi K^*$, hence (xvii) becomes

$$\dot{K}^* + (\psi - \alpha)K^* = -c_2\exp\left\{\left(\frac{\alpha-\psi-r}{v}\right)t\right\},$$

a linear differential equation. Multiplying through by $\exp(\psi-\alpha)t$ gives

$$\frac{\mathrm{d}}{\mathrm{d}t}(K^*\exp\{(\psi-\alpha)t\}) = -c_2\exp\left\{\left(\frac{\alpha-\psi-r}{v} + (\psi-\alpha)\right)t\right\}.$$

Integrating,

$$K^*\exp\{(\psi-\alpha)t\}=c_3+c_4\exp\left\{\left(\frac{\alpha-\psi-r}{v}+\psi-\alpha\right)t\right\}$$

$$\Rightarrow K^*(t)=c_3\exp\{(\alpha-\psi)t\}+c_4\exp\left\{\left(\frac{\alpha-\psi-r}{v}\right)t\right\}.$$

At $t=0$, $K(0)=K_0$, so $c_3+c_4=K_0$;

$$K^*(t)=(K_0-c_4)\exp\{(\alpha-\psi)t\}+c_4\exp\left\{\left(\frac{\alpha-\psi-r}{v}\right)t\right\}. \qquad \text{(xviii)}$$

Since $\alpha-\psi>(\alpha-\psi-r)/v$, if $K_0-c_4\neq 0$, the first term will eventually dominate the second. Thus, since $\lim_{t\to\infty} K(t)\geqslant 0$, we must require that $K_0-c_4\geqslant 0$. Furthermore, we also require $\alpha K^*-I^*=(\alpha-\psi)K^*-\dot{K}^*\geqslant 0$. $\dot{K}^*$ may be obtained by differentiating (xviii). The inequality then reduces to

$$c_4\left\{\alpha-\psi-\left(\frac{\alpha-\psi-r}{v}\right)\right\}\exp\left\{\left(\frac{\alpha-\psi-r}{v}\right)t\right\}\geqslant 0.$$

Hence $c_4\geqslant 0$.

Now, the sufficient conditions of Theorem 7.9 (excepting the transversality conditions) hold since H is concave in (K,I), so the only additional requirement in Theorem 7.10, S(2), is

$$\underline{\text{Lim}}_{t\to\infty}\, \lambda_1(t)[K(t)-K^*(t)]\geqslant 0.$$

This will hold if $\underline{\text{Lim}}_{t\to\infty}\{-\lambda_1(t)K^*(t)\}\geqslant 0$, since $c_1\geqslant\lambda_1(t)\geqslant 0$ and $\underline{\text{Lim}}_{t\to\infty} K(t)\geqslant 0$. Now

$$-\lambda(t)K^*(t)=-c_1\exp\{(\psi-\alpha)t\}$$

$$\times\left[(K_0-c_4)\exp[(\alpha-\psi)t]+c_4\exp\left[\left(\frac{\alpha-\psi-r}{v}\right)t\right]\right]$$

$$=-c_1\left\{(K_0-c_4)+c_4\exp\left\{\left(\frac{\alpha-\psi-r}{v}-(\alpha-\psi)\right)t\right\}\right\}.$$

Hence, for the transversality condition to hold, since $K_0 - c_4 \geqslant 0$, we must require $K_0 = c_4$: Since the coefficient in the exponential is negative (and $c_1 > 0$, $c_4 = K_0$), the condition is then satisfied (that is, for any $\varepsilon > 0$, there exists a t' such that $-\lambda_1(t)K^*(t) \geqslant -\varepsilon$ for $t > t'$). The solution

$$K^*(t) = K_0 \exp\left\{\left(\frac{\psi - \alpha - r}{v}\right)t\right\}$$

is therefore optimal with respect to the catching up criterion.

MATHEMATICAL REVIEW

SETS AND MAPPINGS

A set can be regarded as a collection of objects viewed as a single entity. The objects in the collection are called **elements** or *points* of the set. If a is an element of set A we write $a \in A$; $a \notin A$ denotes that a is not an element of set A. A set can be defined by listing its elements, $A = \{1,2,3\}$, or by stating a common property of its elements; $A = \{x : x \text{ has property } P\}$ denotes the set consisting of all objects x that have property P. When two sets A and B have identical elements they are **equal**, $A = B$: $A \neq B$ then denotes that the elements of A and B are not completely identical with each other. A set B is a **subset** of the set A, denoted $B \subset A$, if for all $b \in B$, $b \in A$. Thus, a set is also defined to be a subset of itself. When $B \subset A$ and $B \neq A$, B is a **proper subset** of A. Clearly, $A = B$ if, and only if, $A \subset B$ and $B \subset A$. The set which contains no elements is called the **null** or **empty set**, denoted $\emptyset$. The null set is a subset of every set. A set can have other sets as its elements. For example, we can consider a set $X = \{A, B, C\}$ whose elements A, B, C are themselves sets. Here we call X a **family** of subsets.

If A and B are two sets, the set of elements that are contained in at least one of the two sets is the **union** of the sets, denoted $A \cup B$; $A \cup B = \{x : x \in A \text{ or } x \in B\}$. If A and B are two sets, the set of elements common to both sets is the **intersection** of the sets, denoted $A \cap B$; $A \cap B = \{x : x \in A \text{ and } x \in B\}$. The union (intersection) can be taken over a finite or infinite collection of sets: $\cup_{i=1}^{n} A_i = A_1 \cup A_2 \cup \ldots \cup A_n$ and $\cap_{i=1}^{n} A_i = A_1 \cap A_2 \cap \ldots \cap A_n$, where $A_1, A_2, \ldots, A_n$ are sets. A and B are **disjoint** sets if, and only if, $A \cap B = \emptyset$. Given two sets A and B the **difference** of A and B (or the **complement** of B relative to A), denoted $A \backslash B$, is the set of all elements of A which do not belong to B; $A \backslash B = \{x : x \in A, x \notin B\}$.

We call (a, b) an **ordered-pair** if, and only if, $(a, b) = (c, d)$ implies $a = c$ and $b = d$. Given two sets A and B the set of all ordered-pairs (a, b), where $a \in A$ and $b \in B$ is the **Cartesian product** of A and B denoted $A \otimes B$; $A \otimes B = \{(a, b) : a \in A, b \in B\}$. We can consider the Cartesian product of more than two sets; $\otimes_{i=1}^{n} A_i = A_1 \otimes A_2 \otimes \ldots \otimes A_n = \{(a_1, a_2, \ldots, a_n) : a_i \in A_i, \quad i = 1, \ldots, n\}$, where $A_1, A_2, \ldots, A_n$ are sets. Let $\mathbb{R}$ denote the set of real numbers. Then, for example, $\mathbb{R} \otimes \mathbb{R} = \mathbb{R}^2$ is the set of all points in the plane.

Given two sets X and Y, a rule which assigns to each element of the set X a non-empty subset of the set Y is a **mapping** or **transformation**. Mappings can be grouped into two broad categories. First, there is the class of mappings in which an element of a set is associated with a single element of another set; secondly, there is the class of mappings in which an element of a set is associated with a non-empty subset of another set, where at least one such subset contains more than one element. Mappings of the first class are **functions** or **point-point mappings**; mappings of the second class are **correspondences** or **point-set mappings**.

It is useful to distinguish among the different components of a function. Let F denote the rule defining the function from the set X into the set Y. X is the **domain** of the function and the unique element of set Y which is associated with the element x of set X is the **image** or **value** of x under f, denoted $f(x)$. The **range** of the function, denoted $f(X)$, is the subset of Y which contains all those $y \in Y$ which appear as images of at least one $x \in X$. The subset of X containing only those elements of X which have as their image the element $y \in Y$ is the **inverse image** of y, denoted $f^{-1}(y) : f^{-1}(y) =$

$\{x{:}x\in X,\ f(x)=y\}$. To symbolically describe a function, it is common to write $f{:}X\rightarrow Y$ together with some statement defining the association rule. Often the latter is given by a formula indicating how the image of $x\in X$, $f(x)$, is to be determined. If $f{:}X\rightarrow X$ where $f(x)=x$ for all $x\in X$, then f is an **identity function** on X. If $f{:}X\rightarrow Y$ is a function and $A\subset X$ then the **image of** A under the function f, denoted $f(A)$, is defined by $f(A)=\{f(x){:}x\in A\}$. If $f{:}X\rightarrow Y$ is a function and $B\subset Y$ then the **inverse image of B** under the function f, denoted $f^{-1}(B)$ is defined by $f^{-1}(B)=\{x{:}x\in X,\ f(x)\in B\}$. Note that while $f^{-1}(B)\subset X$, B is not necessarily a subset of the range of f. Let $f{:}X\rightarrow Y$ be a function. The subset of the Cartesian product $X\otimes Y$ which contains all ordered-pairs $(x, f(x))$ is the **graph** of the function f; the graph of f is the set $\{(x, y){:}\,(x, y)\in X\otimes Y, y=f(x)\}$. If the range of a function f is a subset of $\mathbb{R}$, the set of real numbers, then f is a **real-valued function**. The elements in the domain of a real-valued function can be anything at all. Often the domain of a real-valued function will be a subset of $\mathbb{R}^n$, where $\mathbb{R}^n$ denotes the set of all ordered n-tuples of real numbers.

The various terminologies and concepts introduced for functions can be applied with minor modifications to correspondences. For a correspondence $F{:}X\rightarrow Y$, the domain is simply the set X, the non-empty set $F(x)\subset Y$ is the image set of the element $x\in X$, and the set $\cup_{x\in X}F(x)$ is the range of the mapping. The set $\{(x, y){:}\ (x, y)\in X\otimes Y, y\in F(x)\}$ is the graph of F.

If we take into account the character of the association rule then we can classify functions in various ways. A function $f{:}X\rightarrow Y$ is **one-one** or **injective** if, for any element $y\in Y$, its inverse image $f^{-1}(y)$ contains at most one element; that is, for any element $y\in Y$, $f^{-1}(y)$ is the empty set or a set consisting of a single element. A function $f{:}X\rightarrow Y$ is **onto** or **surjective** if every $y\in Y$ is the image of at least one $x\in X$; that is, the range of f, $f(X)$, is the set Y. We can have functions which are one-one but not onto, and functions which are onto but not one-one. A function $f{:}X\rightarrow Y$ is **invertible** or **bijective** if it is a one-one mapping from X onto Y. If the function $f{:}X\rightarrow Y$ is invertible then the inverse image $f^{-1}(y)$ of any element $y\in Y$ is a non-empty set consisting of a single element (of set X). This allows us to define a new mapping from set Y into set X by defining the association rule 'associate each element $y \in Y$ with the element $f^{-1}(y) \in X$'. This new mapping is clearly a one-one function from set Y onto set X; it is called the **inverse** of the function f and is denoted f^{-1}, $f^{-1}:Y \to X$, where $f^{-1}(y) = \{x : x \in X,\ y = f(x)\}$. If $f:X \to Y$ and $g:W \to Z$, where $f(X) \subset W$, the **composite function of g and f**, denoted $g \circ f$, is the function $g \circ f : X \to Z$, where $(g \circ f)(x) = g(f(x))$. The order in which functions are combined in constructing a composite function is important. We may be able to define $g \circ f$ but not necessarily $f \circ g$. Moreover, even when both $g \circ f$ and $f \circ g$ exist, in general, they will be different functions. Suppose $f:X \to Y$ is invertible and let $f^{-1}:Y \to X$ be its inverse. Then $f^{-1} \circ f : X \to X$, where $(f^{-1} \circ f)(x) = f^{-1}(f(x)) = f^{-1}(y) = x$, so that $f^{-1} \circ f$ is an identity function on X. Similarly, $f \circ f^{-1}$ is an identity function on Y.

Two sets X and Y are **equivalent** if there exists a one-one function from X onto Y. For example, let $X = \{x : x \in \mathbb{R}, 0 \leqslant x \leqslant 1)\}$ and $Y = \{y : y \in \mathbb{R}, 2 \leqslant y \leqslant 5\}$ and let $f:X \to Y$, where $f(x) = 3x + 2$. Here f is a one-one and onto so that X and Y are equivalent. A set X is **finite** if, and only if, it is empty or equivalent to a set $S = \{1, 2, \ldots, n\}$ for some $n \in \mathbb{N}$, where $\mathbb{N}$ is the set of natural numbers. Sets which are not finite are called **infinite sets**. The chief difference between finite and infinite sets is that an infinite set must be equivalent to some proper subset of itself, whereas a finite set cannot be equivalent to any proper subset of itself. A set X is **countable** if it is either finite or equivalent to the set of positive integers. There are infinite sets which are not countable, the classic example being the set of all real numbers, $\mathbb{R}$.

REAL AND COMPLEX NUMBER FIELDS

A **field** is a set F, whose elements are called **scalars**, on which the operations of *addition*, $+$, and *multiplication*, $.$, are defined, where these operations satisfy:

(1) for any $\alpha, \beta, \delta \in F$, $\alpha + \beta \in F$ and $\alpha + \beta = \beta + \alpha$ and $(\alpha + \beta) + \delta = \alpha + (\beta + \delta)$,

(2) for any $\alpha, \beta, \delta \in F$, $\alpha, \beta \in F$ and $\alpha . \beta = \beta . \alpha$ and $(\alpha . \beta) . \delta = \alpha . (\beta . \delta)$,

(3) for any $\alpha, \beta, \delta \in F$, $\alpha . (\beta + \delta) = \alpha . \beta + \alpha . \delta$ and $(\alpha + \beta) . \delta = \alpha . \delta + \beta . \delta$,

(4) there exists a scalar $0 \in F$ such that $0+\alpha=\alpha+0=\alpha$, for any $\alpha \in F$; 0 is the *additive identity*,

(5) for any $\alpha \in F$ there is a scalar $-\alpha \in F$ such that $\alpha+(-\alpha)=(-\alpha)+\alpha=0$; $-\alpha$ is the *additive inverse of* α,

(6) there exists a scalar $1 \in F$ such that $1.\alpha=\alpha.1=\alpha$, for any $\alpha \in F$: 1 is the *multiplicative identity*,

(7) for any $\alpha \in F$, $\alpha \neq 0$, there is a scalar $\alpha^{-1} \in F$ such that $\alpha^{-1}.\alpha=\alpha.\alpha^{-1}=1$; α^{-1} is the *multiplicative inverse of* α.

The set $\mathbb{R}$ of all real numbers and the set of all rational numbers are fields, where the operations of addition and multiplication are defined in the familiar way.

The **open** interval $\{x \in \mathbb{R}: a<x<b\}$ is denoted (a,b) and the **half-open** (half-closed) intervals $\{x \in \mathbb{R}: a \leqslant x<b\}$ and $\{x \in \mathbb{R}: a<x \leqslant b\}$ are denoted $[a,b)$ and $(a,b]$, respectively: the **closed** interval $\{x \in R: a \leqslant x \leqslant b\}$ is denoted $[a,b]$.

Let $X \subset \mathbb{R}$, then X is **bounded above** if there exists $a \in \mathbb{R}$ such that $a \geqslant x$ for all $x \in X$; a is an **upper bound** for the set X. Similarly, X is **bounded below** if there exists a $b \in \mathbb{R}$ such that $b \leqslant x$ for all $x \in X$; b is a **lower bound** for the set X. If X is bounded from above there is a smallest element in the set of upper bounds for X, α; α is called the **supremum** of X and we write $\alpha=\sup\{x \in X\}$. Similarly, if X is bounded below, there is a greatest element in the set of lower bounds for X, β; β is called the **infimum** of X and we write $\beta=\inf\{x \in X\}$. A set $X \subset \mathbb{R}$ may or may not contain its supremum or its infimum. If $\alpha \in X$ and $\alpha=\sup\{x \in X\}$ then α is called the **maximum element of** X: if $\beta \in X$ and $\beta=\inf\{x \in X\}$ then β is called the **minimum element of** X.

The *complex number system* is the set of all ordered-pairs (x,y) of real numbers with the operations of addition and multiplication defined by $(x,y)+(x',y')=(x+x',y+y')$ and $(x,y).(x',y')=(xx'-yy',xy'+x'y)$. The complex number system is a field with $(0,0)$ the additive identity and $(1,0)$ the multiplicative identity; $(-x,-y)$ is the additive inverse of (x,y) and $(x/(x^2+y^2),-y/(x^2+y^2))$ is the multiplicative inverse of (x,y), $(x,y) \neq (0,0)$. The real field is a subset of the complex field. To see this note that $(x,0)+(y,0)=(x+y,0)$ and $(x,0).(y,0)=(xy,0)$, so that complex numbers of the form $(x,0)$ have exactly the same properties as real numbers if we identify the complex number $(x,0)$

with the real number x. Whilst the real number system is a subset of the complex number system, the former has an important property that the latter in general does not possess. We can order real numbers but in general we are unable to order complex numbers.

The complex number $(0, 1)$ is usually denoted by i; $i^2 = (0, 1).(0, 1) = (-1, 0)$. Every complex number $z = (x, y)$ can be expressed in the form $z = x + iy$, where x is termed the **real-part** of z, $Re(z) = x$, and y the **imaginary-part** of z, $\text{Im}(z) = y$. To see this note that $x = (x, 0)$ and $iy = (0, 1).(y, 0) = (0, y)$, so that $x + iy = (x, 0) + (0, y) = (x, y) = z$. If two complex numbers z, z' are expressed in the form $z = x + iy$ and $z' = x' + iy'$ then it is easily verified that $z + z' = (x + x') + i(y + y')$ and $z.z' = (xy - x'y') + i(x'y - xy')$. Let $z = x + iy$ be a complex number; then the complex number $\bar{z} = x - iy$ is called the **complex conjugate** of z. Note that for the complex number $x = (x, 0)$ we have $\bar{x} = x$; that is, the complex conjugate of any real number is that real number. The sum of a complex number and its complex conjugate is a real number; for $z = x + iy$ we have $z + \bar{z} = (x + iy) + (x - iy) = (2x, 0)$. The product of a complex number and its complex conjugate is a non-negative real number; for $z = x + iy$ we have $z.\bar{z} = (x + iy).(x - iy) = (x^2 + y^2, 0)$. It is easy to verify that, for any complex numbers z and w, $\overline{(z + w)} = \bar{z} + \bar{w}$ and $\overline{(z.w)} = \bar{z}.\bar{w}$.

The **absolute value** or **modulus** of a complex number $z = (x, y)$ is the non-negative real number $|z|$ defined by $|z| = (x^2 + y^2)^{1/2}$. Note that $|(x, 0)| = (x^2)^{1/2} = |x|$, so that the definition of the absolute value of a complex number is the same as that for real numbers when the complex number itself is a real number. Being ordered-pairs of real numbers complex numbers can be represented as points in the plane. If we introduce polar coordinates r, θ, by setting $x = r\cos\theta$ and $y = r\sin\theta$, then the complex number $z = x + iy$ can be written as $z = r\cos\theta + ir\sin\theta = r(\cos\theta + i\sin\theta)$. Here $r = |z|$. Complex numbers expressed in the form $z = r(\cos\theta + i\sin\theta)$ are said to be in **polar form**.

We assume the reader is familiar with the exponential function $f: \mathbb{R} \to \mathbb{R}$, where $f(x) = e^x$. We now wish to define e^z, where z is a complex number, in such a way that the principal properties of e^x for x real are preserved; that is the laws of exponents; $e^x e^y = e^{x+y}$ and

$e^\circ = 1$. If we write $z = x + iy$ then for the law of exponents to hold we want $e^{x+iy} = e^x e^{iy}$, so that essentially we need to define e^{iy}. We define e^{iy} by $e^{iy} = \cos y + i \sin y$, so that, for $z = x + iy$, $e^z = e^x(\cos y + i \sin y)$. Note that, for z real, $(z = (x, 0))$, the above definition implies $e^z = e^x$. Now for $z = x + iy$ and $z' = x' + iy'$ we have

$$e^z e^{z'} = e^x e^{x'}[(\cos y \cos y' - \sin y \sin y') + i(\cos y \sin y' + \sin y \cos y')]$$
$$= e^x e^{x'}[\cos(y + y') + i \sin(y + y')] = e^{z+z'}.$$

It is also easily verified that $e^\circ = e^z e^{-z} = 1$. Now $z = x + iy$ can be expressed in polar form $z = r(\cos\theta + i \sin\theta)$, but $\cos\theta + i\sin\theta = e^{i\theta}$, so that z can also be expressed as $z = r\,e^{i\theta}$, where $r = |z|$.

VECTOR SPACES

A non-empty set V, whose elements are called **vectors**, is a **vector space over the field F** (or a **linear space**) if the operations of *vector addition*, $+$, and *scalar multiplication*, $.$, are defined on V and these operations satisfy the following conditions:

(1) for any $x, y, z \in V$, $x + y \in V$ and $x + y = y + x$ and $(x + y) + z = x + (y + z)$,
(2) for any $\alpha, \beta \in F$ and any $x \in V$, $\alpha . x \in V$ and $(\alpha . \beta) . x = \alpha . (\beta . x)$,
(3) for any $x \in V$, $1 . x = x$,
(4) for any $\alpha, \beta \in F$ and any $x, y \in V$, $\alpha . (x + y) = \alpha . x + \alpha . y$ and $(\alpha + \beta) . x = \alpha . x + \beta . x$,
(5) there exists a vector $\theta \in V$ such that $x + \theta = \theta + x = x$, for any $x \in V$; θ is called the *null vector*,
(6) there exists a vector $-x \in V$ such that $x + (-x) = (-x) + x = \theta$, for any $x \in V$.

Note that in the above definition we have used the same symbol, $+$, for addition of scalars and vector addition and the same symbol, $.$, for multiplication of scalars and scalar multiplication. No confusion should result from this double use of symbols; indeed, henceforth, we shall omit the symbol . and simply write $\alpha\beta$ for $\alpha . \beta$ and αx for $\alpha . x$.

From the above definition it is easy to see that, if θ is the null vector in the vector space V over the field F, then for any $x \in V$ and $0, 1, \alpha \in F$ we have $0x = \theta$, $(-1)x = -x$, and $\alpha\theta = \theta$.

The set of ordered n-tuples of real numbers $\mathbb{R}^n = \{x = (x_1, \ldots, x_n) : x_i \in \mathbb{R}, i = 1, \ldots, n\}$ is a vector space over the real field, where vector addition and scalar multiplication are defined respectively by $x + y = (x_1 + y_1, \ldots, x_n + y_n)$ and $\alpha x = (\alpha x_1, \ldots, \alpha x_n)$, for $\alpha \in \mathbb{R}$. The null vector for this vector space is $\theta = (0, \ldots, 0)$. We call this vector space the real vector space $\mathbb{R}^n$. Given two vectors $x, y \in \mathbb{R}^n$ we write

$x \geqslant y$ if $x_i \geqslant y_i$ for all $i = 1, \ldots, n$,

$x > y$ if $x_i \geqslant y_i$ for all $i = 1, \ldots, n$, and $x_i > y_i$ for at least one i, $i = 1, \ldots, n$,

$x \gg y$ if $x_i > y_i$ for all $i = 1, \ldots, n$,

and $\mathbb{R}^n_+ \equiv \{x : x \in \mathbb{R}^n, x \geqslant \theta\}$,

$\mathbb{R}^n_{++} \equiv \{x : x \in \mathbb{R}^n, x \gg \theta\}$.

Any vector $x \in \mathbb{R}$ may be expressed as a column vector

$$x = \begin{bmatrix} x_1 \\ \vdots \\ x_n \end{bmatrix} \text{ or as a row vector } x = (x_1, \ldots, x_n).$$

In this text, unless otherwise stated, all vectors are regarded as column vectors. To indicate that a vector x is a row vector we write x' and call x' the **transpose** of x.

A non-empty subset M of a vector space V over a field F is a **subspace** of V if M is itself a vector space over the field F; that is, if for any $x, y \in M$ and any $\alpha, \beta \in F$, $\alpha x + \beta y \in M$. Note that the null vector θ of V must be contained in M. The subset $\{\theta\}$ and the entire space V are subspaces of the vector space V. If S and M are subspaces of a vector space V then their intersection $S \cap M$ is also a subspace of V and their **sum** defined as $S + M = \{x : x = s + m, s \in S, m \in M\}$ is also a subspace of V.

Let V be a vector space over a field F. Then given r vectors $x_1, \ldots, x_r$ in V the vector x defined by $x = \sum_{i=1}^{r} \lambda_i x_i$, $\lambda_i \in F$, $i = 1, \ldots, r$, is said to be a **linear combination** of these r vectors, and the set of all linear combinations of these vectors constitutes a subspace of V, which is called the **subspace generated** or **spanned** by $x_1, \ldots, x_r$.

Let V be a vector space over the field F and $x_1, \ldots, x_r \in V$. Then the set of vectors $\{x_1, \ldots, x_r\}$ is said to be **linearly independent** if, and only if, $\sum_{i=1}^{r} \lambda_i x_i = \theta$, $\lambda_i \in F$, implies $\lambda_i = 0$ for every i. A set of vectors which is not linearly independent is said to be **linearly dependent**. Any set of vectors which contains the null vector θ is linearly dependent.

Let S be the subspace generated by the vectors $x_1, \ldots, x_r$. If $\{x_1, \ldots, x_r\}$ is linearly independent then the set $\{x_1, \ldots, x_r\}$ is said to be a **basis** for S: that is, a basis for a vector space is a linearly independent subset of that space which generates or spans that space. The set of vectors $\{\mathrm{e}_1, \ldots, \mathrm{e}_n\}$ where $\mathrm{e}_i \in \mathbb{R}^n$, $i = 1, \ldots, n$, with e_i the vector whose ith component is unity and whose other components are all zero, is a basis for the real vector space $\mathbb{R}^n$; it is usually called the **standard basis** for $\mathbb{R}^n$. Let $\{\alpha_1, \ldots, \alpha_n\}$ be a basis for the vector space V over the field F. Then $v \in V$ is represented by a unique linear combination of the basis vectors; that is, $v = \sum_{i=1}^{n} a_i \alpha_i$, where the a_i are the **coordinates of *v*** with respect to the basis $\{\alpha_1, \ldots, \alpha_n\}$.

A vector space is said to be **finite dimensional** if, and only if, it has a finite basis; that is, it is spanned by a finite number of linearly independent vectors. While there can be many different bases for a given finite dimensional vector space V, every basis for V must contain the same number of elements. The number of elements in any basis of a finite dimensional vector space V is called the **dimension of *V***, denoted dim V. If a vector space V has dimension n, then any set containing $n+1$ vectors of V must be linearly dependent, and no set containing $n-1$ vectors of V can generate V. The subspace $\{\theta\}$ has no basis; by convention its dimension is defined to be zero.

A subset L of $\mathbb{R}^n$ is **affine** if $\lambda x + (1-\lambda) y \in L$ whenever $x, y \in L$ and $\lambda \in \mathbb{R}$. This simply asserts that, for every pair of points in L, the straight line through these points lies wholly within L. Every subspace of $\mathbb{R}^n$ is affine and if $\theta \in L \subset \mathbb{R}^n$ and L is affine then L is a subspace of $\mathbb{R}^n$. If $L \neq \emptyset$, $L \subset \mathbb{R}^n$ and L is affine then there is a unique subspace V of $\mathbb{R}^n$ such that $L = b + V$ where $b \in L$ (but b is otherwise arbitrary). The affine subset L is also termed a **linear manifold** or **flat**. Let $S \subset \mathbb{R}^n$; then the unique smallest affine set containing S is the

affine hull of S, denoted aff S. If $S \neq \theta$ and $S \subset \mathbb{R}^n$ then aff S consists of all x of the form $\sum_{i=1}^{k} \lambda_i x_i$, where $x_i \in S$, $\lambda_i \in \mathbb{R}$, $i = 1, \ldots, k$, $\sum_{i=1}^{k} \lambda_1 = 1$. If S is a non-empty subset of $\mathbb{R}^n$ and aff $S = b + V$ where V is a subspace of $\mathbb{R}^n$ then the **dimension of S**, denoted dim S, is defined to be the dimension of V.

An **inner product space** is a vector space V over the complex field C together with an **inner product** defined on $V \otimes V$ as follows: corresponding to each pair of vectors $x, y \in V$, the inner product of x and y, denoted $x.y$, is a scalar satisfying:

(i) $x.y = \overline{x.y}$, where $\overline{x.y}$ is the complex conjugate of $x.y$,
(ii) $(x+y).z = (x.z) + (y.z)$, for $z \in V$,
(iii) $(\alpha x).y = \alpha(x.y)$ for any scalar $\alpha \in C$,
(iv) $x.x \geqslant 0$, where equality holds if, and only if, $x = \theta$.

For the real vector space $\mathbb{R}^n$, $x.y = \sum_{i=1}^{n} x_i y_i$ is an inner product, called the *Euclidean inner product*.

A **normed space** is a vector space V over a field F on which a real-valued function is defined such that it assigns to each vector $x \in V$ a real number, denoted $\|x\|$, called the **norm** of x, which satisfies the following conditions:

(i) $\|x\| \geqslant 0$ for all $x \in V$, where equality holds if, and only if, $x = \theta$,
(ii) $\|x+y\| \leqslant \|x\| + \|y\|$ for all $x, y \in V$ (*triangle inequality*),
(iii) $\|\alpha x\| = |\alpha| . \|x\|$ for all $x \in V$ and scalars $\alpha \in F$, where $|\alpha|$ is the absolute value of α.

Consider the real vector space $\mathbb{R}^n$ and for any $x = (x_1, \ldots, x_n) \in \mathbb{R}^n$ define $\|x\| = (\sum_{i=1}^{n} x_i^2)^{1/2} = (x.x)^{1/2}$, then $\|x\|$ satisfies the conditions for a norm; it is called the *Euclidean norm*. The choice of a norm is not unique; $\|x\|$ defined as $\|x\| = \max_{1 \leqslant i \leqslant n} |x|$ or $\|x\| = \sum_{i=1}^{n} |x_i|$, where $|x_i|$ is the absolute value of x_i, are also norms for the real vector space $\mathbb{R}^n$, as is easily verified. In this text, unless otherwise specified, $\|x\|$ will denote the Euclidean norm. Note that for the Euclidean norm we have $x.y \leqslant \|x\| . \|y\|$ for any $x, y \in \mathbb{R}^n$ (*Schwarz inequality*).

It will be useful on occasion to consider the angle between two vectors $x, y \in \mathbb{R}^n$. We define the **angle**, ψ, between $x, y \in \mathbb{R}^n$, $x, y \neq \theta$, by

$$\cos \psi = x.y / \|x\| . \|y\|, \text{ where } \|.\| \text{ denotes the Euclidean norm.}$$

Non-zero vectors are termed **orthogonal** if the angle ψ between them is $\psi=\pi/2$; that is, if $x.y=\theta$.

Let X be a non-empty set. A real-valued function $d:X\otimes X\to\mathbb{R}$ is a **metric** if, and only if, for every $x, y\in X$:

(i) $d(x,y)\geqslant 0$, and $d(x,y)=0$ if, and only if, $x=y$,
(ii) $d(x,y)=d(y,x)$,
(iii) $d(x,z)\leqslant d(x,y)+d(y,z)$ for any $z\in X$.

A non-empty set together with a metric $d:X\otimes X\to\mathbb{R}$ is a **metric space**. In essence, a metric space is a set of points together with a well-defined quantitative (distance) measure giving the degree of closeness of points in the set. The number $d(x,y)$ is the **distance** between the points x and y. It should be clear that different real-valued functions may satisfy the requirements of the above definition, so that many different metrics may be associated with a given set. The function $d_1:\mathbb{R}^n\otimes\mathbb{R}^n\to\mathbb{R}$ defined by $d_1(x,y)=\max_i|x_i-y_i|$ for $x, y\in\mathbb{R}^n$ is a metric for $\mathbb{R}^n$, as is the function $d:\mathbb{R}^n\otimes\mathbb{R}^n\to\mathbb{R}$ defined by $d(x,y)=[\sum_{i=1}^n(x_i-y_i)^2]^{1/2}$ for $x, y\in\mathbb{R}^n$. The latter metric is known as the *Euclidean metric*. In this text, except where otherwise specified, the metric for $\mathbb{R}^n$ will be the Euclidean metric.

MATRICES

An $m\times n$ **real matrix** A is an array of m rows and n columns of real numbers. The entry in the ith row and jth column, $i=1,\ldots,m$, $j=1,\ldots,n$, is denoted a_{ij}. Often we write $[a_{ij}]$ for the matrix A. With **addition** and **scalar multiplication** defined entrywise by $A+B=[a_{ij}+b_{ij}]$ and $\alpha A=[\alpha a_{ij}]$ the set $M(m,n)$ of all $m\times n$ matrices is a vector space over the real field. The null vector for this vector space is the $m\times n$ matrix whose entries are all zero; we denote this matrix by θ. $M(m,1)$ is identified with $\mathbb{R}^m$: if $A\in M(m,1)$) then A is a column vector with m components. Each member of $M(1,n)$ is a row vector with n components. $M(1,1)$ is identified with $\mathbb{R}$, the set of real numbers.

If $A\in M(m,n)$ and $B\in M(n,p)$ then the product AB is defined to be the $m\times p$ matrix whose ijth entry is $\sum_{k=1}^n a_{ik}b_{kj}$: $AB\in M(m,p)$ where

$AB=[\sum_{k=1}^{n} a_{ik}b_{kj}]$. Note that for AB to be defined A must have the same number of columns as B has rows. If $A \in M(m,n)$ and $x \in \mathbb{R}^n$ then $Ax=[\sum_{k=1}^{n} a_{ik}x_k] \in \mathbb{R}^m$. In general, $AB \neq BA$, even when both products are defined.

If $A=[a_{ij}] \in M(m,n)$ its **transpose** $A' \in M(n,m)$ is defined by $A'=[a_{ji}]$. Thus, to transpose a matrix we simply interchange its rows and columns. $(A+B)'=A'+B'$ and $(AB)'=B'A'$. A matrix $A \in M(n,n)$ is *symmetric* if, and only if, $A=A'$, while it is *skew-symmetric* if, and only if, $A=-A'$.

Recall that for $x, y \in \mathbb{R}^n$ we defined the inner product $x.y$ by $x.y=\sum_{i=1}^{n} x_i y_i$. Now $x, y \in \mathbb{R}^n$ is equivalent to $x, y \in M(n, 1)$, and if $x \in M(n, 1)$ then $x' \in M(1,n)$. We can therefore form the product $x'y$. $x'y \in M(1, 1)$, where $x'y=\sum_{i=1}^{n} x_i y_i \in \mathbb{R}$. Thus, $x.y=x'y$.

If $A \in M(m,n)$ we write a_{i*} for the row vector, $i=1,\ldots,m$, in $M(1,n)$, where $a_{i*}=(a_{i1}, a_{i2},\ldots, a_{in})$ for $j=1,\ldots,n$. Similarly, we write a_{*j} for the column vector, $j=1,\ldots,n$ in $M(m,1)$, where $a_{*j}=(a_{1j}, a_{2j},\ldots, a_{nj})$ for $i=1,\ldots,m$. If $A \in M(m,n)$ then each column a_{*j} for A is in $\mathbb{R}^m$. The subspace of $\mathbb{R}^m$ spanned by these columns a_{*j}, $j=1,\ldots,n$, is the **column space** of A. Its dimension is called the **column rank of** A. The column rank of A is the maximum number of linearly independent columns in A. Similarly, the **row rank of** A is the dimension of the subspace of $M(1,n)$ spanned by the rows a_{i*}, $i=1,\ldots,m$. The row rank of A is the maximum number of linearly independent rows in A. It can be shown that the row rank of A is equal to the column rank of A, so we refer simply to the **rank of** A, denoted $\rho(A)$. Since the column space of A is a subspace of $\mathbb{R}^m$ (which has dimension m), $\rho(A) \leqslant m$. Similarly, $\rho(A) \leqslant n$. It can be shown that $\rho(AB) \leqslant \min\{\rho(A), \rho(B)\}$, $\rho(A)=\rho(A')$ and $\rho(A+B) \leqslant \rho(A)+\rho(B)$. Furthermore, the rank of a matrix is unchanged if one row (column) is multiplied by a non-zero constant or if a multiple of one row (column) is added to another row (column).

If A is square, $A \in M(n,n)$ then A is **non-singular** or **invertible** if there is in $M(n,n)$ a matrix B such that $AB=BA=I$, where I is the $n \times n$ **identity matrix**, defined by $I=[\delta_{ij}]$, where $\delta_{ij}=1$ for $i=j$, $\delta_{ij}=0$ otherwise. If A is invertible, then B is unique and is denoted A^{-1}; A^{-1} is the **inverse** of A. If $A \in M(n,n)$, then A is in invertible if, and only if, $\rho(A)=n$. The rank of a matrix remains unchanged when the matrix is pre- or post-multiplied by a non-singular matrix. $(A')^{-1}=(A^{-1})'$ and $(AB)^{-1}=B^{-1}A^{-1}$. If $A \in M(n,n)$, then A is *orthogonal* if, and only if, $A^{-1}=A'$.

Since the set $M(n,n)$ is a vector space we can define the norm of the real matrix $A \in M(n,n)$. We define the **norm of the $n \times n$ real matrix A** to be the real number $\|A\|$ which satisfies the following conditions:

(i) $\|A\| \geqslant 0$ for all $A \in M(n,n)$, where equality holds if, and only if, $A=\theta$, where θ is the $n \times n$ matrix whose elements are all zero,

(ii) $\|\alpha A\|=|\alpha|\,\|A\|$, for $\alpha \in \mathbb{R}$,

(iii) $\|A+B\| \leqslant \|A\|+\|B\|$ for all $A, B \in M(n,n)$,

(iv) $\|AB\| \leqslant \|A\|\,\|B\|$ for all $A, B \in M(n,n)$, (*Schwarz inequality*).

Examples of matrix norms are (a) $\max_j \sum_i |a_{ij}|$, (b) $\max_i \sum_j |a_{ij}|$, (c) $(\sum_i \sum_j |a_{ij}|^2)^{1/2}$, and (d) $n.\max_{i,j}|a_{ij}|$.

Let V and U be two vector spaces over a field F. A function $T:V \to U$ is a **linear transformation** from V into U if for all $v, w \in V$ and $\alpha, \beta \in F$, $T(\alpha v+\beta w)=\alpha T(v)+\beta T(w)$. The **range space** of the linear transformation $T:V \to U$ is the set $R_T=\{u: u \in U$ and $u=T(v)$, for some $v \in V\}$, and the **null space** of T is the set $N_T=\{v: v \in V$ and $T(v)=\theta\}$. The dimension of R_T is called the **rank of T**, denoted $\rho(T)$; the dimension of N_T is called the **nullity of T**, denoted $n(T)$. If $T:V \to U$ is a linear transformation then $\dim V=\rho(T)+n(T)$. Given two linear transformations $T_1: V \to U$, and $T_2: V \to U$, their **sum** is the mapping $T_1+T_2: V \to U$, defined by $(T_1+T_2)(v)=T_1(v)+T_2(v)$, for all $v \in V$; the **scalar transformation** of T_1 is the mapping $\gamma T_1: V \to U$, defined by $(\gamma T_1)(v)=\gamma T_1(v)$, for all $v \in V$, where γ is a scalar. The mappings T_1+T_2 and γT_1 are easily seen to be linear transformations. Let V, W, U be vector spaces over a field F, with $\dim V=n$, $\dim W=m$ and $\dim U=p$. Let $T_1: V \to W$ and $T_2: W \to U$ be linear transformations. Then the composite mapping $T_1 \circ T_2: V \to U$, defined by $(T_1 \circ T_2)(v)=T_2(T_1(v))$, for all $v \in V$, is a linear transformation. A linear transformation $T:V \to U$ is **non-singular** or **invertible** if it is a one-one mapping from V onto U. If $T:V \to U$ is non-singular then there exists a unique function T^{-1} from U onto V, where T^{-1} is defined by $T^{-1}(u)=\{v: v \in V, u \in T(v)\}$ such that $T^{-1} \circ T=i_v$, where i_v is the identity function on V. The mapping T^{-1} is itself a linear transformation, called the **inverse of T**.

T is non-singular if, and only if, the domain and range space of T have the same dimension. To see this, note that, for any linear transformation $T:V \to U$, $T(\theta)=\theta$, so that T is non-singular if, and only if, $N_T=\{\theta\}$: however, $\dim\{\theta\}=0$, so $n(T)=0$, and, since $\dim V=\rho(T)+n(T)$, it follows that $\dim V=\rho(T)$.

Matrices and linear transformations are intimately related. Any linear transformation T from the n-dimensional vector space V into the m-dimensional vector space U can be represented by a unique matrix A of order $m \times n$, with respect to a given pair of bases for V and U, with $\rho(T)=\rho(A)$. Conversely, any $m \times n$ matrix A, of rank n, represents a unique linear transformation from an n-dimensional vector space V into an m-dimensional vector space U, with respect to some pair of bases for V and U.

If $T_1:V \to U$ and $T_2:V \to U$ and T_1 and T_2 are represented by the $m \times n$ matrices A and B, respectively, with respect to a given pair of bases for V and U, then their sum, T_1+T_2, and the scalar transformation of $T_1, \gamma T_1$, are mappings represented by the matrices $A+B$ and γA, respectively, with respect to the given pair of bases for V and U. Similarly, if $T_1:V \to W$ and $T_2:W \to U$, and T_1 is represented by the $m \times n$ matrix A, with respect to given bases for V and W, and T_2 is represented by the $p \times m$ matrix B, with respect to the given basis for W and a given basis for U, then $T_1 \circ T_2:V \to U$ is uniquely represented by the $p \times n$ matrix BA, with respect to the given bases. Moreover, if $T:V \to U$ is a non-singular linear transformation, uniquely represented by the $n \times n$ matrix A, with respect to a given basis for V, then the inverse of T, T^{-1}, is uniquely represented by the $n \times n$ matrix A^{-1}, with respect to the given basis for V. The matrix which represents a given linear transformation will depend upon the bases chosen for the domain and range space of the transformation. Two matrices A and B are said to be **equivalent** if they represent the same linear transformation with respect to (some) different pairs of bases. It can be shown that A and B are equivalent if $B=EAF$, where E and F are non-singular matrices. Two matrices representing the same linear transformation $T:V \to V$ are said to be **similar**. This is equivalent to saying that A and B are similar if, and only if, there exists a non-singular matrix P such that $B=P^{-1}AP$, in which case $B'=P'A'P$ and $\rho(A)=\rho(B)$.

A rearrangement of the natural order of n integers is called a *permutation*. The number of *inversions* in a permutation of n integers is the number of pairs of elements (not necessarily adjacent) in which a larger integer precedes a smaller one; in the permutation (3, 2, 1, 4) there are three inversions (3, 2), (3, 1) and (2, 1). A permutation is *even* when the number of inversions is even, and *odd* when the number of inversions is odd. The **determinant** of an $n \times n$ matrix A, denoted $|A|$, is defined to be

$$|A| \equiv \sum(\pm)a_{1i}a_{2j}\ldots a_{nr}$$

the sum being taken over all permutations of the second subscripts. A term in the sum is assigned a plus sign if the second subscripts are in natural order or if they form an even permutation; a term in the sum is assigned a minus sign if the second subscripts form an odd permutation. For example, if A is of order 2×2 then $|A| = a_{11}a_{22} - a_{12}a_{21}$. (Note that the context will distinguish whether $|\,|$ denotes determinant or modulus.) The following are the major properties of determinants: (i) $|A| = |A'|$; (ii) $|AB| = |A||B|$; (iii) if A is a diagonal matrix ($a_{ij} = 0$, $i \neq j$) then $|A|$ is the product of its diagonal elements; (iv) if any row (column) of A is a non-trivial linear combination of all the other rows (columns) of A then $|A| = 0$; in particular, if two rows (or columns) of A are identical, or a row (or column) contains only zeros then $|A| = 0$; (v) if B results from A by interchanging two rows (or columns) then $|B| = -|A|$; (vi) if B results from **A** by multiplying one row (or column) by a scalar μ then $|B| = \mu|A|$; (vii) if B is obtained from A by adding a multiple of row j to row i ($j \neq i$) of A (or a multiple of column j to colum i ($i \neq j$) of A) then $|B| = |A|$; (viii) $|A^{-1}| = |A|^{-1}$; (ix) if A and B are similar, then $|A| = |B|$.

The ***i*, *j* minor** of the $n \times n$ matrix A, denoted M_{ij}, is the determinant of the $(n-1) \times (n-1)$ matrix obtained by deleting the ith row and jth column of A. The ***i*, *j* cofactor** of the $n \times n$ matrix A, denoted C_{ij}, is the same as the i, j minor of A if $i + j$ is even, and the negative of the i, j minor of A if $i + j$ is odd: $C_{ij} = (-1)^{i+j}M_{ij}$, $i = 1, \ldots, n, j = 1, \ldots, n$. A determinant can be evaluated by *expansion by its cofactors*; $|A| = \sum_{i=1}^{n} a_{ij}C_{ij}$, any j, $j = 1, \ldots, n$; $|A| = \sum_{j=1}^{n} a_{ij}C_{ij}$, any i, $i=1,\ldots,n$. The **kth order leading principal minor** of the $n\times n$ matrix A, denoted M_k, is the determinant of the $k\times k$ matrix consisting of the first k rows and columns of A. A matrix satisfies the *Hawkins–Simon conditions* if, and only if, all its leading principal minors are positive.

A *permutation matrix* is a square matrix for which each row and each column contain a one, all other elements being zero; for example,

$$\begin{bmatrix}1&0\\0&1\end{bmatrix},\begin{bmatrix}0&1\\1&0\end{bmatrix},\begin{bmatrix}0&1&0\\0&0&1\\1&0&0\end{bmatrix}\text{ are permutation matrices.}$$

There are $n!$ permutation matrices of order n, one of which is an identity matrix. A **kth order principal minor** of the $n\times n$ matrix A, denoted $\tilde{M}_k$, is the kth order leading principal minor of $P'AP$, where P is some permutation matrix. There are $n!/k!(n-k)!$ possible kth order principal minors. The second order principal minors of the 2×2 matrix A are:

$$\left|\begin{bmatrix}1&0\\0&1\end{bmatrix}\begin{bmatrix}a_{11}&a_{12}\\a_{21}&a_{22}\end{bmatrix}\begin{bmatrix}1&0\\0&1\end{bmatrix}\right|=\begin{vmatrix}a_{11}&a_{12}\\a_{21}&a_{22}\end{vmatrix}=|A|,$$

and

$$\left|\begin{bmatrix}0&1\\1&0\end{bmatrix}\begin{bmatrix}a_{11}&a_{12}\\a_{21}&a_{22}\end{bmatrix}\begin{bmatrix}0&1\\1&0\end{bmatrix}\right|=\begin{vmatrix}a_{22}&a_{21}\\a_{12}&a_{11}\end{vmatrix}=|A|.$$

The first order principal minors of A are a_{11} and a_{22}.

The rank of any matrix A, $\rho(A)$, is the size of the largest non-vanishing determinant contained in A. Thus, if A is diagonal, $\rho(A)$ is the number of non-zero elements of A. An $n\times n$ matrix A is of rank n if, and only if, $|A|\neq 0$; thus, A is non-singular if, and only if, $|A|\neq 0$. The inverse of A can be computed as:

$$A^{-1}=\frac{(C_{ij})'}{|A|}=\frac{((-1)^{i+j}M_{ij})}{|A|},$$

where (C_{ij}) is the matrix of cofactors, and $(C_{ij})'$ is the **adjoint matrix**.

APPENDIX

LINEAR EQUATIONS

The system of m linear equations in n unknowns, $x_j, j=1,\ldots,n$, $\sum_{j=1}^{n} a_{ij}x_j=b_i$, $i=1,\ldots,m$, can be expressed as $Ax=b$, where $A=[a_{ij}]$ is the $m\times n$ coefficient matrix, x the vector of unknowns and b the vector of the constants b_i. The system $Ax=b$ can have a unique solution, a non-unique solution, or no solution. Given A and b, the **augmented matrix** $[A,b]$ of the system $Ax=b$ is defined to be

$$[A,b]=\begin{bmatrix} a_{11}\ldots a_{1n} & b_1 \\ \vdots \quad\quad \vdots & \vdots \\ a_{m1}\ldots a_{mn} & b_m \end{bmatrix}$$

The system of linear equations $Ax=b$ has a solution if, and only if, the rank of A, $\rho(A)$, is the same as the rank of the augmented matrix $[A,b]$, $\rho(A,b)$. The solution is unique if, and only if, $\rho(A)=\rho(A,b)=n$. If $\rho(A)=\rho(A,b)=k<m$, then any x which satisfies k of the equations for which the corresponding rows of A are linearly independent satisfies all equations. Furthermore, for $k<n$, $n-k$ of the variables can be assigned arbitrary values and then the remaining k variables can be solved, provided the columns of A associated with the k variables are linearly independent. The **homogeneous system** $Ax=\theta$ always has a solution, since $\rho(A)=\rho(A,\theta)$.

When A is non-singular, the system $Ax=b$ has a unique solution $x=A^{-1}b$, and the corresponding homogeneous system $Ax=\theta$ has the unique solution $x=\theta$. The system $Ax=b$, where A is of order $n\times n$, has a unique solution if, and only if, $|A|\neq 0$. In this case the unique solution can be obtained from **Cramer's Rule**

$$x_k=|A_k|/|A|, \qquad k=1,\ldots,n,$$

where $|A_k|$ denotes the determinant of the matrix obtained from A by replacing the kth column of A with b.

EIGENVALUES

A vector $\gamma\in\mathbb{R}^n$, $\gamma\neq\theta$, and scalar λ are called, respectively, an

eigenvector (characteristic vector) of A and an **eigenvalue (characteristic value)** of A, if $A\gamma=\lambda\gamma$. The system $A\gamma=\lambda\gamma$, $\gamma\neq\theta$, is equivalent to the system $(\lambda I-A)\gamma=\theta$, $\gamma\neq\theta$, where I is the $n\times n$ identity matrix. The latter system has a solution if, and only if, $|\lambda I-A|=0$. Now

$$|\lambda I-A|=\begin{vmatrix}\lambda-a_{11} & -a_{12} & \cdots & -a_{1n}\\ -a_{21} & \lambda-a_{22} & \cdots & -a_{2n}\\ \vdots & \vdots & & \vdots\\ -a_{n1} & -a_{n2} & \cdots & \lambda-a_{nn}\end{vmatrix},$$

and expanding the determinant gives a polynomial of degree n in λ:

$$|\lambda I-A|=\lambda^n+b_{n-1}\lambda^{n-1}+b_{n-2}\lambda^{n-2}+\ldots+b_1\lambda+b_0,$$

where

$$b_{n-1}=-\sum_{i=1}^{n}a_{ii}$$

$$b_{n-2}=(-1)\sum_{\substack{i,j=1\\ i<j}}^{n}\begin{vmatrix}a_{ii} & a_{ij}\\ a_{ji} & a_{jj}\end{vmatrix}$$

$$\vdots$$

$$b_r=(-1)^{n-r}\sum_{\substack{i,j,\ldots,k=1\\ i<j<\cdots<k}}^{n}\underbrace{\begin{vmatrix}a_{ii} & a_{ij} & \cdots & a_{ik}\\ a_{ji} & a_{jj} & \cdots & a_{jk}\\ \vdots & \vdots & & \vdots\\ a_{k1} & a_{kj} & \cdots & a_{kk}\end{vmatrix}}_{n-r}$$

$$\vdots \qquad \vdots$$

$$b_0=(-1)^n|A|.$$

The expressions $|\lambda I-A|$ and $|\lambda I-A|=0$ are called the **characteristic polynomial** of A and the **characteristic equation** of A, respectively. Clearly, the eigenvalues of A are the roots of the characteristic equation of A.

The characteristic equation $|\lambda I-A|=0$, where A is $n\times n$, has exactly n roots in the complex field, provided we count multiple roots according to their multiplicity. Since complex roots appear in conjugate pairs, complex eigenvectors also appear in conjugate pairs. If γ is an eigenvector of A associated with the eigenvalue λ, then the vectors $\alpha\gamma, \alpha \neq 0$, are also eigenvectors of A associated with λ. Thus, each eigenvector has a unique proportionality in its components, but its length can be arbitrary.

If the eigenvalues $\lambda_1, \ldots, \lambda_n$ are all distinct and $\{\gamma_1, \ldots, \gamma_n\}$ is a set of eigenvectors corresponding to the λ_i, the set $\{\gamma_1, \ldots, \gamma_n\}$ is linearly independent. This can be shown as follows. Suppose $\{\gamma_1, \ldots, \gamma_n\}$ is linearly dependent and that the matrix $X = [\gamma_1, \ldots, \gamma_n]$, whose columns are the eigenvectors of A, has rank $r < n$. Without loss of generality we can take the first r eigenvectors to be linearly independent. Any $\gamma_j, j = r+1, \ldots, n$, can be expressed as $\gamma_j = \sum_{i=1}^{r} \alpha_i \gamma_i$, with at least one $\alpha_i \neq 0$, since $\gamma_j \neq \theta$, so that $\lambda_j \gamma_j = \sum_{i=1}^{r} \alpha_i \lambda_j \gamma_i$, $j = r+1, \ldots, n$. But $\lambda_j \gamma_j = A\gamma_j = \sum_{i=1}^{r} \alpha_i A\gamma_i = \sum_{i=1}^{r} \alpha_i \lambda_i \gamma_i$, since γ_i is an eigenvector of A. Hence, $\sum_{i=1}^{r} \alpha_i (\lambda_i - \lambda_j)\gamma_i = 0$. But as $\lambda_i \neq \lambda_j$, $j = r+1, \ldots, n$, and at least one $\alpha_i \neq 0$, this implies $\{\gamma_1, \ldots, \gamma_r\}$ is linearly dependent; a contradiction. Hence, $\{\gamma_1, \ldots, \gamma_n\}$ cannot be linearly dependent.

Similar matrices have the same eigenvalues and eigenvectors. Suppose A and B are similar matrices. Then, by definition, there exists a non-singular matrix P, such that $A = P^{-1}BP$. Hence, we have $\lambda I - A = \lambda I - P^{-1}BP = \lambda P^{-1}IP - P^{-1}BP = P^{-1}(\lambda I - B)P$. This implies, $|\lambda I - A| = |P^{-1}(\lambda I - B)P| = |P^{-1}| . |\lambda I - B| . |P| = |\lambda I - B|$, since $|P^{-1}| . |P| = 1$.

Since the determinant of a diagonal matrix is the product of its diagonal elements, it follows that the eigenvalues of a diagonal matrix are its diagonal elements. From the properties of determinants, it can also be easily established that A and A' have the same eigenvalues, and that, if A^{-1} exists, its eigenvalues are the reciprocals of the eigenvalues of A. Let λ_1, λ_2 be two distinct eigenvalues of A and let v_i and u_i be the eigenvectors of A and A', respectively, associated with λ_i $(i = 1, 2)$. Then it is easily established that $u_1' v_2 = u_2' v_1 = 0$. Moreover, if all the eigenvalues $\lambda_i, i = 1, \ldots, n$, of A are distinct, then the following holds for the associated eigenvectors v_i and u_i, respectively, of A and A': $u_i' v_i \neq 0$, for $i = 1, \ldots, n$.

If A has n linearly independent eigenvectors $\gamma_1,\ldots,\gamma_n$ then A is similar to a diagonal matrix; $P^{-1}AP=D$, where D is a diagonal matrix whose diagonal elements are the eigenvalues of A and P is the matrix whose columns are the n linearly independent eigenvectors. This can be shown as follows. Let γ_j be an eigenvector of A corresponding to the eigenvalue λ_j of A (not all λ_j are necessarily distinct). Let $D\equiv[\lambda_j\delta_{ij}]$, where $\delta_{ij}=1$ if $i=j$, $\delta_{ij}=0$ otherwise, and let $P\equiv[\gamma_1,\ldots,\gamma_n]$, so that P is the matrix whose columns are the n linearly independent eigenvectors of A. Then $PD=[\lambda_1\gamma_1,\ldots,\lambda_n\gamma_n]$ and $AP=[A\gamma_1,\ldots,A\gamma_n]=[\lambda_1\gamma_1,\ldots,\lambda_n\gamma_n]$. Hence, $AP=PD$, so that $D=P^{-1}AP$. In the special case where A has n distinct eigenvalues the above result immediately implies that A is similar to a diagonal matrix. While a matrix A with repeated eigenvalues may not be similar to a diagonal matrix, Bellman (1960, Chapter 11), has shown that there exists a matrix C with distinct eigenvalues whose elements do not differ from those of A by a total of more than ε, where ε is a preassigned quantity as small as we please. In economics there are few, if any, situations where a repetition of eigenvalues would be regarded as more than an accident of numbers.

In many economic problems we are concerned with real symmetric matrices. Such matrices have important special properties. In particular, all the eigenvalues of a real symmetric matrix are real and the eigenvectors of a real symmetric matrix are orthogonal. Moreover, A is similar to a diagonal matrix $D=P^{-1}AP$, where P has the property $P^{-1}=P'$ (P is an **orthogonal matrix**). The rank of a real symmetric matrix is equal to the number of non-zero eigenvalues (counting multiplicities) of the matrix.

For any square matrix $A=[a_{ij}]$ the **trace** of A, denoted tr A, is defined by tr $A=\sum_{i=1}^{n}a_{ii}$. Since tr $AB=\sum_{i=1}^{n}(\sum_{j=1}^{n}a_{ij}b_{ji})=\sum_{j=1}^{n}(\sum_{i=1}^{n}b_{ji}a_{ij})=\text{tr}\,BA$, we have $\text{tr}\,P^{-1}AP=\text{tr}\,AP^{-1}P=\text{tr}\,A$: similar matrices have the same trace. Furthermore, since the coefficient b_{n-1} of λ^{n-1} in the expansion of $|\lambda I-A|$ is $-\sum_{i=1}^{n}a_{ii}$, we have $-b_{n-1}=\text{tr}\,A$. Also tr A is the sum of the eigenvalues of A and $|A|$ is the product of the eigenvalues of A (in both cases each eigenvalue is counted with its multiplicity).

QUADRATIC FORMS

In this section we shall be concerned with a non-linear mapping $q:V \to F$, where V is an n-dimensional real vector space and F is the real field. To be precise we shall be concerned with the function $q:V \to F$ defined by $q(v) = \sum_{i=1}^{n} \sum_{j=1}^{n} a_{ij} v_i v_j = v'Av$, where $A = [a_{ij}]$ is a real matrix of order $n \times n$. A function of this form is called a **real quadratic form**. Our main concern is with the behaviour of q as v varies over the space V or some subset of V. Let $q:V \to \mathbb{R}$, where $q(v) = v'Av$. Then (i) q is **positive (negative) definite** if $q(v) > 0$ ($q(v) < 0$) for all $v \neq \theta$, (ii) q is **positive (negative) semidefinite** if $q(v) \geqslant 0$ ($q(v) \leqslant 0$) for all v with $q(v) = 0$ for some $v \neq \theta$, (iii) q is **indefinite** if $q(v)$ is positive for some v and negative for other v. This classification provides us with a corresponding way of classifying real symmetric matrices. A real symmetric matrix A is said to be **positive (negative) definite (semidefinite)** if the quadratic form $v'Av$ is positive (negative) definite (semidefinite).

If $v'Av$ is positive (negative) definite and $v = Py$, where P is a non-singular nth order square matrix, then $y'P'APy = y'By$ is positive (negative) definite. To see this we know that the range of $y'By$ is the same as that of $v'Av$, as A and B are equivalent since P is non-singular, so that it is only necessary to show that $y = \theta$ is the only y for which $y'By = 0$. Now $v'Av = 0$ only if $v = \theta$, but $y = P^{-1}v$ and $y = \theta$ is the only value of y for which $v = \theta$. Similarly, it can be shown that semidefinite and indefinite forms remain semidefinite and indefinite, respectively, under a non-singular linear transformation. Given the quadratic form $v'Av$, where A is a real symmetric matrix, we know that there exists an orthogonal matrix P, whose columns are the eigenvectors of A, such that $P^{-1}AP = P'AP$ is a diagonal matrix whose diagonal elements are the eigenvalues of A. Hence, letting $v = Py$ we have $v'Av = (Py)'A(Py) = y'P'APy = y'By$, where B is a diagonal matrix whose diagonal elements are the real eigenvalues of A. Now P is orthogonal and hence non-singular, so that $v'Av$ is positive (negative) definite (semidefinite) if and only if $y'By$ is positive (negative) definite (semidefinite). But $y'By = \sum_{i=1}^{n} \lambda_i y_i^2$, so that $q(v)$ is positive (negative) definite if, and only if, $\lambda_i > 0$ ($\lambda_i < 0$) for all i, and $q(v)$ is positive (negative) semidefinite if, and only if, $\lambda_i \geqslant 0$

($\lambda_i \leqslant 0$) with at least one $\lambda_i = 0$. Clearly, $q(v)$ is indefinite if, and only if, $\lambda_i > 0$ for some i and $\lambda_j < 0$ for some j. Thus, we have the following result.

A real symmetric matrix A is positive (negative) definite if, and only if, all eigenvalues of A are positive (negative); A is positive (negative) semidefinite if, and only if, all eigenvalues of A are non-negative (non-positive) and at least one eigenvalue of A is zero; A is indefinite if, and only if, A has both positive and negative eigenvalues.

It is often not convenient to discover the sign of a quadratic form by calculating the eigenvalues of its associated matrix, since the latter are often difficult to determine if the matrix is of order three or more. However, there exist alternative necessary and sufficient conditions for positive (negative) definite (semidefinite) quadratic forms which are more easily applicable.

Let A be a real symmetric matrix of order $n \times n$. Then:

(i) A is positive definite if, and only if, $|A| > 0$ and all other leading principal minors of A are positive,

(ii) A is negative definite if, and only if, $|A|$ has sign $(-1)^n$ and for all other leading principal minors of A, M_k, we have $(-1)^k M_k > 0$,

(iii) A is positive semidefinite if, and only if, all principal minors $\tilde{M}_k$ of A are non-negative,

(iv) A is negative semidefinite if, and only if, for all principal minors $\tilde{M}_k$ of A, we have $(-1)^k \tilde{M}_k \geqslant 0$.

Note that if A is positive (negative) semidefinite then $|A| = 0$, since if A is positive (negative) semidefinite then at least one eigenvalue of A must be zero and hence $|A| = 0$.

In many applications of real quadratic forms we are interested in the sign of the quadratic form for $v \in S$, where S is some subset of V. A particular, and important, case is where S is the set of all solutions to the system of linear equations $B'v = \theta$, where B is an $n \times n$ matrix. The following results due to Debreu (1952) provide us with necessary and sufficient conditions for the real quadratic form $v'Av$ to be positive (negative) definite (semidefinite) subject to $B'v = \theta$.

Let A be a real symmetric matrix of order $n \times n$. Let A_r be the matrix obtained from A by keeping only the elements in the first r

rows and first r columns. Let B be an $n \times m$ matrix and let B_{rm} be the matrix obtained from B by retaining only the elements in the first r rows and the first m columns. Let π denote a permutation of the first n integers and A^{π} the matrix obtained from A by performing the permutation π on its rows and on its columns, and let B^{π} be the matrix obtained from B by performing the permutation π on its rows. Assume $|B_{mm}| \neq 0$ and let θ_{mm} be the $m \times m$ null matrix. Then we have:

(i) $v'Av > 0$ for every $v \neq \theta$ such that $B'v = \theta$ if, and only if,

$$(-1)^m \begin{vmatrix} A_r & B_{rm} \\ B'_{rm} & \theta_{mm} \end{vmatrix} > 0, \quad \text{for } r = m+1, \ldots, n,$$

(ii) $v'Av < 0$ for every $v \neq \theta$ such that $B'v = \theta$ if, and only if,

$$(-1)^r \begin{vmatrix} A_r & B_{rm} \\ B'_{rm} & \theta_{mm} \end{vmatrix} > 0, \quad \text{for } r = m+1, \ldots, n,$$

(iii) $v'Av \geqslant 0$ for every v such that $B'v = \theta$ if, and only if,

$$(-1)^m \begin{vmatrix} A_r^{\pi} & B_{rm}^{\pi} \\ (B_{rm}^{\pi})' & \theta_{mm} \end{vmatrix} \geqslant 0, \quad \text{for } r = m+1, \ldots, n \text{ and any } \pi,$$

(iv) $v'Av \leqslant 0$ for every v such that $B'v = \theta$ if, and only if,

$$(-1)^r \begin{vmatrix} A_r^{\pi} & B_{rm}^{\pi} \\ (B_{rm}^{\pi})' & \theta_{mm} \end{vmatrix} \geqslant 0, \quad \text{for } r = m+1, \ldots, n \text{ and any } \pi.$$

The above results require that A be real and symmetric. An $n \times n$ real matrix A is said to be **positive (negative) quasidefinite** if $x'Ax$ is positive (negative) for every x, $x \neq \theta$, or equivalently, if $(A + A')$ is positive (negative) definite, since $x'Ax = \frac{1}{2}x'(A + A')x$. Clearly, definiteness of a matrix A is the special case of quasidefiniteness when A is symmetric.

FURTHER SET PROPERTIES

The set $\{y \in \mathbb{R}^n : \|x - y\| < r, x \in \mathbb{R}^n, r > 0\}$ is called an **open ball**, denoted $B_r(x)$, with centre x and radius r. The set $\{y \in \mathbb{R}^n : \|x - y\| \leqslant r,$

$x \in \mathbb{R}^n, r > 0\}$ is the corresponding **closed ball**. The definition of a ball is clearly norm dependent.

A set $A \subset \mathbb{R}^n$ is **open** (in $\mathbb{R}^n$) if, for every $x \in A$, there exists an $r > 0$ such that every $y \in \mathbb{R}^n$ satisfying $\|x - y\| < r$ belongs to A. Thus, A is open if every $x \in A$ is the centre of some open ball entirely contained in A. $\emptyset$ and $\mathbb{R}^n$ are both open sets. The union of open sets is open and the intersection of finitely many open sets is open. A set $A \subset \mathbb{R}^n$ is **closed** (in $\mathbb{R}^n$) if $\mathbb{R}^n \backslash A$ is open. $\emptyset$ and $\mathbb{R}^n$ are both closed sets. $\emptyset$ and $\mathbb{R}^n$ are the only sets in $\mathbb{R}^n$ which are both open and closed. The union of a finite number of closed sets is closed and arbitrary intersections of closed sets are closed. If $X \subset \mathbb{R}^n$ and $A \subset X$, a **neighbourhood of A** is a subset of X that contains an open set containing A. In particular, if $x \in \mathbb{R}^n$, then any set which contains an open set containing x is a **neighbourhood of x**, denoted $N(x)$. A set $A \subset \mathbb{R}^n$ is a neighbourhood of a point $x \in \mathbb{R}^n$ if, and only if, there exists an open ball with centre x entirely contained within A. A point $x \in \mathbb{R}^n$ is an **interior point** of $A \subset \mathbb{R}^n$ if there is a $N(x) \subset A$. If $A \subset \mathbb{R}^n$ the following statements are equivalent: (i) A is open, (ii) if $x \in A$, x is an interior point of A, (iii) A is a neighbourhood of each of its points. Given any subset $A \subset \mathbb{R}^n$ the subset which is the union of all open sets contained in A is called the **interior of A**, denoted int A. int A is an open set, and an element $x \in \mathbb{R}^n$ belongs to int A if, and only if, it is an interior point of A. A point $x \in \mathbb{R}^n$ is a **boundary point** of A, $A \subset \mathbb{R}^n$, if every neighbourhood of x contains a point in A and a point not in A. The set of all boundary points of A is the **boundary of A**, denoted bd A. A set $A \subset \mathbb{R}^n$ is closed if, and only if, it contains all its boundary points. A point $x \in \mathbb{R}^n$ is a **cluster point** (or **limit point** or **point of accumulation**) of a subset $A \subset \mathbb{R}^n$ if every neighbourhood $N(x)$ contains at least one point of A distinct from x. A set is closed if, and only if, it contains all its cluster points. Let cp A denote the set of all cluster points of A. The **closure of A**, denoted cl A, is the set cl $A = A \cup$ cp A. Clearly, A is closed if, and only if, $A =$ cl A.

An **open cell** $B \subset \mathbb{R}^n$ is the Cartesian product of n open intervals of real numbers: $B = \{x = (x_1, \ldots, x_n)' \in \mathbb{R}^n : a_i < x_i < b_i, i = 1, \ldots, n\}$. With weak inequalities B is a **closed cell**. A subset $A \subset \mathbb{R}^n$ is **bounded** if it is contained in some cell.

A set A is **compact** if, whenever it is contained in the union of a collection A of open sets, it is also contained in the union of some *finite* number of the sets in A. Compact subsets of $\mathbb{R}^n$ are relatively easy to describe. A set $A \subset \mathbb{R}^n$ is compact if, and only if, it is closed and bounded.

A set $A \subset \mathbb{R}^n$ is **connected** if it cannot be expressed as the union of two disjoint non-empty open subsets of A. If, on the other hand, we can find non-empty open subsets H and K of A with $H \cap K = \emptyset$ and $H \cup K = A$ then A is **disconnected**. A subset $A \subset \mathbb{R}$ is connected if, and only if, it is an interval (open, closed or half-open).

CONTINUITY CONCEPTS

Let $\mathbb{N}$ denote the set of positive integers in increasing order. A **sequence** in a metric space (S, d) is a function $s: \mathbb{N} \to S$. The image of an element $n \in \mathbb{N}$ under the sequence s, we often denote by s_n (or s^n) rather than by $s(n)$, and we often denote the sequence s by $\{s_n\}$ (or $\{s^n\}$). A sequence $\{s_n\}$ in a metric space (S, d) is said to **converge** if there is a point $p \in S$ with the following properties: for every $\varepsilon > 0$ there exists a $k \in \mathbb{N}$ such that $d(s_n, p) < \varepsilon$ whenever $n \geqslant k$. We also say that $\{s_n\}$ *converges to p* and write $s_n \to p$ as $n \to \infty$, or simply $s_n \to p$. If there is no such $p \in S$ then the sequence $\{s_n\}$ is said to **diverge**. A sequence $\{s_n\}$ in a metric space (S, d) can converge to at most one point in S. If $\{s_n\}$ converges, the unique point to which it converges is called the **limit** of the sequence. If $p \in S$ is the limit of the sequence $\{s_n\}$ we write $\lim_{n \to \infty} s_n = p$.

Let $s = \{s_n\}$ be a sequence and let g be a function whose domain is the set of positive integers and whose range is a subset of the positive integers, and assume that $g(m) < g(n)$ if $m < n$. Then the composite function $s \circ g$ is defined for all $n \in \mathbb{N}$ and for each such n we have $(s \circ g)(n) = s(g(n)) \equiv s_{g_n}$. Such a composite function is said to be a **subsequence** of s. For example, we can take a subsequence consisting of every other member of the sequence s, in which case $g_n = 2n$. In a metric space (S, d) a sequence $\{s_n\}$ converges to $p \in S$ if, and only if, every subsequence $\{s_{g_n}\}$ converges to p.

A sequence $\{s_n\}$ in a metric space (S, d) is a **Cauchy sequence** if, for any $\varepsilon > 0$, there is a positive integer k such that $n, m \geqslant k$ implies

$d(s_n, s_m) < \varepsilon$. A metric space (S, d) is **complete** if every Cauchy sequence in it converges to some point in the space.

In this book we shall be concerned mainly with the metric space $(\mathbb{R}^n, d)$, where d is the Euclidean metric. Clearly, if $\{x_n\}$ is a sequence in $\mathbb{R}^m$, then a point $x \in \mathbb{R}^m$ is a limit of $\{x_n\}$ if, and only if, for each $\varepsilon > 0$ there is a positive integer $k(\varepsilon)$ such that for all $n \geqslant k(\varepsilon)$, $\|x_n - x\| < \varepsilon$, where $\| \;\|$ denotes the Euclidean norm. A sequence $\{x_n\}$ in $\mathbb{R}^m$ is **bounded** if there exists $M > 0$ such that $\|x_n\| < M$ for all $n \in \mathbb{N}$. A convergent sequence in $\mathbb{R}^m$ is bounded, and a bounded sequence in $\mathbb{R}^m$ has a convergent subsequence. Furthermore, a sequence in $\mathbb{R}^m$ is convergent if, and only if, it is a Cauchy sequence. If $\{x_n\}$ and $\{y_n\}$ are sequences in $\mathbb{R}^m$ we define their **sum** to be the sequence $\{x_n + y_n\}$, their **difference** to be the sequence $\{x_n - y_n\}$, and their **inner product** to be the sequence $\{x_n' y_n\}$ in $\mathbb{R}$. If $\{x_n\}$ and $\{y_n\}$ converge to $x, y \in \mathbb{R}^m$, respectively, then $\{x_n + y_n\}$, $\{x_n + y_n\}$ and $\{x_n' y_n\}$ converge to $x + y \in \mathbb{R}^m$, $x - y \in \mathbb{R}^m$ and $x' y \in \mathbb{R}$, respectively.

Let f be a function with domain $X \subset \mathbb{R}^n$ and range in $\mathbb{R}^m$. If $x^\circ \in X$ then f is **continuous at x°** if for every neighbourhood V of $f(x^\circ)$ there exists a neighbourhood W of x° such that if x is an element of W then $f(x) \in V$.

Clearly, f is continuous at x° if, and only if, for every neighbourhood V of $f(x^\circ)$ there is a neighbourhood W of x° such that $W \cap X = f^{-1}(V)$. The following statements are equivalent: (i) f is continuous at x°, (ii) for any $\varepsilon > 0$, there exists a number $\delta > 0$, such that, if $x \in X$ is any element such that $\|x - x^\circ\| < \delta$, then $\|f(x) - f(x^\circ)\| < \varepsilon$; in general δ will depend on both ε and x°, (iii) if $\{x_n\}$ is a sequence of elements of X which converge to x°, then the sequence $\{f(x_n)\}$ converges to $f(x^\circ)$.

Let f and g be functions with domains $X \subset \mathbb{R}^n$, $Y \subset \mathbb{R}^n$, respectively, and ranges in $\mathbb{R}^m$. Then we can define the functions $f + g$, $f - g$, fg for each $x \in X \cap Y$ by $(f + g)(x) = f(x) + g(x)$, $(f - g)(x) = f(x) - g(x)$ and $(fg)(x) = f(x)' g(x)$, respectively: The ranges of $f + g$ and $f - g$ are subsets of $\mathbb{R}^m$, while fg is a real-valued function. If $\alpha \in \mathbb{R}$, we can also define the function αf for $x \in X$ by $(\alpha f)(x) = \alpha f(x)$; the range of αf is a subset of $\mathbb{R}^m$. Suppose h is a real-valued function with domain $Z \subset \mathbb{R}^n$. Then we can also define the function hf for $x \in X \cap Z$ by $(hf)(x) = h(x) f(x)$; the range of hf is a subset of $\mathbb{R}^m$. If $h(x)\neq 0$ for $x\in Z^\circ\subset Z$, then we can also define the function f/h for $x\in X\cap Z^\circ$ by $(f/h)(x)=f(x)/h(x)$; the range of f/h is a subset of $\mathbb{R}^m$. If f and g are continuous at a point then the functions $f+g, f-g, fg, \alpha f, hf$ and f/h are also continuous at that point. Let f be a function on $X\subset\mathbb{R}^n$ with range $f(X)\subset W\subset\mathbb{R}^m$, and let g be a function with domain W and range $g(W)\subset\mathbb{R}^p$. Then the composite function $g\circ f$ exists. If f is continuous at $x^\circ\in X$, and g is continuous at $f(x^\circ)$, then the composite function $g\circ f$ is continuous at x°.

Continuity is a 'local' concept in that the basic definition refers to a function being continuous at a point in its domain. If f is continuous at all points in its domain then we say that f is a **continuous function**. If f has domain $X\subset\mathbb{R}^n$ and f is continuous at every point of $A\subset X$, then f is **continuous on A**. If f is a function with domain $X\subset\mathbb{R}^n$ and range in $\mathbb{R}^m$ then the following statements are equivalent: (i) f is a continuous function, (ii) if A is any open set in $\mathbb{R}^m$ then there exists an open set A° in $\mathbb{R}^m$ such that $A^\circ\cap X=f^{-1}(A)$, (iii) if B is any closed set in $\mathbb{R}^m$ then there exists a closed set B° in $\mathbb{R}^m$ such that $B^\circ\cap X=f^{-1}(B)$. Note that the above does *not* say that if f is continuous and A is an open set in $\mathbb{R}^m$ then the image $f(A)=\{f(x):x\in A\}$ is open in $\mathbb{R}^m$. (Recall that while $f^{-1}(A)$ is automatically a subset of X, A is not necessarily a subset of the range of f.)

The property of a set being open or closed is not necessarily preserved under mapping by a continuous function. However, there are important properties of a set which are preserved under continuous mappings. In particular, the properties of connectedness and compactness have this character. If f is a function with domain $X\subset\mathbb{R}^n$ and range in $\mathbb{R}^m$ and if $A\subset X$ is connected in $\mathbb{R}^n$ and f is continuous on A then $f(A)$ is connected in $\mathbb{R}^m$. If f is a function with domain $X\subset\mathbb{R}^n$ and range in $\mathbb{R}^m$ and if $A\subset X$ is compact and f is continuous on A then $f(A)$ is compact (i.e., $f(A)$ is closed and bounded in $\mathbb{R}^m$). Let X be a compact subset of $\mathbb{R}^n$ and let f be a continuous one-one function with domain X and range $f(X)$ in $\mathbb{R}^m$. Then the inverse function f^{-1} is continuous with domain $f(X)$ and range X.

The following result is used frequently throughout the text.

(Weierstrass Theorem). Let $A\subset X$ be compact in $\mathbb{R}^n$ and let f be a continuous real-valued function with domain X. Then there are points x° and x_0 in A such that $f(x^\circ)=\sup\{f(x): x \in A\}$ and $f(x_0)=\inf\{f(x): x \in A)$; that is, a continuous real-valued function on a compact set attains its maximum and minimum values.

Let f be a function with domain $X \in \mathbb{R}^n$ and range in $\mathbb{R}^m$. Then f is **uniformly continuous** on a set $A \subset X$ if for any $\varepsilon > 0$ there is a $\delta > 0$, δ depending on ε but not on x°, such that if x and x° belong to A and $\|x - x^\circ\| < \delta$ then $\|f(x) - f(x^\circ)\| < \varepsilon$. Clearly, uniform continuity implies continuity, but not vice versa. If f is a continuous function with domain $X \subset \mathbb{R}^n$ and range in $\mathbb{R}^m$ and if $A \subset X$ is compact, then f is uniformly continuous on A. Thus, a continuous function is automatically uniformly continuous on any compact subset of its domain.

If f is a function with domain X in $\mathbb{R}^n$ and range in $\mathbb{R}^m$ then f is said to satisfy a **Lipschitz condition** (or be **Lipschitz continuous**) if there exists a $k > 0$ such that $\|f(x) - f(x^\circ)\| \leqslant k\|x - x^\circ\|$ for all $x, x^\circ \in X$. If the inequality holds with $k < 1$ the function is termed a **contraction**. If f satisfies a Lipschitz condition then f is uniformly continuous; the converse, however, is not true. Let f be a function with domain $X \subset \mathbb{R}^n$ and range in $\mathbb{R}^m$. An element b of $\mathbb{R}^m$ is termed the **limit of f at x°** if, for every neighbourhood V of b, there is a neighbourhood W of x° such that, if $x \in W \cap X$ and $x \neq x^\circ$, then $f(x) \in V$; we then write $b = \lim_{x \to x^\circ} f(x)$. The following statements are equivalent: (i) $b = \lim_{x \to x^\circ} f(x)$, (ii) for any $\varepsilon > 0$ there is a $\delta > 0$ such that, if $x \in X$ and $\|x - x^\circ\| < \delta$, then $\|f(x) - b\| < \varepsilon$, (iii) if $\{x_n\}$ is any sequence in X such that $x_n \neq x^\circ$ and $x^\circ = \lim\{x_n\}$, then $b = \lim\{f(x_n)\}$.

The relationship between the concept of the limit of a function and the continuity of a function at a point are given by the following. If $x^\circ \in X$ is a cluster point then the following statements are equivalent: (i) f is continuous at x°, (ii) $\lim_{x \to x^\circ} f(x)$ exists and is equal to $f(x^\circ)$. If f and g are two functions whose limits exist at a cluster point x° of the domain of $f+g$ then $\lim_{x \to x^\circ}(f+g)(x)$ exists and $\lim_{x \to x^\circ}(f+g)(x) = \lim_{x \to x^\circ} f(x) + \lim_{x \to x^\circ} g(x)$. Similar relationships hold for other algebraic combinations of functions, as well as for composite functions.

We shall often be concerned with functions whose domain X is a subset of $\mathbb{R}$ and whose range is a subset of $\mathbb{R}^m$. In defining the

$\lim_{x\to x^\circ} f(x)$ we allow $x \to x^\circ$ from either side—that is, from values greater than x° and from values less than x°. By restricting the direction in which $x \to x^\circ$, we obtain the concepts of the **right (left)-limit of f at x°**. We define the right-hand limit of f at x° to be b, if for any $\varepsilon > 0$ there is a $\delta > 0$ such that, if $x \in X$ and $x - x^\circ < \delta$, then $\|f(x) - b\| < \varepsilon$. We denote the right-hand limit of f at x° by $\lim_{x\to x^\circ +} f(x)$ or $\lim_{x\to x^\circ} f(x^\circ +)$ or simply $f(x^\circ +)$. The left-hand limit of f at x° is defined in a similar way, and is denoted by $\lim_{x\to x^\circ -} f(x)$ or $\lim_{x\to x^\circ} f(x^\circ -)$ or $f(x^\circ -)$. Note that $\lim_{x\to x^\circ} f(x)$ exists if, and only if, $\lim_{x\to x^\circ +} f(x) = \lim_{x\to x^\circ -} f(x)$ and that f is continuous at x° if, and only if, $\lim_{x\to x^\circ +} f(x) = \lim_{x\to x^\circ -} f(x) = f(x^\circ)$. If $\lim_{x\to x^\circ +} f(x)$ and $\lim_{x\to x^\circ -} f(x)$ both exist but are not equal then f is said to have a **jump discontinuity at x°**.

Let $f: X \to \mathbb{R}^n$, where $X = [a, b]$, and f is a function. Then f is said to be **piecewise continuous** over the interval $[a, b]$ if (i) there exists not more than a finite number of points in the interval where f is not continuous, (ii) at any point where f is not continuous, f has a jump discontinuity.

Note that the definition of a piecewise continuous function does not exclude the possibility that it is continuous over the entire interval.

Let f be a real-valued function defined on a subset $X \subset \mathbb{R}$. Then f is said to be **increasing** on X if, for every pair of points $x, y \in X$, $x < y$ implies $f(x) \leqslant f(y)$; if $x < y$ implies $f(x) < f(y)$ then f is said to be **strictly increasing** on X. **Decreasing** and **strictly decreasing** functions are similarly defined. A function f is said to be **monotone** on X if it is increasing on X or decreasing on X. Functions which are monotone on compact intervals always have finite right- and left-hand limits, so their discontinuities (if any) must be jump discontinuities. If f is strictly increasing (strictly decreasing) on $X \subset \mathbb{R}$ then f^{-1} exists and is strictly increasing (strictly decreasing) on $f(X)$.

The following generalization of the notion of continuity will prove useful. Let $f: X \to \mathbb{R}$, where $X \subset \mathbb{R}^n$, and let $x^\circ \in X$. Then the function f is **uppersemicontinuous at x°** if, and only if, for any given $\varepsilon > 0$, there exists a $\delta > 0$ such that $f(x^\circ) \geqslant f(x) - \varepsilon$ for all $x \in X$ with $\|x - x^\circ\| < \delta$. Similarly, f is **lowersemicontinuous at x°** if, and only if, for any given $\varepsilon > 0$ there is a $\delta > 0$ such that $f(x^\circ) \leqslant f(x) + \varepsilon$ for all $x \in X$ with

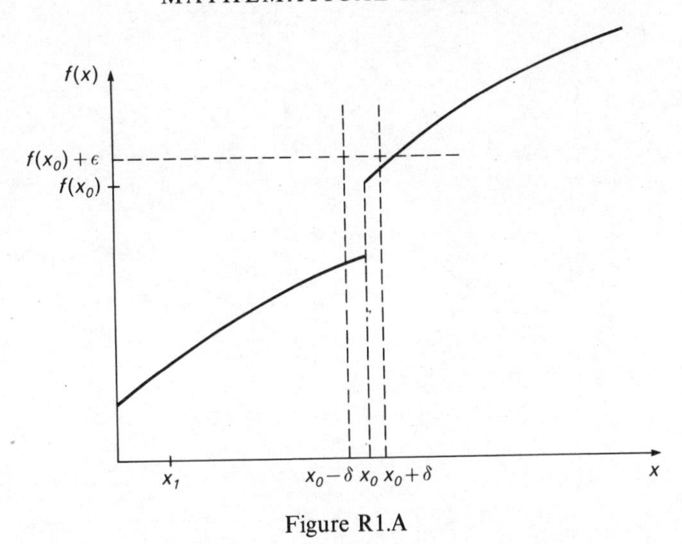

Figure R1.A

$\|x-x^\circ\|<\delta$. The function f is **upper (lower) semicontinuous** if it is upper (lower) semicontinuous at each point in its domain. The function, illustrated in Figure R1.A is uppersemicontinuous at x°, since for any given $\varepsilon>0$ there exists an open ball $B_\delta(x^\circ)$ such that $f(x^\circ)+\varepsilon \geqslant f(x)$ for all $x \in B_\delta(x^\circ)$. However, the function illustrated in Figure R1.B is not uppersemicontinuous at x°, since it is clear that for the given $\varepsilon>0$ there does not exist any open ball with centre x° such that $f(x^\circ)+\varepsilon \geqslant f(x)$ for all x in the open ball. This function is actually lowersemicontinuous at x°, since for any given $\varepsilon>0$ it can be seen that an open ball $B_\delta(x^\circ)$ exists such that for all $x \in B_\delta(x^\circ)$, $f(x^\circ)-\varepsilon \leqslant f(x)$. Note that, at the point x, both functions are lower and uppersemicontinuous. The following are important properties of upper (lower) semicontinuous functions. Let f be a real-valued function defined on a subset X of $\mathbb{R}^n$. Then, (i) f is continuous at $x^\circ \in X$ if, and only if, f is both upper and lowersemicontinuous at x°, (ii) f is lowersemicontinuous on X if, and only if, the sets $\{x: x \in X, f(x) \leqslant \alpha\}$ are closed for all $\alpha \in \mathrm{R}$, (iii) f is uppersemicontinuous on X if, and only if, the sets $\{x: x \in X, f(x) \geqslant \alpha\}$ are closed for all $\alpha \in \mathbb{R}$.

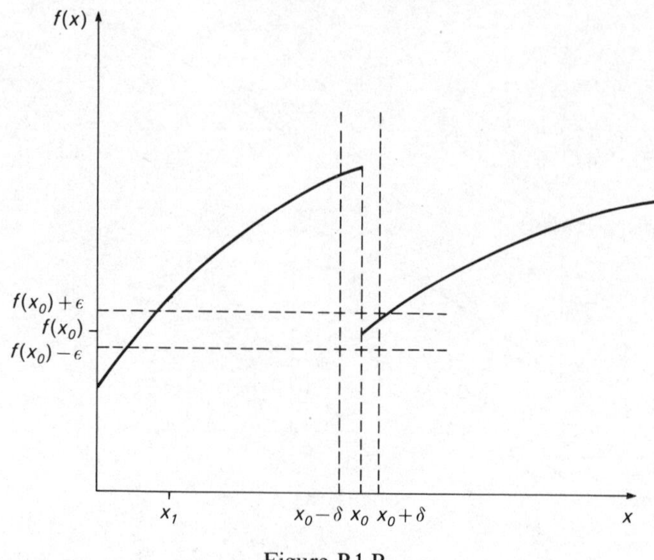

Figure R1.B

The following result generalizes the Weierstrass Theorem. Let f be a real-valued function defined on a compact subset X of $\mathbb{R}^n$. If f is lower (upper) semicontinuous on X then f has a minimum (maximum) on X.

DIFFERENTIABILITY

Let $f: X \rightarrow \mathbb{R}$, where $X \subset \mathbb{R}^n$. Then f is **differentiable at x°**, where x° is an interior point of X, if there exists an $a \in \mathbb{R}^n$, which in general depends on x°, such that

$$\lim_{\|h\| \rightarrow 0} \frac{f(x^\circ + h) - f(x^\circ) - a'h}{\|h\|} = 0,$$

where $h \in \mathbb{R}^n$ and $\|h\|$ denotes the Euclidean norm; or equivalently,

$$f(x^\circ + h) - f(x^\circ) = a'h + o(\|h\|),$$

where the notation $o(\|h\|)$ denotes that for any $\varepsilon > 0$ there exists a

$\delta>0$ such that $\|h\|<\delta$ implies $o(\|h\|)<\varepsilon\|h\|$. The vector a' is called the **derivative of f at x°**; we shall denote it by $f_x(x^\circ)$. If it exists $f_x(x^\circ)$ is unique. If f is differentiable at every point of X then f is said to be a **differentiable function**. The term $a'h=f_x(x^\circ)h$ in the above definition, termed the **differential of f at x°**, is easily seen to be a linear transformation from $\mathbb{R}^n\to\mathbb{R}$, for given x°. If $f:X\to\mathbb{R}$, where $X\subset R^n$, is differentiable at x° then f is continuous at x°: the converse is not true in general. The function $f:X\to\mathbb{R}$, where $X=\mathbb{R}$, defined by $f(x)=|x|$ is continuous on X, but is not differentiable at $x=0$.

If $f:X\to\mathbb{R}$, where $X\subset\mathbb{R}^n$, then we can also introduce the concept of a partial derivative. Let $e_i\in\mathbb{R}^n$ with its ith component unity and all other components equal to zero. Let $f:X\to\mathbb{R}$, where $X\subset\mathbb{R}^n$. Then f is said to have a **partial derivative with respect to x_i at x°**, where x° is an interior point of X, if there exists an $a_i\in\mathbb{R}$, which in general depends on x°, such that

$$\lim_{h\to 0}\frac{f(x^\circ+he_i)-f(x^\circ)-a_ih}{h}=0$$

or equivalently,

$$f(x^\circ+he_i)-f(x^\circ)=a_ih+o(|h|).$$

The scalar a_i is called the **partial derivative of f at x° with respect to x_i**, and is denoted by $f_i(x^\circ)$ or $\partial f(x^\circ)/\partial x_i$. If it exists, $f_i(x^\circ)$ is unique. It is important to note that $f:X\to\mathbb{R}$, with $X\subset\mathbb{R}^n$, can have partial derivatives at x° with respect to each x_i and yet not be continuous at x°. For example the function $f:\mathbb{R}^2\to\mathbb{R}$ defined by $f(x)=x_1+x_2$ if $x_1=0$ or $x_2=0$, and $f(x)=1$ otherwise, is not continuous at $x=(0,0)'$, but $f_1(0,0)=f_2(0,0)=1$.

Since $f_i(x^\circ)$, if it exists, is unique, we can construct a new function f_i, for each i, $i=1,\ldots,n$, which assigns to each point $x^\circ\in X$, at which f has a partial derivative with respect to x_i, the scalar $f_i(x^\circ)$. The f_i's are called the **first-order partial derivatives of f**: note that $f_i:S\to\mathbb{R}$, with $S\subset X\subset\mathbb{R}^n$. Since the f_i's are themselves real-valued functions defined on a subset S of $\mathbb{R}^n$ we can consider their associated first-order partial derivative functions. Clearly, we can associate with each f_i, n first-order partial derivatives: the first-order partial derivatives of f_i are called the **second-order partial derivatives of** f, and are denoted f_{ij} or $\partial^2 f/\partial x_j \partial x_i \equiv \partial(\partial f/\partial x_i)/\partial x_j$. If $f:X \rightarrow \mathbb{R}$ with $X \subset \mathbb{R}^n$, then f will have n^2 second-order partial derivatives. The matrix of second-order partial derivatives of f at x° is termed the **Hessian matrix of** f at x°; we denote it by $f_{xx}(x^\circ)$:

$$f_{xx}(x^\circ) \equiv \begin{bmatrix} f_{11}(x^\circ) & \ldots & f_{1n}(x^\circ) \\ \vdots & & \vdots \\ f_{n1}(x^\circ) & \ldots & f_{nn}(x^\circ) \end{bmatrix}.$$

Higher-order partial derivatives of f can be defined in a similar manner. If all first-order partial derivatives of f are themselves continuous on X then f is said to be **continuously differentiable** and we write $f \in C_1$; if all second-order partial derivatives of f are themselves continuous on X then f is said to be **twice-continuously differentiable** and we write $f \in C_2$. If $f \in C_2$ at $x^\circ \in X$ then $f_{ij}(x^\circ) = f_{ji}(x^\circ)$ for all i and j. $f \in C_0$ denotes that f is a continuous function. If $f \in C_0$ then the graph of f is called a **hypersurface** or **manifold** of dimension n in $\mathbb{R}^{n+1}$. If $f \in C_1$ the hypersurface or manifold is termed **smooth**.

If $f:X \rightarrow \mathbb{R}$, with $X \subset \mathbb{R}^n$, it is natural to enquire into the relationship between the derivative of f at x°, $f_x(x^\circ)$, and its first-order partial derivatives f_i at x°, $f_i(x^\circ)$. If f is differentiable at x° then for any scalar λ, $f(x^\circ + \lambda e_i) - f(x^\circ) = f_x(x^\circ)(\lambda e_i) + o(\|\lambda e_i\|)$. But $f_x(x^\circ)(\lambda e_i)$ is a linear transformation, so that $f_x(x^\circ)(\lambda e_i) = \lambda f_x(x^\circ) e_i$, and thus $[f(x^\circ + \lambda e_i) - f(x^\circ)]/\lambda = f_x(x^\circ) e_i + o(\|\lambda e_i\|)$. Taking the limit as $\lambda \rightarrow 0$ we have $\lim_{\lambda \rightarrow 0}[f(x^\circ + \lambda e_i) - f(x^\circ)]/\lambda = f_x(x^\circ) e_i$, so $f_i(x^\circ) = f_x(x^\circ) e_i$: that is, the ith component of the vector $f_x(x^\circ)$ is $f_i(x^\circ)$. Thus, $f_x(x^\circ) = (f_1(x^\circ), \ldots, f_n(x^\circ))$. The vector of first-order partial derivatives of at x° is sometimes called the *gradient vector of* f *at* x° and denoted $\nabla f(x^\circ)$ or Grad $f(x^\circ)$.

We have already noted that, if f is differentiable at x°, it is also continuous at x°, but that f need not be continuous at x° even if all partial derivatives of f exist at x°. It thus follows that the mere existence of all partial derivatives of f at x° is not sufficient to guarantee that f is differentiable at x°. It can be shown that f is differentiable at x° if, and only if, f has continuous first-order partial derivatives at x° with respect to each variable.

In the case where $f:X \to \mathbb{R}$ with $X \subset \mathbb{R}$ we have that f is differentiable at x°, where x° is an interior point of X, if there exists an $a \in \mathbb{R}$, which in general depends on x°, such that $\lim_{h\to 0}[f(x^\circ+h)-f(x^\circ)-ah]/h=0$. In this case we denote the derivative of f at x° by $f'(x^\circ)$ or $\mathrm{d}f(x^\circ)/\mathrm{d}x$. Since x° is an interior point of X, this ensures that $x^\circ+h \in X$ for sufficiently small h, where h can be either positive or negative. If X is a closed set and x° is a boundary point of X then we cannot ensure that $x^\circ+h \in X$ for h positive or negative, no matter how small we allow h to be. To deal with such situations we introduce the concepts of right- and left-hand derivatives. Let $f:X \to \mathbb{R}$, where $X \subset \mathbb{R}$. Then f has a **right-hand** (respectively, **left-hand**) **derivative at $x^\circ \in X$**, if there exists an $a^+ \in \mathbb{R}$ (respectively $a^- \in \mathbb{R}$), generally depending on x°, such that $\lim_{h\to 0^+}[f(x^\circ+h)-f(x^\circ)-a^+h]/h=0$ (respectively, $\lim_{h\to 0^-}[f(x^\circ+h)-f(x^\circ)-a^-h]/h=0$): the scalars a^+ and a^-, if they exist, are the **right-hand derivative** and the **left-hand derivative** of f at x°, respectively, and are denoted by $f'_+(x^\circ)$ and $f'_-(x^\circ)$, respectively, or by $f'(x^{\circ+})$ and $f'(x^{\circ-})$, respectively. If $X=[a,b]$ then at $x=a$ only $f'_+(a)$ may be defined, and at $x=b$ only $f'_-(b)$ may be defined. If x° is an interior point of X, then $f'(x^\circ)$ exists if, and only if, $f'_+(x^\circ)=f'_-(x^\circ)$, in which case $f'(x^\circ)=f'_+(x^\circ)=f'_-(x^\circ)$. In the definition of partial derivatives of $f:X\to\mathbb{R}$, where $X\subset\mathbb{R}^n$, f is essentially being treated as a function of a single variable, x_i, with the remaining $n-1$ variables, x_j, $j=1,\ldots,n$, $j\neq i$, being held constant; this allows us to extend the notion of right- and left-hand derivatives to partial derivatives in an obvious way.

So far our concepts of differentiability have been restricted to real-valued functions defined on subsets X of $\mathbb{R}^n$. We now consider functions which map points in $\mathbb{R}^n$ into points in $\mathbb{R}^m$. A function $f:X\to\mathbb{R}^m$, with $X\subset\mathbb{R}^n$, can be regarded as a set of m real-valued functions $\{f^1:\mathbb{R}^n\to\mathbb{R},\ldots,f^m:\mathbb{R}^n\to\mathbb{R}\}$, where the functions $f^j:\mathbb{R}^n\to\mathbb{R}$ provide the rule for determining the jth component of the image of x, $f(x)$. Such a function f is often denoted by the vector whose components are the functions f^j. Let $f:X\to\mathbb{R}^m$, where $X\subset\mathbb{R}^n$. Then f is **differentiable at x°**, where x° is an interior point of

X, if there exists an $m \times n$ real matrix A, which in general depends on x°, such that

$$\lim_{\|h\| \to 0} [\| f(x^\circ + h) - f(x^\circ) - Ah \|]/\|h\| = 0,$$

or equivalently,

$$f(x^\circ + h) - f(x^\circ) = Ah + o(\|h\|),$$

where $h \in \mathbb{R}^n$ and $\| \ \|$ is the Euclidean norm. The matrix A, if it exists, is called the **derivative of f at x°**, and we denote it by $f_x(x^\circ)$. If $f_x(x^\circ)$ exists, it is unique. If f is differentiable at every point in X then f is a **differentiable function**. If f is differentiable at x° then f is continuous at x°. Let f be represented by its components $f^1, \ldots, f^m$, where $f^j: \mathbb{R}^n \to \mathbb{R}$ provides the rule for determining the jth component for the image of x, $f(x)$. Then we have $f(x) = \sum_{j=1}^m f^j(x) u_j$, where $\{u_1, \ldots, u_m\}$ is the standard basis for $\mathbb{R}^m$. Let $\{e_1, \ldots, e_n\}$ be the standard basis for $\mathbb{R}^n$. Then, if f is differentiable at x°, we have, for any scalar λ, $f(x^\circ + \lambda e_i) - f(x^\circ) = A\lambda e_i + o(\|\lambda e_i\|)$, so that $\lim_{\lambda \to 0} \sum_{j=1}^m [(f^j(x^\circ + \lambda e_i) - f^j(x^\circ))/\lambda] u_j = Ae_i$. But $\lim_{\lambda \to 0} \sum_{j=1}^m [f^j(x^\circ + \lambda e_i) - f^j(x^\circ))/\lambda] = f_i^j(x^\circ) = \partial f^j(x^\circ)/\partial x_i$, so that $\sum_{j=1}^m f_i^j(x^\circ) u_j = Ae_i$. Now Ae_i is the ith column of A, so that $f_i^j(x^\circ)$ must be the element in the jth row and ith column of A. Hence,

$$f_x(x^\circ) = A = \begin{bmatrix} \partial f^1(x^\circ)/\partial x_1 & \ldots & \partial f^1(x^\circ)/\partial x_n \\ \vdots & & \vdots \\ \partial f^m(x^\circ)/\partial x_1 & \ldots & \partial f^m(x^\circ)/\partial x_n \end{bmatrix}.$$

This matrix is called the **Jacobian matrix of f at x°**. We shall often denote the Jacobian matrix of f at x° by $[\partial f^i(x^\circ)/\partial x_j]$, where $\partial f^i(x^\circ)/\partial x_j$ is the element in the ith row and jth column of the matrix, or by $f_x(x^\circ)$.

We assume that the reader is familiar with the basic rules concerning derivatives of functions. The 'chain rule' for differentiable functions $f: X \to \mathbb{R}$, where $X \subset \mathbb{R}$, can be generalized to functions $f: X \to \mathbb{R}^m$, where $X \subset \mathbb{R}^n$. Let $f: X \to \mathbb{R}^m$, where X is an open subset of $\mathbb{R}^n$. Suppose that f is differentiable at $x^\circ \in X$, and let $g: Y \to \mathbb{R}^p$, where Y is a subset of the range of f. If g is differentiable at $y = f(x^\circ) \in Y$ then the composite function $h \equiv g \circ f$, $h: X \to \mathbb{R}^p$, defined by $h(x)=g(f(x))$, is differentiable at x°, and its derivative at x° is

$$h_x(x^\circ)=g_x(f(x^\circ))f_x(x^\circ)$$
$$=\begin{bmatrix} g_1^1(f(x^\circ))\ldots g_m^1(f(x^\circ)) \\ \vdots \qquad\qquad \vdots \\ g_1^p(f(x^\circ))\ldots g_m^p(f(x^\circ)) \end{bmatrix}\begin{bmatrix} f_1^1(x^\circ)\ldots f_n^1(x^\circ) \\ \vdots \qquad\qquad \vdots \\ f_1^m(x^\circ)\ldots f_n^m(x^\circ) \end{bmatrix}.$$

TAYLOR'S THEOREM

Let $f:X\to\mathbb{R}$, where $X\subset\mathbb{R}$, and suppose that f is differentiable at x°. Then there exists a linear function $T_{x^\circ}:X\to\mathbb{R}$ such that $f(x^\circ+h)=f(x^\circ)+T_{x^\circ}(h)+o(h)=f(x^\circ)+f'(x^\circ)h+o(h)$. Thus, $f(x^\circ+h)\simeq f(x^\circ)+f'(x^\circ)h$, for small h. Taylor's Theorem tells us that, more generally, f can be approximated by a polynomial function of degree $n-1$ if f has derivatives of order n. Moreover, Taylor's Theorem gives a useful expression for the error involved in this approximation.

(Taylor's Theorem). Let $f:X\to\mathbb{R}$, where $X\subset\mathbb{R}$ have an nth derivative $f^{(n)}$ in the open inerval (a,b) and assume that $f^{(n-1)}$ is continuous on the closed interval $[a,b]$. Assume that $x^\circ\in(a,b)$. Then, for every $h\neq 0$, $x^\circ+h\in[a,b]$, there exists a point $x_1=x^\circ+\psi h$, $0<\psi<1$, such that

$$f(x^\circ+h)=f(x^\circ)+\textstyle\sum_{k=1}^{n-1}f^{(k)}(x^\circ)h^k/k!+f^{(n)}(x_1)h^n/n!.$$

Here $f(x^\circ)+\sum_{k=1}^{n-1}f^{(k)}(x^\circ)h^k/k!$ is the approximating polynomial function (in terms of h) of degree $n-1$ and $f^{(n)}(x_1)h^n/n!$ is the error involved in using this approximation. Letting $x=x^\circ+h$ we have

$$f(x)=f(x^\circ)+\textstyle\sum_{k=1}^{n-1}f^{(k)}(x^\circ)(x-x^\circ)^k/k!+f^{(n)}(x_1)(x-x^\circ)^n/n!,$$

where $x_1=x^\circ+\psi(x-x^\circ)$, $0<\psi<1$.

Taylor's theorem can be extended to real-valued functions defined on subsets of $\mathbb{R}^n$ and can be expressed in several forms. We shall find the following forms most suitable for our purposes.

(Taylor's Theorem). Let $f:X\to\mathbb{R}$, where X is an open subset of $\mathbb{R}^n$ with $f\in C_2$ on X. Assume $x^\circ\in X$. Then for every $h\neq\theta$, $x^\circ+h\in X$, there exists a point x_1 such that:

(i) $f(x^\circ+h)=f(x^\circ)+f_x(x_1)h$, where $x_1=x^\circ+\psi h$ for some ψ, $0<\Psi<1$.

(ii) $f(x^\circ+h)=f(x^\circ)+f_x(x^\circ)h+\frac{1}{2}h'f_{xx}(x_1)h$, where $f_{xx}(x_1)$ is the Hessian matrix of f at x_1 where $x_1=x^\circ+\gamma h$ for some γ, $0<\gamma<1$.

or equivalently, letting $x=x^\circ+h$:

(i′) $f(x)=f(x^\circ)+f_x(x_1)(x-x^\circ)$,

(ii′) $f(x)=f(x^\circ)+f_x(x^\circ)(x-x^\circ)+\frac{1}{2}(x-x^\circ)'f_{xx}(x_1)(x-x^\circ)$.

Note that $h'f_{xx}(x_1)h$ and $(x-x^\circ)'f_{xx}(x_1)(x-x^\circ)$ are quadratic forms in the variables h and $(x-x^\circ)$, respectively.

HOMOGENEOUS FUNCTIONS

A function $f:X\to\mathbb{R}$, where $X\subset\mathbb{R}^n$, is **positively homogeneous of degree *k*** if for every $x\in X$, $f(\lambda x)=\lambda^k f(x)$ for all $\lambda>0$. If f is positively homogeneous of degree $k=1$ then we simply say that f is **positively homogeneous**. If f is differentiable and positively homogeneous of degree k then: (i) **(Euler's Theorem)** $kf(x)=\sum_{i=1}^n f_i(x)x_i$, (ii) the first-order partial derivatives of f are themselves homogeneous of degree $k-1$, (iii) $k(k-1)f(x)=x'f_{xx}(x)x=\sum_{i=1}^n\sum_{j=1}^n f_{ij}(x)x_ix_j$. A **homothetic function** h is a monotone increasing transform of a positively homogeneous of degree k function: $h(x)=g(f(x))$, where g is a monotone increasing function and f is homogeneous of some degree $k>0$.

IMPLICIT FUNCTION THEOREM

Suppose we represent points in $\mathbb{R}^{n+k}$ by (x,b), where $x=(x_1,\ldots,x_n)'\in\mathbb{R}^n$ and $b=(b_1,\ldots,b_k)'\in\mathbb{R}^k$. Let $f:S\to\mathbb{R}^n$, $S\subset\mathbb{R}^{n+k}$, where $f(x,b)=\theta$ for all $(x,b)\in S$. Then we can ask, does there exist a function $g:\mathbb{R}^k\to\mathbb{R}^n$, where $g(b)=x$ and $f(g(b),b)=\theta$, for all $b\in\mathbb{R}^k$?

An important special case of the above problem is that of solving systems of n linear equations in n unknowns, $Ax=b$, where $A=[a_{ij}]$ is $n\times n$. The system $Ax-b=\theta$ can be described by a function

$f:\mathbb{R}^n \to \mathbb{R}^n$, where f is such that $f(x)=b$, with its component functions $f^1,\ldots,f^n$ given by $f^i(x)=\sum_{j=1}^n a_{ij}x_j=b_i$, $i=1,\ldots,n$. We know that the system $Ax=b$ has a unique solution if, and only if, A is non-singular. Note that since $f^i(x)=\sum_{j=1}^n a_{ij}x_j$ we have $f_j^i(x)=a_{ij}$, so that A is the Jacobian matrix of f. Expressing the unique solution as $x=g(b)$, we have $f(g(b))=b$ if, and only if, the Jacobian matrix of f is non-singular.

In the general case, the equations defined by $f(x,b)=\theta$ are not restricted to be linear. Nevertheless, the non-singularity of the Jacobian matrix of f continues to play a vital role.

Implicit Function Theorem. Let $f:S\to\mathbb{R}^n$, where S is an open subset of $\mathbb{R}^{n+k}$, and suppose $f\in C_m$ on $S, m\geqslant 1$. Let (x°, b°) be a point in S for which $f(x^\circ,b^\circ)=\theta$ and for which the Jacobian matrix of f is non-singular. Then there exists an open set $Y\subset\mathbb{R}^k$ containing b° and a unique function $g:Y\to\mathbb{R}^n$ such that:

(i) $g\in C_m$ on Y,

(ii) $g(b^\circ)=x^\circ$,

(iii) $f(g(b),b)=\theta$ for every $b\in Y$.

Since f is continuously differentiable on S and g is continuously differentiable on Y we can apply the chain rule to $f(g(b),b)=\theta$, for $b\in Y$, to obtain

$$\begin{bmatrix} f_1^1 & \cdots & f_n^1 \\ \vdots & & \vdots \\ f_1^n & \cdots & f_n^n \end{bmatrix}\begin{bmatrix} g_p^1 \\ \vdots \\ g_p^n \end{bmatrix} = -\begin{bmatrix} f_p^1 \\ \vdots \\ f_p^n \end{bmatrix} \quad \text{for } p=1,\ldots,k,$$

where the f^i and g^i are the ith component functions of f and g.

INTEGRATION

We require only the concept and certain properties of the Riemann Integral in the text. For a bounded closed interval $[a,b]$, a finite subset $P=\{x_0,x_1,\ldots,x_n\}$ of $[a,b]$, where $a=x_0<x_1<\ldots<x_n=b$, is called a *partition* of $[a,b]$. Given a bounded function f on $[a,b]$ and writing

$$S_k(f)\equiv\sup\{f(x):x\in[x_{k-1},x_k]\}$$

and

$$s_k(f) \equiv \inf\{f(x): x \in [x_{k-1}, x_k]\}$$

we define

$$U(P, f) \equiv \sum_{k=1}^{n} S_k(f)(x_k - x_{k-1}),$$

$$L(P, f) \equiv \sum_{k=1}^{n} s_k(f)(x_k - x_{k-1}),$$

as the *upper and lower Riemann sums of f with respect to the partition P*. We say that f is **Riemann integrable** if

$$\inf\{U(P, f): \text{for all partitions } P\} = \sup\{L(P, f): \text{for all partitions } P\},$$

and the common value is denoted $\int_a^b f(\tau)\mathrm{d}\tau$. Note that the numerical value of $\int_a^b f(\tau)\mathrm{d}\tau$ depends on f, a and b, and that τ is a 'dummy variable' and may be replaced by any other symbol. Every continuous function on a bounded closed interval is Riemann integrable.

The following results, known as the Fundamental Theorem of Calculus, provide the link between differentiable and integral calculus.

Fundamental Theorem of Calculus. Consider an integrable function f on $[a, b]$:

(i) if F is continuous on $[a, b]$ and differentiable on (a, b) and $F'(x) = f(x)$ for all $x \in (a, b)$ then $\int_a^b f(\tau)\mathrm{d}\tau = F(b) - F(a)$,

(ii) if f is continuous at $c \in (a, b)$ then the function F on $[a, b]$ defined by $F(x) = \int_a^x f(\tau)\mathrm{d}\tau$ is differentiable at c and $F'(c) = f(c)$.

If f and g have continuous derivatives on $[a, b]$, then

$$\int_a^b f(\tau)g'(\tau)\mathrm{d}\tau = f(b)g(b) - f(a)g(a) - \int_a^b g(\tau)f'(\tau)\mathrm{d}\tau.$$

This result is usually referred to as **Integration by Parts**.

We shall need to consider integrals in which the integrand

$f(x,y)$ depends on a parameter y, and where the parameter y may also enter in the limits of integration. Let $D \equiv \{(x,y): a \leqslant x \leqslant b, c \leqslant y \leqslant d\}$ and suppose that f and f_y are continuous on D and that α and β, the limits of integration, are functions of y which are differentiable on the interval $[c,d]$ with values in $[a,b]$. If J is defined on $[c,d]$ by $J(y)=\int_{\alpha(y)}^{\beta(y)} f(\tau,y)\mathrm{d}\tau$, then J has a derivative for each $y \in [c,d]$, which is given by

$$J'(y)=f(\beta(y),y)\beta'(y)-f(\alpha(y),y)\alpha'(y)+\int_{\alpha(y)}^{\beta(y)} f_y(\tau,y)\mathrm{d}\tau.$$

In the case where the limits of integration do not depend on the parameter y then $J'(y)=\int_a^\beta f_y(\tau,y)\mathrm{d}\tau$.

Let f be defined on $(a,b]$ and suppose the integral $J(c) \equiv \int_c^b f(\tau)\mathrm{d}\tau$, $c \in (a,b]$, exists. Then we define the **improper integral** of f over $[a,b]$ to be $\lim_{c\to a} J(c)$, if this limit exists. Clearly, if f is bounded on $(a,b]$ the limit exists.

In the text we need to consider integrals defined on unbounded sets, usually of the form $[a,\infty)$. Suppose f is defined on $[a,\infty)$ and, for $c \geqslant a$, define $J(c) \equiv \int_a^c f(\tau)\mathrm{d}\tau$. Then the **infinite integral** of f is defined as $\lim_{c\to\infty} J(c)$, if the limit exists; if so it is denoted $\int_a^\infty f(\tau)\mathrm{d}\tau$ and we say that the integral is **convergent**. It is important to know whether the limit exists. The following results often prove useful in this regard. If $f(x) \geqslant 0$ for all $x > a$ and if f is integrable over $[a,c]$ for all $c \geqslant a$, then $\int_a^\infty f(\tau)\mathrm{d}\tau$ exists if, and only if, the set $\{J(c): c \geqslant a\}$ is bounded, in which case $\int_a^\infty f(\tau)\mathrm{d}\tau = \sup\{\int_a^c f(\tau)\mathrm{d}\tau : c \geqslant a\}$. If f is continuous for $c \geqslant a$ and $\{J(c): c \geqslant a\}$ is bounded and if g is monotone decreasing to zero as $x \to \infty$, then the infinite integral $\int_a^\infty f(\tau)g(\tau)\mathrm{d}\tau$ exists.

In some applications it is important to consider infinite integrals in which the integrand depends on a parameter. In such situations the notion of uniform convergence of the integral relative to the parameter is of importance. Let f be a real-valued function defined on $\{(x,y): x \geqslant a, \alpha \leqslant y \leqslant \beta\}$. Suppose that for each $y \in [\alpha,\beta]$ the infinite integral $J(y)=\int_a^\infty f(\tau,y)\mathrm{d}\tau$ exists. We say that this convergence is **uniform** on $[\alpha,\beta]$ if, for every $\varepsilon>0$, there exists a number $k(\varepsilon)$ such that, if $c \geqslant k(\varepsilon)$ *and* $y \in [\alpha,\beta]$, then $|J(y)-\int_a^c f(\tau,y)\mathrm{d}\tau| < \varepsilon$. The following often provides a useful test for uniform convergence of the indefinite integral. Let f be continuous in (x, y) for $x \geqslant a$ and $y \in [\alpha, \beta]$ and suppose there exists a constant b such that $\left|\int_a^c f(\tau, y)\mathrm{d}\tau\right| \leqslant b$ for $x \geqslant a$, $y \in [\alpha, \beta]$. Suppose that, for each $y \in [\alpha, \beta]$, the function $g(x, y)$ is monotone decreasing for $x \geqslant a$ and converges to zero as $x \to \infty$ uniformly for $y \in [\alpha, \beta]$. Then the integral $J(y) = \int_a^\infty f(\tau, y)g(\tau, y)\mathrm{d}\tau$ converges uniformly on $[\alpha, \beta]$. This result is often used in economic applications where it is common to encounter $\int_a^\infty f(x, y)\mathrm{e}^{-yx}\mathrm{d}x$. Here e^{-yx} monotonically decreases to zero as $x \to \infty$, for $y > 0$.

Finally, for this case, we consider differentiation under the integral sign when the integrand depends on a parameter y. Suppose $f(x, y)$ is a continuous function defined for $x \geqslant a$ and for $y \in [\alpha, \beta]$. Suppose $J(y) = \int_a^\infty f(\tau, y)\mathrm{d}\tau$ exists for all $y \in [\alpha, \beta]$. Then we have: (i) if the convergence is uniform on $[\alpha, \beta]$, then J is continuous on $[\alpha, \beta]$, and (ii) if f and f_y are continuous in (x, y) for $x \geqslant a$ and $y \in [\alpha, \beta]$ and $G(y) = \int_a^\infty f_y(\tau, y)\mathrm{d}\tau$ is uniformly convergent on $[\alpha, \beta]$, then J is differentiable on $[\alpha, \beta]$ and $J'(y) = G(y)$; that is, $J'(y) = \int_a^\infty f_y(\tau, y)\mathrm{d}\tau$.

GLOSSARY

SET NOTATION

$\in, \notin$ belongs (does not belong) to	358
$\subset, \not\subset$ is (is not) a subset of	358
$\cup$ union	359
$\cap$ intersection	359
$\emptyset$ null set	358
$\{x : x \text{ satisfies } P\}$, the set of x which satisfy property P	358
$A \backslash B$ difference of A and B	359
$A \otimes B$ cartesian product of A and B	359
int A interior of A	381
cl A closure of A	381
bd A boundary of A	381
$B_r(x)$ open ball with centre x and radius r	381
$N(x)$ a neighbourhood of x	381
dim A dimension of A	367

REAL AND COMPLEX NUMBERS

$\mathbb{R}$ set of real numbers	362
$\mathbb{N}$ set of natural numbers	382
$\text{Re}(z)$, $\text{Im}(z)$ real (imaginary) part of the complex number z	363
$\lvert\ \rvert$ absolute value or modulus	363
sup supremum	362
inf infimum	363

GLOSSARY

MAPPINGS

$f: X \to Y$ f is a mapping from X to Y	360
$f(X)$ range of mapping f	359
$f^{-1}(A)$ inverse image of A under mapping f	360
f^{-1} inverse of mapping f	361
$g \circ f$ composition of functions g and f	384
$\{s_n\}$ or $\{s^n)$ sequence whose nth term is s_n or s^n	382

VECTORS AND MATRICES

vectors are understood to be column vectors	
x', A' transpose of vector x (or matrix A)	365, 369
$\mathbb{R}^n$ set of all ordered n-tuples of real numbers	365
For $x, y \in \mathbb{R}^n$,	
$x \geqslant y$ if $x_i \geqslant y_i$ for all i	365
$x > y$ if $x_i \geqslant y_i$ and for at least one i $x_i > y_i$	365
$x \gg y$ if $x_i > y_i$ for all i	365
θ null vector or null matrix	364
$\mathbb{R}^n_+ = \{x : x \in \mathbb{R}^n, x \geqslant \theta\}$	365
$\mathbb{R}^n_{++} = \{x : x \in \mathbb{R}^n, x \gg \theta\}$	365
$\lVert \rVert$ norm of vector or matrix	367, 370
$\rho(A)$ rank of matrix A	369
$\lvert A \rvert$ determinant of matrix A	372
A^{-1} inverse of matrix A	369
tr A trace of matrix A	377
M_k the kth order leading principal minor of an $n \times n$ matrix	373
$\tilde{M}_k$ a kth order principal minor of an $n \times n$ matrix	373

DERIVATIVES

$\mathrm{d}f/\mathrm{d}x$ or f' derivative of a function of a single variable	389
$\partial f/\partial x_i$ or f_i partial derivative with respect to x_i	389
$\partial^2 f/\partial x_i \partial x_j$ or f_{ij} second order partial derivative of f	390
$f \in C_r$ f has continuous partial derivatives up to order r	390
f_x derivative of f	
for $f: X \to \mathbb{R}$, where $X \subset \mathbb{R}^n$, $\nabla f = f_x \equiv (f_1, \ldots, f_n)$	390
f_{xx} matrix of second order partial derivatives of f	390
$f'_-(c)$ or $f'(c^-)$ left-hand derivative of f at c	391
$f'_+(c)$ or $f'(c^+)$ right-hand derivative of f at c	391
$\lim_{x \to c^+} f$, $\lim_{x \to c^-} f$ limit of f as x approaches c from right (left)	386
$\dot{x} = \mathrm{d}x/\mathrm{d}t$, $x^{(n)} = \mathrm{d}^n x/\mathrm{d}t^n$	393
$o(.)$ or $\mathcal{O}(.)$	388

BIBLIOGRAPHY

ARAUJO A. AND SCHEINKMAN J. A. (1983) Maximum principle and transversality conditions for concave infinite horizon economic models, *Journal of Economic Theory*, Vol. 28, pp. 1–16.

ARROW K. J. (1968) Applications of control theory to economic growth, in Dantzig G. B. and Veinott A. F.

ARROW K. J., BLOCK H. D. AND HURWICZ L. (1959) On the stability of the competitive equilibrium, II, *Econometrica*, Vol. 27, pp. 82–109.

ARROW K. J. AND DEBREU G. (1954) Existence of an equilibrium for a competitive economy, *Econometrica*, Vol. 22, pp. 265–90.

ARROW K. J. AND ENTHOVEN A. C. (1961) Quasi-concave programming, *Econometrica*, Vol. 29, pp. 779–800.

ARROW K. J. AND HURWICZ L. (1958) On the stability of competitive equilibrium, I, *Econometrica*, Vol. 26, pp. 522–52.

ARROW K. J., HURWICZ L. AND UZAWA H. (1961) Constraint qualifications in maximisation problems, *Naval Logistics Quarterly*, Vol. 8, pp. 175–86.

ARROW K. J. AND HAHN F. H. (1971) *General competitive analysis*, Holden-Day Inc., San Francisco.

ARROW K. J. AND INTRILIGATOR M. D. (1982) (EDS.) *Handbook of Mathematical economics*, 3 vols, North-Holland, Amsterdam.

ARROW K. J. AND KURZ M. (1970) *Public investment, the rate of return and optimal fiscal policy*, Johns Hopkins, Baltimore.

ATHANS M. AND FALB P. L. (1966) *Optimal control: an introduction to the theory and its applications*, McGraw-Hill, New York.

BIBLIOGRAPHY

ATKINSON A. B. (1971) Capital taxes, the redistribution of wealth and individual savings, *Review of Economic Studies*, Vol. 38, pp. 209–27.

AUMANN R. J. (1964) Markets with a continuum of traders, *Econometrica*, Vol. 32, pp. 39–50.

AUMANN R. J. AND PELEE B. (1960) Von Neumann-Morgenstern solutions to cooperative games without side payments, *Bulletin of the American mathematical society*, Vol. 66, pp. 173–9.

AVRIEL M. (1976) *Non-linear programming*, Prentice Hall, Englewood Cliffs N.J.

BARNETT S. AND STOREY C. (1970) *Matrix methods in stability theory*, Nelson, London.

BARTLE R. G. (1976) *The elements of real analysis*, Wiley, New York.

BAZARAA M., GOODE J. J. AND SHETTY C. (1972). Constraint qualifications revisited, *Management Science*, Vol. 18, pp. 567–73.

BAZARAA M. AND SHETTY C. (1976) *Foundations of optimisation* (Lecture notes in economics and mathematic systems, No. 122), Springer, Berlin.

BEGG D. K. (1982) *The rational expectations revolution in macroeconomics*, Philip Allan, Oxford.

BELLMAN R. E. (1953) *Stability theory of differential equations*, McGraw-Hill, New York.

BELLMAN R. E. (1960) *Introduction to matrix analysis*, McGraw-Hill, New York.

BELLMAN R. E. AND COOKE K. L. (1963) *Differential-difference equations*, Academic Press, New York.

BENVENISTE L. M. AND SCHEINKMAN J. (1982) Duality theory for dynamic optimisation models of economics: the continuous time case, *Journal of Economic Theory*, Vol. 27, pp. 1–19.

BILLINGSLEY P. (1968) *Convergence of probability measures*, Wiley, New York.

BLACKORBY C., PRIMONT D. AND RUSSELL R. R. (1978) *Duality separability and functional structure: theory and economic applications*, Elsevier, New York.

BORDER K. (1985) *Fixed point theorems with applications to economics and the theory of games*, Cambridge University Press.

BOTHWELL F. E. (1952) The method of equivalent linearisation, *Econometrica*, Vol. 20, pp. 269–83.

BRØNSTED A. (1983) *An introduction to convex polytopes*, Springer-Verlag, New York.

BUITER W. H. AND MILLER M. (1981) Monetary policy and international competitiveness: the problems of adjustment, *Oxford Economic Papers*, Vol. 33, pp. 143–75.

BURMEISTER E. (1980) *Capital theory and dynamics*, Cambridge University Press.

CAMERON N. (1985) *Introduction to linear and convex programming*, Cambridge University Press.

CHAMPSAUR P., DREZE J. AND HENRY C. (1977) Stability theorems with economic applications, *Econometrica*, Vol. 45, pp. 273–94.

CHIPMAN J. S. (1950) The multi-sector multiplier, *Econometrica*, Vol. 18, pp. 355-74.

CLARKE F. H. (1976) The maximum principle under minimal hypotheses, *SIAM Journal of Control*, Vol. 14, pp. 1078–91.

CLARKE F. H. (1979) Optimal control and the true Hamiltonian, *SIAM Review*, Vol. 21, pp. 157–66.

CODDINGTON E. A. AND LEVINSON N. (1955) *Theory of differential equations*, McGraw-Hill, New York.

COOTER R. (1978) Optimal tax schedules: Mirrlees and Ramsey, *American Economic Review*, Vol. 68, pp. 756–68.

COURANT R. AND JOHN F. (1965) *Introduction to calculus and analysis*, Vol. 1, Wiley, New York.

DANTZIG G. B. AND VEINOTT A. F. (EDS) (1968) Mathematics of the decision sciences, *American Mathematical Society*, Providence, R. I.

DEBREU G. (1952) Definite and semi-definite quadratic forms, *Econometrica*, Vol. 20, pp. 295–300.

DEBREU G. (1959) *Theory of value*, Wiley, New York.

DEBREU G. (1974) Excess demand functions, *Journal of Mathematical Economics*, Vol. 1, pp. 15–23.

DEBREUG, (1982) Existence of competitive equilibrium, in the *Handbook of Mathematical Economics*, Chapter 15, edited by Arrow K. J. and Intriligator M. D., North-Holland, Amsterdam.

DIERKER E. (1970) Two remarks on the number of equilibria in an economy, *Econometrica*, Vol. 40, pp. 951–3.

DIERKER E. (1982) Regular economies, in *Handbook of Mathematical Economics*, Vol. II, Chapter 17, edited by Arrow K. J. and Intriligator M. D., North-Holland, Amsterdam.

DIEWERT W. E. (1974) Applications of duality theory, in *Frontiers of quantitative economics*, Vol. 2, edited by Intriligator M. D. and Kendrick D. A., North-Holland, Amsterdam.

DIEWERT W. E. (1982) Duality approaches to microeconomic theory, in *Handbook of mathematical economics*, Vol. 2, Chapter 12, edited by Arrow K. J. and Intriligator M. D., North-Holland, Amsterdam.

DREYFUS S. E. (1965) *Dynamic programming and the calculus of variations*, Academic Press, New York.

ECKALBAR J. C. (1985) Inventory fluctuations in a disequilibrium macro model, *Economic Journal*, Vol. 95, pp. 976–91.

EICHHORN W. AND OETTLI W. (1972) A general formulation of the Le-Chatelier-Samuelson principle, *Econometrica*, Vol. 40, pp. 711–17.

EPSTEIN L. G. (1981) Generalized duality and integrability, *Econometrica*, Vol. 49, pp. 655–78.

FAREBROTHER R. W. (1973) *Simplified Samuelson conditions for cubic and quartic equations*, The Manchester School of Economic and Social Studies, Vol. 41, pp. 396–400.

FEINSTEIN C. H. (ED.) (1967) *Socialism, capitalism and economic growth*, (Essays presented in honour of Maurice Dobb), Cambridge University Press.

FIACCO A. V. (1968) Second order sufficient conditions for weak and strict constrained minima, *SIAM Journal of Applied Mathematics*, Vol. 16, pp. 105–8.

FORSYTH P. J. AND KAY J. A. (1980) The economic implications of North Sea oil revenues, *Fiscal studies*, Vol. 1, pp. 1–28.

FRIEDMAN B. (1979) Optimal expectations and the extreme information assumptions of 'rational expectations macro models', *Journal of Monetary Economics*, Vol. 5, pp. 23–41.

FRIEDMAN J. W. (1977) *Oligopoly and the theory of games*, North-Holland, Amsterdam.

FUSS M. AND McFADDEN D. (EDS.) (1978) *Production economics: a dual approach to theory and applications*, Vol. 1, North-Holland, Amsterdam.

GALE D. (1955) The law of supply and demand, *Mathematica Scandinavia*, Vol. 3, pp. 155–89.

GALE D. (1960) *The theory of linear economic models*, McGraw-Hill, New York.

GANDOLFO G. (1980) *Economic dynamics: methods and models*, North-Holland, Amsterdam.

GANTMACHER F. R. (1959) *The theory of matrices*, Chelsea Pub. Co., New York.

GOODWIN R. M. (1967) A growth cycle, in Feinstein C. (ed.), pp. 54–8.

GOULD F. J. AND TOLLE J. W. (1971) A necessary and sufficient qualification for constrained optimisation, *SIAM Journal of Applied Mathematics*, Vol. 20, pp. 168–72.

GUILLEMIN V. AND POLLAK A. (1974) *Differential topology*, Prentice-Hall, Englewood Cliffs, N.J.

GUINN T. (1965) The problem of bounded space coordinates as a problem of Hestenes, *SIAM Journal of Control*, Series A, Vol. 3, pp. 181–90.

HADLEY G. (1967) *Linear programming*, Addison-Wesley, Reading, Massachusetts.

HADLEY G. AND KEMP M. C. (1971) *Variational methods in economics*, North-Holland, Amsterdam.

HAHN F. H. (1958) Gross substitutability and the dynamic stability of general equilibrium, *Econometrica*, Vol. 26, pp. 169–70.

HAHN F. H. (1982) Stability, *The handbook of mathematical economics*, Chapter 16, edited by Arrow K. J. and Intriligator M. D., North-Holland, Amsterdam.

HAHN W. (1967) *Stability of motion*, Springer Verlag, New York.

HALE J. K. (1969) Ordinary differential equations, *Interscience*, New York.

HALKIN H. (1974) Necessary conditins for optimal control problems with infinite horizons, *Econometrica*, Vol. 42, pp. 267–72.

HARTMAN P. (1961) On stability in the large for systems of ordinary differential equations, *Canadian Journal of Mathematics*, Vol. 13, pp. 480–92.

HARTMAN P. (1964) *Ordinary differential equatins*, Wiley, New York.

HARTMAN P. AND OLECH C. (1962) On the global stability of solutions of differential equations, *Trans. Am. Math. Society*, Vol. 104, pp. 154–78.

HENRY C. (1972) Differential equations with discontinuous right hand sides for planning procedures, *Journal of Economic Theory*, Vol. 4, pp. 545–51.

HENRY C. (1973) An exxistence theorem for a class of differential equations with multi-valued right hand side, *Journal of Mathematical Analysis and Applications*, Vol. 41, pp. 179–86.

HENRY C. (1974) *Problemes d'éxistence et de stqabilité pour des processes dynamique considere en economie mathematique*, Compte Rendue de L'Academie des Sciences, Paris, 278 (series A), pp. 97–100.

HESTENES M. R. (1966) *Calculus of variations and optimal control theory*, Wiley, New York.

HESTENES M. R. (1965) On variational theory and optimal control theory, *Journal of SIAM* Series A, Control, Vol. 3, pp. 23–48.

HILDENBRAND W. (1982) Core of an economy, in *Handbook of Mathematical Economics*, Chapter 18, Vol. II, edited by Arrow K. J. and Intriligator M. J., North-Holland, Amsterdam.

HILDENBRAND W. AND KIRMAN A. P. (1976) *Introduction to Equilibrium Analysis*, North-Holland, Amsterdam.

HIRSCH M. AND SMALE S. (1974) *Differential equations, dynamical systems and linear algebra*, Academic Press, New York.

JOHN F. (1948) Extremum problems with inequalities as subsidiary conditions, in *Studies and essays*, Courant Anniversary volume, edited by Friedricks K. D. *et al.*, Interscience Publishers, New York.

KAKUTANI S. (1941) A generalisation of Brower's fixed point theorem, *Duke Mathematical Journal*, Vol. 8, pp. 457–59. Reprinted in Newman P. (ed.) (1968) *Readings in Mathematical Economics*, Vol. 1, John Hopkins Press, Baltimore.

KALMAN R. E. AND BERTRAM J. E. (1960) Control system analysis and design via the 'second method' of Liapunov, *Journal of Basic Engineering*, June, pp. 371–400.

KAMIEN M. AND SCHWARTZ N. (1981) *Dynamic optimisation: calculus of variations and optimal control in economics and management*, North-Holland, New York.

KIRMANA A. P. (1982) Measure theory with applications to economics, in *Handbook of Mathematical Economics*, Chapter 5, Vol. 1, edited by Arrow K. J. and Intriligator M., North-Holland, Amsterdam.

KLEIN E. (1973) *Mathematical methods in theoretical economics*, Academic Press, New York.

LEE E. B. AND MARKUS L. (1967) *Foundations of optimal control theory*, Wiley, New York.

LETSCHETZ S. (1957) Differential equations: geometric theory, *Interscience*, New York.

MANGASARIAN O. L. (1966) Sufficient conditions for the optimal control of non-linear systems, *SIAM Journal of Control*, Vol. 4, pp. 139–52.

MANGASARIAN O. L. (1969) *Nonlinear programming*, McGraw-Hill, New York.

MAS COLELL A. (1985) *The theory of General Economic Equilibrium: A Differentiable Approach*, Cambridge University Press.

MASSERA J. L. (1956) Contributions to stability theory, *Annals of Mathematics*, Vol. 64, pp. 182–206.

McCORMICK G. P. (1967) Second order conditions for constrained minima, *SIAM Journal of Applied Mathematics*, Vol. 15, pp. 641–52.

McFADDEN D. (1978) Cost, revenue and profit functions, in *Production economics: a dual approach to theory and applications*, edited by Fuss M. and McFadden D., North-Holland, Amsterdam.

McKENZIE L. W. (1959) On the existence of general equilibrium for a competitive market, *Econometrica*, Vol. 27, pp. 54–71.

McKENZIE L. W. (1981) The classical theorem on existence of competitive equilibrium, *Econometrica*, Vol. 49, pp. 819–41.

MILNE F. (1977) The adjustment problem with jumps in the state variable, Chapter 5, in Pitchford and Turnovsky (eds.).

MILNOR J. (1965) *Topology from the differentiable viewpoint*, University of Virginia Press, Charlottesville, VA.

MINFORD P. AND PEEL D. A. (1983) *Rational expectations and the new macroeconomics*, Martin Robertson, Oxford.

MURATA Y. (1977) *Mathematics for stability and optimisation of economic systems*, Academic Press, New York.

NEWMAN P. K. (1959) Some notes on stability conditions, *Review of Economic Studies*, Vol 27, pp. 1–9.

NEWMAN P. K. (1961) Approaches to stability analysis, *Economica*, Vol. 28, pp. 12–29.

NICKEL S. J. (1978) Fixed costs, employment and labour demand over the cycle, *Economica*, Vol. 45, pp. 329–45.

NIKAIDO H. (1956) On the classical multilateral exchange problem, *Metroeconomica*, Vol. 8, pp. 135–45.

NIKAIDO H. AND UZAWA H. (1960) Stability and non-negativity in Walrasian Tatonnement processes, *Internatinal Economic Review*, Vol. 1,

OLECH C. (1963) On the global stability of an autonomous system on the plane, in *Contributions to Differential Equations*, Vol. 1, No. 3, pp. 389–400.

PITCHFORD J. D. (1977) Two state variable problems, Chapter 6 in Pitchford J. and Turnovsky S. (eds.).

PITCHFORD J. D. and TURNOVSKY S. (EDS.) (1977) *Applications of control theory to economic analysis*, North-Holland, Amsterdam.

PONSTEIN J. (1967) Seven kinds of convexity, *SIAM Review*, Vol. 9, pp. 115–19.

PONTRYAGIN, L. S. *ET AL.* (1962) The mathematical theory of optimal processes, *Interscience*, New York.

ROCKAFELLAR R. T. (1970) *Convex Analysis*, Princeton University Press, N.J.

BIBLIOGRAPHY

SAMUELSON P. A. (1947) *Foundations of economic analysis*, Harvard University Press.

SCHUR J. (1917) Uber Potenzreihen, die im Innern des Einheitskreises beschrankt sind, *Journal fur Mathematik*, Vol. 147, pp. 205–32.

SEIERSTAD A. AND SYDSAETER K. (1977) Sufficient conditions in optimal control theory, *International Economic Review*, Vol. 18, pp. 367–91.

SEIERSTAD A. AND SYDSAETER K. (1987) *Optimal Control Theory with Economic Applications*, North-Holland, Amsterdam.

SHAPLEY L. S. (1973) On balanced games without side payments, in *Matnematical Programming*, edited by Hu T. C. and Robinson S., Academic Press, New York.

SILBERBERG E. (1971) The Le-Chatelier principle as a corollary to a generalized envelope theorem, *Journal of Economic Theory*, Vol. 3, pp. 146–55.

SONNENSCHEIN H. F. (1977) Some recent results on the existence of equilibrium in finite purely competitive economies, in *Frontiers of quantitative economics*, Vol. IIIA, edited by Intriligator M. D., North-Holland, Amsterdam.

SPANIER E. (1966) *Algebraic topology*, McGraw-Hill, New York.

TAKAYAMA A. (1985) *Mathematical economics*, 2nd edn., Cambridge University Press.

TOMPKINS C. B. (1964) Sperner's lemma and some extensions, in *Applied combinatorial mathematics*, edited by Beckenbach E., Wiley, New York. Reprinted in Newman P. (ed.) (1968), *Readings in Mathematical Economics*, Vol. 1, John Hopkins Press, Baltimore.

UZAWA H. (1961) The stability of dynamic processes, *Econometrica*, Vol. 29,

VAN LONG N. AND VOUSDEN N. (1977) Optimal control theorems, in *Applications of Control Theory to Economic Analysis*, Chapter 1, edited by Pitchford J. D. and Turnovsky S., North-Holland, Amsterdam.

VARIAN H. R. (1982) Dynamical systems with applications in economics, in *Handbook of Mathematical Economics*, Vol. 1, Chapter 3, edited by Arrow K. J. and Intriligator M., North-Holland, Amsterdam.

VIND K. (1967) Control systems with jumps in the state variables, *Econometrica*, Vol. 35, pp. 273–7.

AUTHOR INDEX

INDEX

SUBJECT INDEX

INDEX